Advances in STEM Education

Advances in STEM Education is a book series focusing on cutting-edge research and knowledge development in science, technology, engineering and mathematics (STEM) education from pre-college through continuing education around the world. It is open to all topics in STEM education, both in and outside of classrooms, including innovative approaches and perspectives in promoting and improving STEM education, and the processes of STEM instruction and teacher education. This series values original contributions that view STEM education either in terms of traditionally defined subject-based education or as an educational undertaking involving inter-connected STEM fields. The series is open to new topics identified and proposed by researchers internationally, and also features volumes from invited contributors and editors. It works closely with the *International Journal of STEM Education* and *Journal for STEM Education Research* to publish volumes on topics of interest identified from the journal publications that call for extensive and in-depth scholarly pursuit.

Researchers interested in submitting a book proposal should contact the Series Editor: Yeping Li (yepingli@tamu.edu) or the Publishing Editor: Melissa James (Melissa.James@springer.com) for further information.

More information about this series at http://www.springer.com/series/13546

Yeping Li • Roger E. Howe • W. James Lewis
James J. Madden
Editors

Developing Mathematical Proficiency for Elementary Instruction

Springer

Editors
Yeping Li
Department of Teaching, Learning
& Culture
Texas A&M University
College Station, TX, USA

Roger E. Howe
Department of Teaching, Learning
& Culture
Texas A&M University
College Station, TX, USA

W. James Lewis
Department of Mathematics
University of Nebraska-Lincoln
Lincoln, NE, USA

James J. Madden
Department of Mathematics
Louisiana State University
Baton Rouge, LA, USA

ISSN 2520-8616 ISSN 2520-8624 (electronic)
Advances in STEM Education
ISBN 978-3-030-68955-1 ISBN 978-3-030-68956-8 (eBook)
https://doi.org/10.1007/978-3-030-68956-8

This Springer imprint is published by the registered company Springer Nature Switzerland AG
The registered company address is: Gewerbestrasse 11, 6330 Cham, Switzerland

Acknowledgments

This book is a scholarly product that is derived from national workshops on mathematical training for elementary school teachers, held in 2017 and 2018 at Texas A&M University, College Station, USA. The gathering of many mathematicians, mathematics educators, and other scholars made both workshops memorable and productive. We would like to acknowledge and thank all workshop participants for their participation and support.

We also take this opportunity to acknowledge and thank all those who have been involved in preparing and contributing to this volume. This book would not have been possible without the dedicated group of 20 contributors from 13 universities or institutes across the United States, many of whom also volunteered their time to review chapters. These collective efforts help ensure this book's quality.

We want to thank Melissa James (a senior editor at Springer) and other staff at Springer for their support. Working with Melissa has been a productive experience, with this book's publication as a part of the book series on Advances in STEM Education.

Finally, we want to express our deep appreciation to the workshop's host institution, the College of Education and Human Development at Texas A&M University and especially the Department of Teaching, Learning, and Culture, for their support that made both workshops possible. The selfless support from many dedicated staff and graduate students was beyond the call of duty.

Contents

Editors and Contributors

About the Editors

Yeping Li is Professor of Mathematics Education at Texas A&M University, USA, where he served as the Head of the Department of Teaching, Learning, and Culture from 2011 to 2015. His research interests include mathematics education, STEM education, and teacher education. He is the founding editor-in-chief of the *International Journal of STEM Education* and *Journal for STEM Education Research,* both published by Springer, and is also the editor of several monograph series including, *Advances in STEM Education* also published by Springer. In addition to publishing over 15 books and special journal issues in recent years, he has published more than 150 articles in topic areas that he is interested in. He has also organized and chaired many group sessions at various national and international professional conferences, such as International Congress on Mathematical Education-10 (ICME-10) (2004), ICME-11 (2008), ICME-12 (2012), and American Educational Research Association. He received his Ph.D. in Cognitive Studies in Education from the University of Pittsburgh, USA.

Roger E. Howe is Wm. Kenan Jr. Professor of Mathematics, emeritus, from Yale University, and the Curtis D. Roberts Professor of Mathematics Education, and University Distinguished Professor at Texas A&M University, USA. His scholarly interests include mathematics and mathematics education, with a focus on conceptual understanding and teacher preparation. He served on the Study Committee of the National Research Council that produced *Adding It Up*. He was Chair of the Committee on Education for the American Mathematical Society, 2000–2004. He was a member of the reaction group for the creation of the Common Core State Standards in Mathematics for the United States. He served on the U.S. National Commission on Mathematics Instruction from 2008 to 2016, and on Executive Committee of the International Commission on Mathematics Education of the International Mathematical Union, 2010–2016. He helped to plan ICMI Study 23, on whole number arithmetic. He has been a member of the Editorial Board of the

Springer Undergraduate Texts in Mathematics since 2012. He received his Ph.D. in Mathematics from the University of California, Berkeley.

W. James Lewis is Aaron Douglas Professor of Mathematics at the University of Nebraska-Lincoln, USA, where he has served as chair of the Department of Mathematics and currently serves as Director of the Center for Science, Mathematics and Computer Education and Director of STEM Education Research Initiatives. Lewis was chair of the writing team for *The Mathematical Education of Teachers II*, and a member of the writing team for the *Standards for Mathematics Teacher Preparation* published by the Association for Mathematics Teacher Educators. From 2015 to 2018 he worked at the National Science Foundation (NSF), serving first as Deputy Assistant Director and then as Acting Assistant Director for NSF's Education and Human Resources Directorate. His research interests include mathematics education and teacher education.

James J. Madden is Patricia Hewlett Bodin Distinguished Professor of Mathematics at Louisiana State University (LSU), USA, Director of Special programs in the LSU College of Science and Associate Director of the Gordon A. Cain Center for STEM Literacy. In 2007, Madden launched a new track of the College of Science Master of Natural Science (MNS) Program for secondary STEM teachers, now called LaMSTI, or the Louisiana Math and Science Teacher Institute. This program has served nearly 200 Louisiana teachers, of whom over 150 have earned the MNS degree. Madden has over 20 years of experience in developing curricula for professional development. He advised the secondary Eureka Math curriculum and he has published scholarly work the history of the mathematics curriculum.

About the Authors

Sybilla Beckmann is Meigs Distinguished Professor of Mathematics, Emeritus, at the University of Georgia. She has done research in Arithmetic Geometry, but is currently interested in mathematics teacher education and how people think about and learn the interrelated ideas of multiplication, division, fractions, ratio, and proportional relationships. Beckmann is author of *Mathematics for Elementary and Middle School Teachers*, whose 6th edition is forthcoming. She has been on numerous national and international committees concerning mathematics education, including as a member of the writing teams for the *Common Core State Standards for Mathematics* and the *Mathematical Education of Teachers II*. Beckmann has won a number of awards for her work in mathematics education, including the Louise Hay Award from the Association for Women in Mathematics, the Mary P. Dolciani Award from the Mathematical Association of America, and the Award for Excellence in Mathematics Education from Texas A&M University.

Laura Kyser Callis is an Assistant Professor in the Department of Natural Science and Mathematics at Curry College in Milton, Massachusetts, where she has taught Mathematics for Teachers, Statistics, Abstract Algebra, and Mathematical Modeling and served as the coordinator for faculty professional development. She holds a doctorate from Boston University's School of Education. She has taught in a variety of settings, including Jiangsu College for International Education, a college for Chinese students intending to study in English-speaking countries, Boston Public Schools, and both public and private colleges. She is currently researching the impact of simulation-based curriculum on students' understanding of statistical inference and methods for reducing prerequisite barriers to upper-level college mathematics courses, particularly with neurodiverse populations.

Cynthia J. Carter has served as a teacher, a professional developer, and a coach in mathematics and has just retired from teaching middle school mathematics. She especially loved the teaching and particularly loved showing both how "bottom level students" can become "top level" and how nearly *all* students, even in middle school, will develop fascination with mathematics and its logic, and will appreciate its surprises and beauty if given the chance. She studied mathematics and economics at Brandeis and psychology at Harvard. She has published on mathematical thinking in classrooms and co-authored books including *Computers, Education, and Special Needs* and *Making Sense of Algebra*.

Suzanne H. Chapin is Professor Emerita at Boston University's Wheelock College of Education & Human Development where she has prepared future educators and conducted research for over 30 years. She is the principal investigator of the Elementary Mathematics Project (NSF, 2011–2021). Her interests include gifted education, curriculum development, and professional development. Her research has focused on undergraduates' mathematical knowledge for teaching and classroom discussions.

Douglas H. Clements, Distinguished University Professor, Kennedy Endowed Chair in Early Childhood Learning, and Executive Director of the Marsico Institute for Early Learning at the University of Denver, is regarded as a major scholar in the field of early childhood mathematics and STEM education, one with equal relevance to the academy, to the classroom, and to the educational policy arena. He has published over 166 refereed research studies, 27 books, 100 chapters, and 300 additional works, including the development of new mathematics curricula, teaching approaches, teacher training initiatives, and models of "scaling up" interventions. He has served on the U.S. President's National Mathematics Advisory Panel, the *Common Core State Standards* committee of the National Governor's Association and the Council of Chief State School Officers, the National Research Council's Committee on Early Mathematics, the National Council of Teachers of Mathematics national standards committees, and is and co-author of each of their reports. A prolific and widely cited scholar, he has earned external grant support totaling over $25 million, including 36 major grants from the National Science Foundation, the

National Institutes of Health, and the Institute of Education Sciences of the US Department of Education. http://www.researchgate.net/profile/Douglas_Clements/

Albert A. Cuoco is Distinguished Scholar, mathematics, at Education Development Center (EDC), in curriculum and professional development. He was lead author on *The CME Project*, a 4-year high-school curriculum and collaborated with Paul Goldenberg, June Mark, and others to develop *Beauty and Joy of Computing*, an AP computer science curriculum that evolved from work at Berkeley. For the past three decades, Al and Glenn Stevens collaborated on Boston University's *PROMYS for Teachers*, a professional development program based on a mathematics immersion experience. Prior to EDC, from 1969 until 1993, Al taught mathematics to a wide range of students in Woburn (MA) High School. A student of Ralph Greenberg at Brandeis, Al's thesis and research were in Iwasawa theory. Recent books include *Learning Modern Algebra* (with Joseph Rotman), *Applications of Algebra and Geometry to the Work of Teaching* (with others), and *Mathematical Connections: A Companion for Teachers and Others*, all published by MAA/AMS.

Ziv Feldman has been a Clinical Associate Professor of Mathematics Education at Boston University Wheelock College of Education & Human Development for 8 years. For the last 2 years, he also served as chair of the Teaching and Learning department. Dr. Feldman taught mathematics content courses for prospective elementary teachers and methods courses for secondary math teachers. As a co-PI on the Elementary Mathematics Project (EMP), he examined the role of the EMP curriculum in developing prospective elementary teachers' mathematical knowledge for teaching and mathematical identities. Dr. Feldman was also the PI of an NSF Noyce grant titled Enacting Curriculum for Engaging Discourse, which prepares secondary math teachers for work in high need schools. Since June 2020, Dr. Feldman has served as the head of instructor development at Wall Street Prep, an educational company focused on financial modeling training.

Hala N. Ghousseini is the John Grover Harvey Professor of Mathematics Education at the University of Wisconsin-Madison. Her scholarship focuses on understanding how teachers learn complex interactive practice and how the designs of professional education can support them in developing knowledge, skills, and commitments that center students as sensemakers and resist oppressive practices. Her work is also concerned with supporting practitioners to learn collaboratively in the contexts of their own work.

Lynsey K. Gibbons is currently an Assistant Professor of Mathematics Education at the University of Delaware. She started her career in education as an elementary teacher mathematics coach in Lexington, KY before pursuing doctoral studies at Vanderbilt University. Her scholarship seeks to understand how teachers and teacher educators can be provided with rich opportunities to learn how to teach in ways that respond to their students' brilliance and position them as capable sense makers. In her research, Gibbons research focuses on designing for, facilitating, and examining teacher learning; specifying the practices of those who support teacher learning, such as teacher educators and instructional leaders; and identifying tools and resources to support teacher educators and instructional leaders. She has written numerous articles about these topics in a number of journals, including *The Elementary School Journal, Journal of Mathematical Behavior, Journal of Teacher Education, Journal of School Leadership*, and *Mathematics Teacher Education and Development*. Gibbons has also published numerous articles in *Teaching Children Mathematics, Educational Leadership*, and other journals whose primary audience are teachers and school leaders.

E. Paul Goldenberg has been at Education Development Center (EDC) for 34 years and is a Distinguished Scholar, Mathematics and Computer Science. Prior to EDC, he worked at the MIT Logo Laboratory and has taught every age: in early elementary school self-contained classes, middle school mathematics, high school computer science, and university mathematics and education. At EDC, his work focuses on curriculum development in mathematics at all levels, with a particular emphasis on developing curiosity, logic, and the habits of a mathematical mind. He has written several books including *Making Sense of Algebra* (co-authored with June Mark, Cynthia J. Carter, and others), *Developing Essential Understanding of Geometry and Measurement for PreK–Grade2* (with Douglas Clements), *Computers, Education, and Special Needs* (with Susan Jo Russell and Cynthia J. Carter), *Exploring Language with Logo* (with Wallace Feurzeig), and *Special Technology for Special Children*.

Michelle Homp earned her Ph.D. from the University of Nebraska-Lincoln in applied mathematics. Having earned a B.S. in secondary mathematics education, her efforts focus on designing and teaching courses for mathematics teachers. Homp joined the Center for Science, Mathematics and Computer Education (CSMCE) faculty in 2005 and has played a lead role in several NSF grants which partner with K–12 schools. She coordinates the online Master of Arts for Teachers (MAT) degree program at UNL and directs the Nebraska Math and Science Summer Institutes, a summer graduate program for teachers. She currently holds a joint appointment with the CSMCE and the Department of Mathematics. Recent department efforts include the promotion of active student engagement in class instruction; recent outreach efforts have led to professional development partnerships in Senegal, Africa.

Roger E. Howe is Wm. Kenan Jr. Professor of Mathematics, emeritus, from Yale University, and the Curtis D. Roberts Professor of Mathematics Education, and University Distinguished Professor at Texas A&M University. His scholarly interests include mathematics and mathematics education, with a focus on conceptual understanding and teacher preparation. He served on the Study Committee of the National Research Council that produced *Adding It Up*. He was Chair of the Committee on Education for the American Mathematical Society, 2000–2004. He was a member of the reaction group for the creation of the Common Core State Standards in Mathematics for the United States. He served on the US National

Commission on Mathematics Instruction from 2008 to 2016, and on Executive Committee of the International Commission on Mathematics Education of the International Mathematical Union, 2010–2016. He helped to plan ICMI Study 23, on whole number arithmetic. He has been a member of the Editorial Board of the Springer Undergraduate Texts in Mathematics since 2012. He received his Ph.D. in Mathematics from the University of California, Berkeley.

Andrew Izsák is Professor of Education at Tufts University. He has conducted research on teachers' and students' cognition around topics related to multiplication, including multiplication and division with whole numbers, multiplication and division with fractions, proportional relationships, and linear functions. He has received multiple grants from the National Science Foundation to support his research, has published regularly in leading mathematics education and learning sciences journals, and has served on the editorial board for *Journal for Research in Mathematics Education* and as an editor for *Cognition and Instruction*.

W. James Lewis is Aaron Douglas Professor of Mathematics at the University of Nebraska-Lincoln, where he has served as chair of the Department of Mathematics and currently serves as Director of the Center for Science, Mathematics, and Computer Education and Director of STEM Education Research Initiatives. Lewis was chair of the writing team for *The Mathematical Education of Teachers II*, and a member of the writing team for the *Standards for Mathematics Teacher Preparation* published by the Association for Mathematics Teacher Educators. From 2015 to 2018 he worked at the National Science Foundation, serving first as Deputy Assistant Director and then as Acting Assistant Director for NSF's Education and Human Resources Directorate. His research interests include mathematics education and teacher education.

Yeping Li is Professor of Mathematics Education at Texas A&M University, USA, where he served as the Head of the Department of Teaching, Learning, and Culture from 2011 to 2015. His research interests include mathematics education, STEM education, and teacher education. He is the founding editor-in-chief of the *International Journal of STEM Education* and *Journal for STEM Education Research*, both published by Springer, and is also the editor of several monograph series including, *Advances in STEM Education* also published by Springer. In addition to publishing over 15 books and special journal issues in recent years, he has published more than 150 articles in topic areas that he is interested in. He has also organized and chaired many group sessions at various national and international professional conferences, such as ICME-10 (2004), ICME-11 (2008), ICME-12 (2012), and AERA. He received his Ph.D. in Cognitive Studies in Education from the University of Pittsburgh, USA.

James J. Madden is Patricia Hewlett Bodin Distinguished Professor of Mathematics at LSU, Director of Special programs in the LSU College of Science, and Associate Director of the Gordon A. Cain Center for STEM Literacy. In

2007, Madden launched a new track of the College of Science Master of Natural Science (MNS) Program for secondary STEM teachers, now called LaMSTI, or the Louisiana Math and Science Teacher Institute. This program has served nearly 200 Louisiana teachers, of whom over 150 have earned the MNS degree. Madden has over 20 years of experience in developing curricula for professional development. He advised the secondary Eureka Math curriculum, and he has published scholarly work the history of the math curriculum.

Alejandra Salinas is a Clinical Associate Professor of mathematics education where she teaches both undergraduate and graduate-level courses in mathematics content and methods. She is also the Practicum Director for the Mathematics Education program. Her broad research interest is the integration of equitable and socially just teaching in mathematics. As the mathematics education and professional development specialist on PBS's Ready to Learn initiative, she has developed materials to help early childhood educators effectively implement technology such as video games, multimedia, and interactive whiteboards in their classrooms. Lastly, Dr. Salinas is the PI for an NSF-funded project aimed at preparing mathematics teachers who are able to facilitate meaningful mathematical discourse in high-need secondary classrooms.

Julie Sarama is the Kennedy Endowed Chair in Innovative Learning Technologies and Distinguished University Professor at the University of Denver. She conducts research on young children's development of mathematical concepts and competencies, implementation and scale-up of curricula, professional development models and their influence on student learning, and implementation and effects of learning technologies (including those she has created) in mathematics classrooms. These studies have been published in more than 77 refereed articles, 6 books, 55 chapters, and over 80 additional publications. Dr. Sarama has directed over 25 projects funded by the National Science Foundation (NSF), the U.S. Department of Education's Institute of Education Sciences (IES) and the National Institute of Health (NIH), the Bill and Melinda Gates Foundation and the Heising-Simons Foundation. She is currently conducting research on learning trajectories and developing tools and activities for teachers and caregivers who want to implement learning trajectories in early mathematics. She has taught secondary mathematics and computer science, gifted math at the middle school level, preschool and kindergarten mathematics enrichment classes, and mathematics methods and content courses for elementary and secondary teachers. She designed and programmed over several computer programs and apps for children birth-grade 6 and is interested in the unique role technology can play in learning.

Eva Thanheiser is a mathematics teacher educator at Portland State University, in Portland OR. Eva has studied prospective teachers' content knowledge, the development thereof, as well as their motivation to learn mathematics. Elementary mathematics is often perceived as easy but is complex and as such it takes deep understanding to make sense of it. Eva works with her students to develop such deep understanding so that the prospective teachers learn the math to make sense of their world and be able to foster their own students' sense making.

Hung-Hsi Wu is Professor Emeritus of Mathematics at the University of California at Berkeley where he taught from 1965 to 2009. He got into mathematics education in 1992 in response to his observation of serious problems in the teaching of mathematics in schools, including the school mathematics curriculum, textbooks, assessment, and professional development. He worked with the State of California in 1997–2005 in all aspects of mathematics education. From 2000 to 2013, he gave 3-week summer professional development institutes every year for teachers in K–8. He was on the committee that wrote the volume *Adding It Up*; he was also a member of the National Mathematics Advisory Panel in 2006–2008 and the TIMSS 2011 Science and Mathematics Item Review Committee. He is the author of six volumes, totaling 2500 pages, that give a detailed exposition of the school mathematics curriculum from K to grade 12 that consistently respects mathematical integrity. Their titles can be obtained from his homepage http://math.berkeley.edu/~wu, where one can also find his articles on mathematics education.

Contributors

Sybilla Beckmann Department of Mathematics, University of Georgia, Athens, GA, USA

Laura Kyser Callis Curry College, Milton, MA, USA

Cynthia J. Carter The Rashi School, Dedham, MA, USA

Suzanne H. Chapin Boston University, Boston, MA, USA

Douglas H. Clements Marsico Institute for Early Learning and Literacy, Morgridge College of Education, University of Denver, Denver, CO, USA

Albert A. Cuoco Education Development Center, Waltham, MA, USA

Ziv Feldman Boston University, Boston, MA, USA

Hala N. Ghousseini Department of Curriculum and Instruction, University of Wisconsin-Madison, Madison, WI, USA

Lynsey K. Gibbons University of Delaware, Newark, DE, USA

E. Paul Goldenberg Education Development Center, Waltham, MA, USA

Michelle Homp Department of Mathematics, Center for Science, Mathematics and Computer Education, University of Nebraska-Lincoln, Lincoln, NE, USA

Roger E. Howe Department of Teaching, Learning & Culture, Texas A&M University, College Station, TX, USA

Andrew Izsák Department of Education, Tufts University, Medford, MA, USA

W. James Lewis Department of Mathematics, Center for Science, Mathematics and Computer Education, University of Nebraska-Lincoln, Lincoln, NE, USA

Yeping Li Department of Teaching, Learning & Culture, Texas A&M University, College Station, TX, USA

James J. Madden Department of Mathematics, Louisiana State University, Baton Rouge, LA, USA

Alejandra Salinas Boston University, Boston, MA, USA

Julie Sarama Marsico Institute for Early Learning and Literacy, Morgridge College of Education, University of Denver, Denver, CO, USA

Eva Thanheiser Portland State University, Portland, OR, USA

Hung-Hsi Wu Department of Mathematics, University of California at Berkeley, Berkeley, CA, USA

Part I
Introduction

Chapter 1
Developing Mathematical Proficiency for Elementary Instruction: An Introduction

Yeping Li, Roger E. Howe, W. James Lewis, and James J. Madden

1.1 Introduction

This volume grew out of two national workshops held in 2017 and 2018 on the campus of Texas A&M University, College Station, as part of a workshop series on "Mathematics Matters in Education." The workshop series has been developed to provide a national forum for mathematicians, mathematics educators, teachers, and scholars from different disciplines to come together to discuss issues important in school education. The series started in 2015, with one volume that grew out of the first workshop published in 2018 (Li, Lewis, & Madden, 2018a). In that volume, as a dedication to Roger E. Howe, a group of 20 scholars from more than 15 universities or institutions across the United States contributed chapters in three topic areas: knowledge of mathematics for teaching and teacher education, core ideas and practices in K-12 mathematics, and supporting and engaging mathematicians in K-12 education. The selection of these three topic areas aimed to highlight what Roger Howe has been working on over the years together with mathematicians, mathematics educators, and other scholars (Li, Lewis, & Madden, 2018b). The volume was a great success in terms of not only what it intended to accomplish but also what it generated for further inquiries needed in various aspects of mathematics

Y. Li (✉) · R. E. Howe
Department of Teaching, Learning & Culture, Texas A&M University, College Station, TX, USA
e-mail: yepingli@tamu.edu; roger.howe@yale.edu; rogerhowe@tamu.edu

W. J. Lewis
Department of Mathematics, Center for Science, Mathematics and Computer Education, University of Nebraska-Lincoln, Lincoln, NE, USA
e-mail: jlewis@unl.edu

J. J. Madden
Department of Mathematics, Louisiana State University, Baton Rouge, LA, USA
e-mail: madden@math.lsu.edu

Y. Li et al. (eds.), *Developing Mathematical Proficiency for Elementary Instruction*, Advances in STEM Education, https://doi.org/10.1007/978-3-030-68956-8_1

education. One of the important ideas arising from the first workshop was to take a deeper look at mathematical training needed for elementary school teachers. This became the focus of the follow-up two national workshops, in 2017 and 2018, respectively.

What teachers need to know and learn for teaching mathematics effectively has been a long-standing question in mathematics education. As reported in the previous volume (Li et al., 2018a), the question has multi-facets that call for extensive research from different perspectives such as the importance of identifying specific aspects of mathematics that are pertinent to its teaching for understanding (Howe, 2018), mathematical content knowledge that teachers need to have from a disciplinary perspective (Wu, 2018), the need of knowing about mathematical content topic development from a historical perspective (Madden, 2018), and the importance of identifying and responding to preservice teachers' difficulties in learning mathematics (Beckmann & Kulow, 2018). Although the question itself is relevant to all K-12 mathematics teachers, the specifics would understandably be different across school levels. The situation is even more complicated for elementary school teachers as they are typically trained as generalists in the United States and other education systems (e.g., Li, 2008). What elementary teachers need to know and learn in mathematics and how to provide them adequate mathematical training thus call for special attention and dedicated research.

The 2017 workshop took a special focus on preservice elementary teachers. Although there has been work toward providing improved mathematics training to preservice elementary teachers, we lack a clear understanding of how experiences in college courses influence teaching practice, how these courses might be made more effective, and how other forms of professional support might add value to preservice training. How to better provide preservice elementary teachers with solid mathematics training remains a big challenge to mathematics educators, mathematicians, and teacher educators alike. The workshop was structured with the following three questions:

1. What do we know about the mathematical proficiency preservice teachers need to start teaching mathematics effectively, and what do we know about providing it?
2. How are elementary teacher preparation programs structured and what are the core ideas and practices in the mathematical preparation of elementary teachers?
3. How can partnerships between Mathematics Departments and Colleges of Education (and other agencies) be built for effective mathematics education of preservice teachers?

The 2018 workshop was a sequel to the 2017 workshop with a focus on in-service elementary teachers. Different from preservice elementary teachers, in-service teachers are often overwhelmed with daily instructional responsibilities in schools. There are not only limited opportunities made available for their professional development but also a lack of consensus for what in-service teachers need and how to assist them. Those variables make teachers' professional development challenging and also call for further efforts in developing, identifying, and assessing effective models for teachers' professional development in mathematics, especially for elementary school teachers. In particular, three questions were identified and used to help structure the workshop:

1. What do we know about the needs of in-service elementary teachers in professional development, and possible challenges in designing and assessing effective professional development programs?
2. What do we know about the nature of subject content and pedagogy training typically provided through elementary teachers' professional development in mathematics and how has such focus on content and pedagogy evolved over the years?
3. What role may Mathematics Departments play in providing elementary teachers' professional development in mathematics, and what successful collaborations may be developed among mathematicians, educators, and teachers?

Information about these two workshops can help provide general background information about this volume. The volume also goes beyond these two workshops in three important ways:

1. Having a theme across these two workshops. On one hand, we originally thought it would not be difficult to identify a theme, as both workshops focused on elementary school teachers. However, these two workshops dealt with aspects where elementary school teachers have very different professional experiences in two separate periods of time. Attempts to cover those experiences in a teacher's life-long professional career would hinder making connections between these two workshops. On the other hand, both workshops were devoted to understanding what elementary teachers need in mathematics and how to support them in gaining the needed training. Thus we believe that the theme of examining mathematical proficiency needed for elementary instruction and improving needed training for elementary teachers should be at the core of such a volume. In fact, the nature of mathematical proficiency needed should stay the same for both preservice and in-service elementary school teachers, but teachers can experience different needs and challenges over their professional lifetime.
2. Making cross-workshop communications possible. As these two workshops were organized with different foci, participants varied in terms of their professional interest and work. There were hardly further communications across these two workshops, as with many other workshops in the field. This volume is positioned to bridge these two workshops with contributions beyond the scope of each individual workshop by itself.
3. Sharing the knowledge available to a broader international community. Although the vast majority of participants in these two workshops were from the United States, we recognized that the ideas being shared have value for a broader international community. Further work in refining and presenting ideas as formal scholarly articles with expanded contributions beyond these two workshops help ensure its intended quality and contribution to the field. Similar to the nature of the previous volume (Li et al., 2018a), readers can easily find in this volume

the value and importance of joint efforts of mathematicians and mathematics educators in addressing this important topic.

In the following section, we provide a brief summary of each chapter as structured in three different parts.

1.2 Structure of the Book

This book is designed to offer mathematicians and mathematics educators with the opportunity to share their work and ideas on mathematical training for preservice and in-service elementary school teachers to a broader international community. It is organized in three parts with the last two parts to parallel the sequence of the two workshops: (I) introduction, (II) mathematical training for preservice elementary school teachers, and (III) professional development in mathematics for in-service elementary school teachers.

Readers can certainly feel free to read through the book in whatever chapter sequence is preferred. Everyone is also encouraged to reflect on these chapters and make possible connections and differentiations in ideas and suggestions conveyed across different chapters.

Part I contains one chapter. The present introductory chapter provides readers with the background of this book development, and an overview of related perspectives and content.

Part II focuses on the nature of mathematical training that would be important to have for preservice elementary school teachers (PSTs). A total of seven chapters (Chaps. 2–8) are included in this part. Readers can find quite a range of perspectives and approaches from these chapters in answering questions such as: what is important in mathematical training needed for PSTs? How to provide mathematical training so that elementary teachers can be well prepared for teaching in the future? While some chapters focus mainly on one aspect, others make points related to both aspects.

In Chap. 2, Li and Howe (2021) first provide a review of relevant research about difficulties elementary teachers may have in mathematics and existing approaches to provide mathematical training to them. Then they build upon the review to highlight the need of emphasizing and developing a thinking-oriented training in mathematics for elementary teachers. Goldenberg, Cuoco, and Carter (2021) in Chap. 3 emphasize the importance for elementary teachers to experience and learn what mathematics is beyond knowledge as given. Through highlighting the need of *doing* mathematics, they point out a key idea that may be missing in instruction that focuses on knowledge acquisition. In the next chapter, Chapin, Gibbons, Feldman, Callis, and Salinas (2021) report their approach, through developing curricula and instructor support materials, to deepen PSTs' understanding of mathematical knowledge for teaching. In their approach, problem solving and discussions are used to immerse students in the mathematical practices of sensemaking, justification, communication, and generalization. It appears that all three chapters (Chaps. 2, 3, and 4) focus on the question of what elementary teachers would need to learn beyond the knowledge itself. Although different approaches in these three chapters are used in developing specific arguments, they share a similar perspective of emphasizing the thinking development for elementary teachers.

In Chap. 5, Homp and Lewis (2021) share the course structure in mathematics and mathematics education for elementary teacher preparation at the University of Nebraska-Lincoln. They highlight the emphasis given in their use of homework and writing assignments to help develop PSTs' habits of mind as mathematical thinkers. Moreover, the partnership between mathematicians and mathematics educators is also recognized as a key aspect contributing to their success. Beckmann and Izsák (2021) in Chap. 6 present and discuss their design of mathematical courses for preservice teachers with a primary focus on mathematical structure and sensemaking. By focusing on the multiplicative conceptual field, they share how several interrelated topics of multiplication, division, fractions, proportional relationships, and linear relationships can be treated in a logical and coherent way, while balancing on PSTs' own sensemaking abilities. The success and challenges of implementing such courses are also discussed with illustrative examples of PSTs' work. In the next chapter, Ghousseini (2021) examines how PSTs can learn mathematical knowledge for teaching (MKT) through the interactive work of teaching, rather than having MKT treated as isolated from the reality of teaching and learning. Focusing on the pedagogy of rehearsals of teaching, Ghousseini explores what the teacher educator does to enlist MKT, using four types of interventions, to help PSTs learn MKT as knowledge-in-action. In Chap. 8, Thanheiser (2021) shares an approach to help PSTs recognize and appreciate the complexity of elementary mathematics and what they learned from a mathematics content course. Specifically, she videotaped interviews with PSTs before taking the course, and then let them watch the videotaped interviews at the end of the course. The contrast helped PSTs to realize that they have learned content from the course.

Part III takes on professional development in mathematics that would be important for in-service elementary school teachers (ISTs). The three chapters (Chaps. 9–11), contributed by scholars from different disciplines, treat topics spanning from mathematics, curriculum, textbooks, to students' learning trajectories in mathematics.

In Chap. 9, Howe (2021) describes the Yale Teachers Institute (YTI), a distinctive form of professional development for in-service teachers. By focusing on some carefully selected key topics, the YTI helps teachers to develop a conceptual understanding of the standard curriculum beyond textbooks. Howe shares and discusses some topics in elementary mathematics covered in mathematical seminars in recent years. In Chap. 10, Clements and Sarama (2021) describe and discuss the Technology-enhanced, Research-based, Instruction, Assessment, and professional Development, or TRIAD model for teacher professional development (PD). Research-validated students' learning trajectories in mathematics are at the core of TRIAD. They present specific PD strategies used and the empirical evidence supporting the approach's efficacy and equity in mathematics PD. Finally, Wu

(2021) in Chap. 11 presents and argues the importance for ISTs to learn and teach correct mathematics. Specifically, Wu discusses the problematic nature of textbook school mathematics, and highlights the importance for ISTs to learn and teach principle-based mathematics through PD.

References

Beckmann, S., & Izsák, A. (2021). Focusing on mathematical structure *and* sense-making in courses for future teachers. In Y. Li, R. E. Howe, W. J. Lewis, & J. J. Madden (Eds.), *Developing mathematical proficiency for elementary instruction*. Cham, Switzerland: Springer.

Beckmann, S., & Kulow, T. K. (2018). How future teachers reasoned with variable parts and strip diagrams to develop equations for proportional relationships and lines. In Y. Li, W. J. Lewis, & J. J. Madden (Eds.), *Mathematics matters in education—Essays in honor of Roger E. Howe* (pp. 117–148). Cham, Switzerland: Springer.

Chapin, S. H., Gibbons, L., Feldman, Z., Callis, L. K., & Salinas, A. (2021). The elementary mathematics project: Supporting pre-service teachers' content knowledge for teaching mathematics. In Y. Li, R. E. Howe, W. J. Lewis, & J. J. Madden (Eds.), *Developing mathematical proficiency for elementary instruction*. Cham, Switzerland: Springer.

Clements, D. H., & Sarama, J. (2021). Sustainable, scalable development in early mathematics: Strategies, evaluation, and tools. In Y. Li, R. E. Howe, W. J. Lewis, & J. J. Madden (Eds.), *Developing mathematical proficiency for elementary instruction*. Cham, Switzerland: Springer.

Ghousseini, H. N. (2021). The work of coaching in rehearsals to enlist mathematical knowledge for teaching. In Y. Li, R. E. Howe, W. J. Lewis, & J. J. Madden (Eds.), *Developing mathematical proficiency for elementary instruction*. Cham, Switzerland: Springer.

Goldenberg, E. P., Cuoco, A. A., & Carter, C. J. (2021). The missing element for teachers: Learning what mathematics *is*. In Y. Li, R. E. Howe, W. J. Lewis, & J. J. Madden (Eds.), *Developing mathematical proficiency for elementary instruction*. Cham, Switzerland: Springer.

Homp, M., & Lewis, W. J. (2021). Mathematics and mathematics education at the University of Nebraska-Lincoln: Partnering to implement core practices in elementary teacher preparation. In Y. Li, R. E. Howe, W. J. Lewis, & J. J. Madden (Eds.), *Developing mathematical proficiency for elementary instruction*. Cham, Switzerland: Springer.

Howe, R. E. (2018). Cultural knowledge for teaching mathematics. In Y. Li, W. J. Lewis, & J. J. Madden (Eds.), *Mathematics matters in education—Essays in honor of Roger E. Howe* (pp. 19–39). Cham, Switzerland: Springer.

Howe, R. E. (2021). Mathematics in the Yale Teachers Institute. In Y. Li, R. E. Howe, W. J. Lewis, & J. J. Madden (Eds.), *Developing mathematical proficiency for elementary instruction*. Cham, Switzerland: Springer.

Li, Y. (2008). Mathematical preparation of elementary school teachers: Generalists versus content specialists. *School Science and Mathematics, 108*, 169–172.

Li, Y., & Howe, R. E. (2021). Towards a thinking-oriented training in mathematics for elementary teachers. In Y. Li, R. E. Howe, W. J. Lewis, & J. J. Madden (Eds.), *Developing mathematical proficiency for elementary instruction*. Cham, Switzerland: Springer.

Li, Y., Lewis, W. J., & Madden, J. J. (Eds.). (2018a). *Mathematics matters in education—Essays in honor of Roger E. Howe*. Cham, Switzerland: Springer.

Li, Y., Lewis, W. J., & Madden, J. J. (2018b). Mathematics matters in education—Essays in honor of Roger E. Howe: An introduction. In Y. Li, W. J. Lewis, & J. J. Madden (Eds.), *Mathematics matters in education—Essays in honor of Roger E. Howe* (pp. 3–7). Cham, Switzerland: Springer.

Madden, J. J. (2018). Knowing ratio and proportion for teaching. In Y. Li, W. J. Lewis, & J. J. Madden (Eds.), *Mathematics matters in education—Essays in honor of Roger E. Howe* (pp. 93–115). Cham, Switzerland: Springer.

Thanheiser, E. (2021). How do I know I learned something? Reflecting on learning by using video-recorded interviews to battle hindsight ("I-knew-it-all-along") bias. In Y. Li, R. E. Howe, W. J. Lewis, & J. J. Madden (Eds.), *Developing mathematical proficiency for elementary instruction*. Cham, Switzerland: Springer.

Wu, H.-H. (2018). The content knowledge mathematics teachers need. In Y. Li, W. J. Lewis, & J. J. Madden (Eds.), *Mathematics matters in education—Essays in honor of Roger E. Howe* (pp. 43–91). Cham, Switzerland: Springer.

Wu, H.-H. (2021). In-service mathematics professional development: The hard work of learning mathematics. In Y. Li, R. E. Howe, W. J. Lewis, & J. J. Madden (Eds.), *Developing mathematical proficiency for elementary instruction*. Cham, Switzerland: Springer.

Part II
Mathematical Training for Pre-service Elementary School Teachers

Chapter 2
Toward a Thinking-Oriented Training in Mathematics for Elementary School Teachers

Yeping Li and Roger E. Howe

2.1 Introduction

There have been consistent concerns about the quality and outcomes of mathematics education in the United States, based on what we can learn from the results of large-scale international studies in mathematics over the years (National Mathematics Advisory Panel, 2008). Recent results from the Program for International Student Assessment (PISA) tests given in 2018 further suggested that the achievement of American 15-year-olds in reading and mathematics has been stagnant since 2000 (Goldstein, 2019). While there are continued debates about possible ways to improve student performance, the quality of classroom instruction and effective teacher training remain important to many professionals and policy makers. The existence of the example of Massachusetts (Howe, 2020) shows that this is possible in a US context. Unfortunately, Massachusetts is close to being the unique state that has succeeded in broad improvement of mathematics achievement.

To empower students mathematically would require first to make mathematical ideas accessible and meaningful to students. By making mathematics accessible, we do not mean simply presenting mathematical content to students as "given" (Li & Schoenfeld, 2019), nor do we mean making mathematics easy for students-that can actually contribute to low achievement and diminish their self-esteem (e.g., Rattan, Good, & Dweck, 2012). We mean providing well-designed and structured challenges that can engage students and develop their understanding and ability to make sense of mathematical concepts and solve mathematical problems. Achieving this ambitious goal places a high demand on teachers, who need to possess a solid understanding of the mathematics they teach and the ability of making feasible

Y. Li (✉) · R. E. Howe
Department of Teaching, Learning & Culture, Texas A&M University, College Station, TX, USA
e-mail: yepingli@tamu.edu; rogerhowe@tamu.edu; roger.howe@yale.edu

Y. Li et al. (eds.), *Developing Mathematical Proficiency for Elementary Instruction*, Advances in STEM Education, https://doi.org/10.1007/978-3-030-68956-8_2

connections between mathematics and students. Although having highly competent teachers at the high school and middle school levels is important, so is having mathematically competent teachers in elementary schools. This is particularly true at the stage of elementary education (kindergarten–grade 6) when students' learning of mathematics should lay a vital foundation for their science, technology, engineering and mathematics (STEM) education in the future (NAEYC & NCTM, 2010; NRC, 2011).

The need for improving mathematical preparation for elementary preservice teachers (PSTs) in the United States is clearly acknowledged (e.g., AMTE, 2017; CBMS, 2001, 2012). Although there have been works toward providing improved mathematics training to elementary PSTs, we lack generally recognized, replicable, and widely adopted models for elementary teacher training in mathematics. How to better provide elementary PSTs with solid mathematical training remains a big challenge to mathematics educators, mathematicians, and teacher educators alike. In this chapter, we aim to provide an overview of relevant research about the possible difficulties in mathematics elementary PSTs may have and existing approaches to provide mathematical training to elementary PSTs. Building upon existing research, we propose an alternative approach to provide mathematical training to elementary PSTs, and conclude with discussion about the need of developing new research and scholarship.

2.2 Elementary PSTs' Knowing and Understanding of Mathematics and Mathematics Instruction

Recognizing the need of improving mathematical training for elementary PSTs leads to the questions of what elementary PSTs need to know and how they can learn it. In fact, elementary PSTs are college students who already went through K–12 education. By assumption, they already learned and passed whatever was required of them in elementary mathematics. Are they ready to teach elementary mathematics then? If not, what may they not know or not know well enough? What may be good ways to help them learn or relearn elementary mathematics as future teachers?

Existing research over the past several decades provided rich information about what PSTs know and do not know about school mathematics. The following three sections are organized to highlight relevant research on

1. PSTs' knowledge and understanding of elementary mathematics.
2. PSTs' performance in problem-solving and posing.
3. PSTs' beliefs and conceptions of mathematics and mathematics teaching.

2.2.1 Elementary PSTs' Knowing and Understanding of Mathematics

2.2.1.1 Studies of Elementary PSTs in the United States

In examining the mathematical knowledge and understanding that PSTs bring to teacher education, Ball (1990a) surveyed 252 PSTs (217 elementary, 35 secondary) and interviewed 35 of these PSTs: 25 elementary and 10 secondary. The secondary PSTs were mathematics majors. Ball found that the majority of PSTs have difficulty in generating, and even in selecting, an appropriate representation for division with fractions. For example, all of these PSTs during interviews showed that they know how to calculate $1\frac{3}{4} \div \frac{1}{2}$, predominantly based on the remembered procedure "invert and multiply." However, only about 30% of 252 PSTs could identify a correct story problem that represented this fraction division. At the same time, among those 30% PSTs that chose the correct story problem, many of them also selected one or more incorrect responses. The result became even worse when they were asked to generate appropriate representations (e.g., pictures, models, stories, or real-world representations) for this fraction division expression. In all the interviews, no elementary PSTs and only four secondary PSTs were able to describe a mathematically appropriate representation. The majority (19, 15 elementary and 4 secondary) of these 35 PSTs were unable to generate a representation at all for this division. The remaining 12 PSTs (34%) generated inappropriate representations, with the majority representing a division by 2 instead of division by ½. For example, a secondary PST came up with a story problem using pizza: "If we had one and three-quarters pizzas left and there were two of us dying to split it, then how would we be able to split that?" (p. 455). The PST explained further conceptually how to do the computation to come up with an answer that each person would get 7/8 of a pizza altogether (i.e., $1\frac{3}{4} \div 2$). Although Ball did not explain why these PSTs did not cross-check or verify their answer with directly calculating $1\frac{3}{4} \div \frac{1}{2}$, these PSTs' explanation suggests that one main difficulty for them is that they perceived the question as one about *fractions* (i.e., having ½ means to divide by 2) instead of *division*. And in real life, people tend to think about such questions using pizza and sharing in terms of whole numbers, not fractions.

The findings led Ball (1990a) to argue that PSTs brought with them, from their precollege and college mathematics experiences, a conception of mathematics as *rule-bound*, *thin*, and *compartmentalized*. Their performance in answering questions related to fraction division indicated that PSTs have a narrow understanding of division in multiple aspects:

1. They can do the fraction division computation by performing the memorized procedure of "invert and multiply".
2. Training in formal mathematics helped PSTs being more successful at computing correctly, but not at providing conceptual explanation (Ball, 1990a).

3. They would perceive and interpret a fraction division computation in terms of fractions but not division (Ball, 1990a), indicating a limited understanding of how the division operation is affected by the type of numbers involved.
4. They tended to think division in partitive (or sharing) terms only, and not in terms of measurement. Thus, it was difficult for them to make meaning out of $1\frac{3}{4} \div \frac{1}{2}$ (Ball, 1990b).

The limited knowledge and narrow understanding of elementary mathematics of U.S. PSTs as revealed by Ball (1990a, 1990b) are well supported by many more recent studies (e.g., Li & Kulm, 2008; Lo, Grant, & Flowers, 2008; Mewborn, 2001; Simon, 1993; Thanheiser et al., 2014), and also by cross-national studies (Ma, 1999; Schmidt et al., 2007). Several review articles have summarized what we can learn from existing research on PSTs' knowledge and understanding of mathematics (e.g., Browning et al., 2014; Mewborn, 2001). For example, in 2014, a special issue of the *Mathematics Enthusiast* focused on elementary PSTs' content knowledge. This issue includes a set of review articles which provide summaries of the peer-reviewed research literature from 1978 to 2012 on PSTs' content knowledge of five key mathematics topics (Browning, Thanheiser, et al., 2014; Thanheiser, Browning, et al., 2014). Each article summarizes relevant research on one of: whole-number concepts and operations (Thanheiser, Whitacre, & Roy, 2014); fractions (Olanoff, Lo, & Tobias, 2014); decimals (Kastberg & Morton, 2014); geometry and measurement (Browning, Edson, Kimani, & Aslan-Tutak, 2014); and algebra (Strand & Mills, 2014). A common theme identified throughout all these review articles is that PSTs' understanding of each topic is limited to using algorithms, and they have difficulties in justifying why the algorithms work (Browning, Thanheiser, et al., 2014). This is consistent with what Ball (1990a) concluded from her studies.

Across different content topics in elementary mathematics, there are unequal distributions of research, with some topics receiving more attention (e.g., fractions, esp. fraction division, see the above Ball's studies, 1990a, 1990b) than others (e.g., decimals). On the topic of whole numbers and operations, as an example, Thanheiser, Whitacre, and Roy (2014) summarized that research in this topic area is scarce with 7 journal articles published from 1978 to 1998, 18 journal articles from 1998 to 2011, and 1 journal article, and 2 peer-reviewed conference proceedings in 2011–2012. Existing research revealed that PSTs' knowledge of whole numbers and operations is insufficient and needs improvement. In particular, PSTs can perform whole-number computation using the standard algorithms but struggle to explain why these algorithms work. For example, Thanheiser (2009) reported that a PST can perform the following computational procedure for 527 − 135 to get the correct answer of 392.

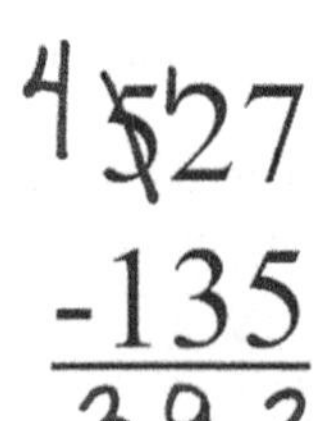

But the PST had difficulty to explain why the procedure works, "You put a 1 over next to the number and that gives you 10 ... I don't get how the 1 can become a 10. One and 10 are two different numbers. How can you subtract 1 from here and then add 10 over here? Where did the other 9 come from? ... I have no idea of why it works. Just principles that you had learned throughout school" (Thanheiser, 2009, pp. 269–270).

Some researchers have noted that some PSTs are aware of their lack of conceptual understanding of mathematics, such as the one noted above in Thanheiser's study (2009). Ball (1990a) also reported that a PST noted, "I am really worried about teaching something to kids I may not know. Like long division—I can *do* it—but I don't know if I could really *teach* it because I don't know if I really *know* it or know how to *word* it" (Ball, 1990a, p. 449). It would certainly be great if all PSTs realized their need for greater understanding as they could then be expected to know the need of taking mathematics courses as an important part of teacher preparation for their mathematics teaching in the future.

However, it is reported in other studies that many PSTs are not aware of what they do not know and what else they need to know (e.g., Li & Kulm, 2008; Li & Smith, 2007). For example, in a study of PSTs' knowledge of fraction division (Li & Kulm, 2008), 46 elementary and middle-school PSTs were asked about their own perceptions about knowledge preparation and were also assessed with a mathematics test on the topic of fraction division. The results from the study show two different and seemingly contradictory sides of PSTs' knowledge in mathematics and pedagogy for teaching with further information provided as follows.

One side is quite positive, as revealed by these PSTs' responses to selected survey questions. For example, one question asked: considering your training and experience in both mathematics and instruction, how ready do you feel you are to teach the topic of "Number—Representing and explaining computations with fractions using words, numbers, or models?" On a scale of three (Very ready, Ready, Not ready), 60% PSTs thought they are ready, while 38% chose "very ready," and only 2% "not ready." Although the positive results are only perceptions and self-assessment of these PSTs' knowledge preparation, the results likely suggest that this group of PSTs is confident in their knowledge preparation for teaching fraction computations, including fraction division.

On the other side, these PSTs performed well in fraction computation but not on other items that require conceptual understanding. For example, only 59% of these PSTs provided a correct answer to the item, "Tell whether $\frac{9}{11} \div \frac{2}{3}$ is greater than or less than $\frac{9}{11} \div \frac{3}{4}$ without evaluating. Explain your reasoning." Many PSTs were confused about which approach to use for making a justification. If a problem is presented not in simple computation but requiring justification with conceptual understanding, these PSTs' performance was even worse as shown in solving the following problem (adapted from Tirosh, 2000).

> You are discussing operations with fractions in your class. During this discussion, John says:
>
> *It is easy to multiply fractions; you just multiply the numerators and the denominators. I think that we should define the other operations on fractions in a similar way:*

$$\textit{Addition}\ \frac{a}{b}+\frac{c}{d}=\frac{(a+c)}{(b+d)}$$
$$\textit{Subtraction}\ \frac{a}{b}-\frac{c}{d}=\frac{(a-c)}{(b-d)}$$
$$\textit{Division}\ \frac{a}{b}\div\frac{c}{d}=\frac{(a\div c)}{(b\div d)}$$

How will you respond to John's suggestions? (Deal with each separately.)

Ninety percent of these PSTs indicated that the given computations for fraction addition and subtraction are not correct, but only 2 out of the 46 PSTs stated that the given computation for fraction division is correct. The majority of others stated that the fraction division should be done as "flip and multiply" or "KFC" (i.e., keep the first, flip the second, and change the computation). No further justification was provided. These PSTs' performance in answering these questions involving fraction division reveals their limited knowledge about fraction computation and is consistent with what was reported by Ball (1990a, 1990b) and others (e.g., Olanoff et al., 2014).

The contrasting results suggest that these PSTs' positive self-perceptions are likely built upon their insufficient (or limited) content knowledge of fraction division, mainly in using algorithms for computation. In fact, with their high confidence in knowledge preparation, these PSTs would not even realize that they have deficiencies in their mathematical understanding. It thus becomes necessary for US teacher educators to identify what PSTs know and do not know and what they need to learn through their program of study. This may be even more important than simply adding more mathematics content courses.

2.2.1.2 Elementary PSTs in the United States in Comparison to Their Counterparts in Other Education Systems

The above results about U.S. PSTs' content knowledge and understanding of mathematics present a picture that calls for changes and improvement. At the same time, we wonder how elementary teachers in other education systems perform in mathematics. Through a cross-national lens, we can learn if the situation in the United States is unique and what may be the possible extent of improvement if PSTs, or teachers, in other countries perform better in mathematics.

A well-publicized study by Ma (1999) revealed that sampled Chinese in-service elementary teachers had substantially stronger understanding of the mathematics they teach than the US teacher participants surveyed in her study. As argued by Ma, the results provided a partial explanation for what contributes to Chinese students' achievement in school mathematics. Paradoxically, according to Ma, Chinese elementary teachers actually received less formal education, in terms of the number of years spent before entering the teaching profession, than their US counterparts in her study. Ma's findings suggested the possibility of improving mathematical training for elementary teachers. At the same time, the findings also raised more questions about what elementary PSTs can and should know and how to help them achieve that through teacher education.

Efforts to provide a better mathematical preparation for PSTs also led to large-scale international comparative studies to investigate what PSTs around the world learn in and through their program studies, including the project of Mathematics Teaching in the Twenty-first Century (MT21, see Schmidt et al., 2007) that involved six education systems (Bulgaria, Chinese Taipei, Germany, Mexico, South Korea, and the United States), and Teacher Education and Development Study—Learning to Teach Mathematics (TEDS-M, see Blömeke, Hsieh, Kaiser, & Schmidt, 2014) that involved 17 education systems. These studies revealed many more cross-system differences than similarities in PSTs' course taking, knowledge of mathematics, and mathematics pedagogy, and beliefs about mathematics at the end of their program studies. Based on the study's findings, for example, Schmidt et al. (2007) indicated that there is a preparation gap in mathematical training provided to PSTs between the United States and two participating education systems in East Asia (i.e., Chinese Taipei and South Korea).

Li and his colleagues also did a series of studies about elementary teacher preparation in mathematics in East Asia and the United States (e.g., Li, Ma, & Pang, 2008; Li, Pang, Zhang, & Song, 2019; Li, Zhao, Huang, & Ma, 2008). In particular, Li, Ma, and Pang (2008) did a review of the practices in mathematical preparation for elementary PSTs in six selected education systems in East Asia (i.e., Hong Kong, Japan, mainland China, Singapore, South Korea, and Taiwan) and revealed that, although there are many variations in program setup and curriculum structure both across and within the six education systems, mathematics is seen as an important subject in all programs. The performance of PSTs sampled in mainland China and South Korea also supported the perception that elementary PSTs in East Asia have a strong preparation in mathematics content knowledge, as assessed in 2007 (Li, Ma, & Pang, 2008) and again in 2018 (Li, Pang, et al., 2019), and better than a group of PSTs' performance on the same items in the United States (Li & Kulm, 2008).

At the same time, researchers reported that PSTs in other education systems, including those high-achieving education systems, also experienced difficulties and challenges in mathematics (e.g., Yang, 2007; Yang, Reys, & Reys, 2009). For example, Yang et al. (2009) reported their investigation of 280 elementary PSTs in Taiwan in solving 12 open-ended questions on the use of number sense. They found that over 60% of all the responses provided by these PSTs used rule-based written approaches (e.g., using computation algorithms), even though they were suggested not to use paper-and-pencil, and less than one-third of these PSTs used aspects of number sense (e.g., benchmarks, estimation, and reasonableness) in their explanation. These results are similar to what were reported and found with the performance of many middle school students in multiple education systems including Australia, Sweden, United States, and Taiwan (e.g., Reys et al., 1999). The similarities in performance naturally led to the question about possible impact of mathematics learning in K–12 schools and colleges on PSTs' mathematical performance. Yang et al. (2009) hypothesized some possible reasons, including the school textbook emphases on rule-written methods and relative lack of attention to aspects important for number sense. They also suggested that teacher education in Taiwan needs changes to shift away from rule-based written computation to the development of number sense.

Taken together, the above results suggest that US elementary teachers' knowledge and understanding in mathematics can and should be improved. At the same time, PSTs in other education systems share a range of similar difficulties in mathematics, albeit to a different extent. The need to improve mathematical training provided to elementary PSTs is felt not only for teacher preparation programs in the United States but also those in other education systems. Thus, it is important for us to examine possible efforts toward improving mathematical training in other education systems, when considering possible approaches to take in the United States.

2.2.2 Elementary PSTs' Performance in Problem-Solving and Posing

One important point from the above review of PSTs' performance in mathematics is that PSTs are good at using algorithms for computation, but they have limited understanding of the ideas of mathematics. They struggle in providing explanation, in making connections across concepts, in reasoning, and fail to make needed distinctions. All of these are important parts of mathematical thinking. In other words, PSTs have long been accustomed to memorize and use algorithms and formulas, but struggle in learning how to think and reason mathematically. For example, Lo et al. (2008) documented a course of providing 38 elementary PSTs with explicit instruction on developing and justifying reasoning strategies, first for addition and subtraction for several weeks, and then for multiplication for 2 weeks. These PSTs were assessed with both a writing assignment, collected the day after the completion of the multiplication unit, and then an exam, given 2 weeks after the multiplication unit. Their analyses of PSTs' work revealed six main challenges that these PSTs encountered in developing and justifying reasoning strategies for multiplication: two for developing reasoning strategies and four for developing a justification for reasoning strategies. One of the tasks was to ask PSTs to evaluate 24×38 by demonstrating a reasoning procedure. It started with 20×40, followed by needed steps of the reasoning procedure, and then justified why it works with the equal groups interpretation of multiplication. Lo et al. noted that PSTs struggle with providing justifications in the following four ways:

1. PSTs confused justification of explaining "why" with a description of steps ("how").
2. PSTs took justification as just drawing a picture.
3. PSTs struggled in relating a picture with their reasoning strategy, especially when using the area model.
4. PSTs struggled in coordinating the equal groups interpretation with their reasoning strategy.

The results illustrate various challenges in helping PSTs to learn mathematics with conceptual understanding that requires thinking, reasoning, and justification.

To better understand how comfortable PSTs are with mathematical thinking, we often take PSTs' performance in problem-solving and posing as an important indicator. In fact, if we expect students to develop problem-solving skills and ability in and through school mathematics, we would certainly expect PSTs to be good at solving and posing problems, not just those procedure-oriented computations but also complex and open-ended problems that require serious efforts and thinking. What can we learn from existing research about elementary PSTs' performance in mathematical problem-solving and posing?

Students in elementary schools often find word problems difficult, in comparison to comparable problems presented in numeric format (Cummins, Kintsch, Reusser, & Weimer, 1988). To help students improve their performance in solving arithmetic word problems, it is critical that teachers be well prepared in understanding and solving such word problems. Carpenter, Fennema, Peterson, and Carey (1988) found that selected first-grade teachers knew well about their students, their achievement, and main strategies that their students would use to solve different problems. They could also identify many of the critical distinctions between problems when asked to write up six word problems to best represent six different numerical expressions ($5 + 7 =?$, $6 + ? = 11$, $? + 4 = 12$, $13 - 4 = ?$, $15 - ? = 9$, and $? - 3 = 9$) that correspond to the six "join" and "separate" problem types, but typically their knowledge was not organized into a coherent framework. These results suggest the need and importance of helping teachers to learn explicitly about arithmetic word problems from a systematic and coherent viewpoint, instead of conceiving them simply as a vast collection of problems differentiated simply in terms of the computation needed for solutions.

Going beyond these relatively simple one-step word problems, researchers found that both elementary PSTs and in-service teachers in the United States experience difficulties in solving and posing word problems (Li & Kulm, 2008; Ma, 1999). For example, Li and Kulm (2008) reported that only 39% of 46 PSTs in their study solved the following problem correctly.

> Johnny's Pizza Express sells several different flavors of large-size pizza. One day, it sold 24 pepperoni pizzas. The number of plain cheese pizzas sold on that day was $\frac{3}{4}$ of the number of pepperoni pizzas sold, and was $\frac{2}{3}$ of the number of deluxe pizzas sold. How many deluxe pizzas did the pizza express sell on that day?

In contrast, Li, Pang, et al. (2019) reported that 94% of 319 elementary PSTs sampled from multiple teacher preparation programs in China and 93% of 221 elementary PSTs sampled from multiple teacher preparation programs in South Korea solved the problem correctly. Ma's study (1999) revealed a similar picture of in-service elementary teachers' performance in solving and posing mathematics problems, which included in-service teacher participants from the United States and China. Although different factors can contribute to difficulties in solving and posing problems, typical analyses in research would focus on specific mistakes being made and different strategies or representations being used in the problem-solving process. The strategies and representations used in US elementary teachers' performance are shown as less diverse and effective than their counterparts in China in Ma's study.

In considering possible contributing factors, one important factor should be the training for preservice and in-service teachers to learn how to think in analyzing and solving a problem. Since problem-solving is a core activity in mathematics, the required thinking habits in problem-solving is consistent with what would be needed for developing conceptual and structural understanding of mathematics knowledge.

Existing research revealed that PSTs can be taught to change their beliefs, thinking, and performance in their problem posing and solving (e.g., Crespo, 2003; Crespo & Sinclair, 2008; Ellerton, 2013; Leavy & Hourigan, 2020). For example, Leavy and Hourigan (2020) provided 3-week instruction to 415 elementary PSTs on problem-solving and posing. They reported that PSTs' pre- and post-instruction responses reflect changes in understanding about problems as resulted from PSTs' experiences in solving and posing diverse types of problems through the instruction, such as ill-structured problems and multistep problems. Moreover, in answering the question of making a math problem for children at a class level (first to fourth grade), these PSTs' performance shows changes in the types of problems being posed. Prior to the instruction, the majority of problems posed were traditional one-step and one-solution arithmetic problems typically seen in textbooks, such as "George had 17 Moshi Monsters. He gave 9 of them to Joanne. How many did he have left?" (p. 353). The finding is consistent with what other researchers reported about both elementary in-service and elementary PSTs' performance (e.g., Carpenter et al., 1988; Crespo, 2003; Crespo & Sinclair, 2008). After the instruction, PSTs posed many different types of problems with most notable changes in a decrease in one-step problems together with the increase of puzzle-type problems and problems with more than one correct solution, such as "The children in 1st Class are making Play-Doh caterpillars. They have four colors to choose from. How many different ways can they make a caterpillar that is four sections long? They can only use each color once" (p. 353). The result is also consistent with findings from other studies (e.g., Crespo, 2003; Crespo & Sinclair, 2008; Ellerton, 2013).

In addition to investigating PSTs' changes in beliefs and performance, Leavy and Hourigan (2020) also included questions that require PSTs to reflect on the problems they posed. PSTs' explanations allow the researchers to gain insights about PSTs' development of understanding. For example, Leavy and Hourigan (2020) indicated that PSTs may pay close attention to the inclusion of problem features as requested (e.g., multiple solution paths, multiple steps) when posing problems, but not taking themselves as problem solvers to view the solvability of problems posed. Thus, some problems were not suitable for the intended students. At the same time, some problems posed may have looked simple, but PSTs' explanations pointed out their richness or acknowledged certain limitations. Such justifications suggest that PSTs did have their own thinking and reasoning behind their work, which reflected the development of their understanding about problem posing in this study.

2.2.3 Elementary PSTs' Knowing About and Beliefs in Mathematics and Mathematics Teaching and Learning

Just as teachers' knowledge of mathematics has long been recognized as important for effective teaching (Ball, 1991; Ball, Thames, & Phelps, 2008), researchers have documented that teachers' beliefs and conceptions of mathematics and mathematics teaching also affects their professional practice (e.g., Cooney, 1988; Philipp, 2007; Rattan et al., 2012; Thompson, 1984, 1992). For example, Thompson (1984) conducted case studies of three junior high school teachers to examine their conceptions of mathematics and mathematics teaching, and the relationship between their conceptions and instructional practices. Data were collected through classroom observation, interviews, and written responses to six tasks on teachers' views about various aspects of mathematical teaching. With these data and cross-sectional analyses, Thompson found that teachers' beliefs, views, and preferences about mathematics and its teaching played a significant, though subtle, role in shaping their instructional behavior.

Rattan et al. (2012) found that college students placed in a teaching role with different perceptions of mathematics teach differently. Specifically, Rattan et al. looked at these student-teachers holding an entity (fixed) theory of mathematics intelligence (T1) versus incremental theory (T2). Through their studies, Rattan and colleagues found that T1 teachers more readily judged students to have low ability, comforted students with perceived low mathematical ability, and used "kind" strategies (e.g., assigning less homework) unlikely to promote their engagement with the subject than T2 teachers. Students who received comfort-oriented feedback sensed their teachers' entity theory and low expectations and reported with lowered motivation and expectations for their own performance. The results suggest how teachers' inadequate perceptions of mathematics and beliefs about the nature of students' mathematical intelligence contributed to low achievement, diminished self-esteem, and to the attitude that mathematics is only a set of static facts and procedures. These results suggest that how mathematics is taught has impact beyond affecting students' proficiency with mathematics content. Sun (2018) made a similar argument after synthesizing existing literature and analyzing classroom observation data.

Along with the recognized importance of teachers' beliefs and conceptions of mathematics and mathematics teaching, it is thus important for us to know about teachers' beliefs and conceptions, and if their beliefs and conceptions can be improved through training. Researchers have found that a vast majority of mathematics teachers from primary to high school levels in the United States emphasized pedagogical practices of "give" and "present" definitions. The results present a picture of instructional practice that has been virtually unchanged for several decades (Banilower et al., 2013; Cooney, 1988), and also reflect teachers' conceptions of mathematics as a collection of "given" and fixed products (Li & Schoenfeld, 2019). As identified and summarized by Cooney (1999), teachers' beliefs and conceptions of mathematics and mathematics teaching are often defined by

1. Perceiving a right/wrong answer orientation with mostly single procedures to arrive at the correct answer.
2. Teaching as telling, with teachers giving clear and concise presentation and students taking good notes and doing the assigned problems.
3. Teachers being fair with students and providing a comfortable classroom with reduced challenge in tests.

Mathematics classroom instruction provides students opportunities of not only learning mathematics but also experiencing how mathematics is taught and learned, over the course of about 1500 h for most high school graduates in the United States (Larson, 2017). Such long-term experiences can foster powerful beliefs and conceptions about mathematics and mathematics teaching that PSTs will bring with them when entering teacher preparation programs. It is thus not surprising that researchers found that making changes in teachers' mathematical conceptions and practices is very difficult (e.g., Gregg, 1995; Heyd-Metzuyanim, Tabach, & Nachlieli, 2016), although not impossible (e.g., Anhalt & Cortez, 2016; Carpenter, Fennema, Franke, Levi, & Empson, 1999; Copur-Gencturk & Papakonstantinou, 2016; Fennema et al., 1996). For example, the Cognitively Guided Instruction (CGI) program (Carpenter et al., 1999; Fennema, Carpenter, & Franke, 1992) provided professional development training to elementary school teachers to learn about children's mathematical thinking as a means for changing teachers' beliefs and instructional practices. Researchers documented such changes possible with the use of CGI for both in-service elementary teachers (Fennema et al., 1996) and elementary PSTs (Vacc & Bright, 1999), but would need to take years time and specific efforts.

In order to make possible changes, the question of which (practice vs. beliefs) comes first to bring expected changes has been a topic of interest in education research (Philipp, 2007). Through reviewing relevant studies, Philipp showed that the relationship between changes in beliefs and changes in instructional practice is complex and could vary from teacher to teacher, as found for example in a longitudinal CGI study by Fennema et al. (1996). At the same time, in a follow-up to an original CGI study (Peterson, Fennema, Carpenter, & Loef, 1989), Knapp and Peterson (1995) phone-interviewed 20 of the original 40 CGI teachers 3–4 years after their initial introduction to CGI. They found that the relationship between these changes was often interactive and that it often happened with beliefs to follow practices. Specifically, these teachers developed initial beliefs in CGI principles during the original workshops, and then deepened over time when their students' performance demonstrated changes in solving complex mathematics problems. The results suggest that in mathematical preparation for PSTs, developing initial beliefs in the value of thinking and sensemaking in mathematics needs to be reinforced and strengthened repeatedly through experiencing its value in mathematics activities and teaching.

2.2.4 Summary

The review above helps outline what we can learn from the literature about PSTs' knowledge in elementary mathematics, their performance in problem-solving and posing, and their beliefs and conceptions of mathematics and mathematics teaching that often play an important role for understanding their instructional practices.

In general, the results from existing studies do not provide us a rosy picture. PSTs in the United States are good at performing numerical computation with memorized formulas or procedures learned through K–12 school education. Their knowledge is rule-based: a collection of isolated concepts and procedures. They are not used to thinking and understanding, and problem-solving (often limited to one-step with one correct solution) is a process of applying what they know to compute and get a quick answer. Their beliefs and conceptions of school mathematics are thus restricted to what they know and can do. Their mathematics teaching would be guided by practice of "give" and "present" different concepts and procedures.

By comparing to PSTs' performance in some other education systems, we learned that there is a preparation gap in mathematics teacher education. The mathematical training for PSTs in the United States can and should be improved, before improving students' mathematics performance in schools can be expected. At the same time, since PSTs in other education systems also experience some difficulties in mathematics, it is also important for us to learn about possible approaches and practices developed and practiced in different education systems.

A positive takeaway from the literature is that improvement is possible, although it may take years of effort and support. A simple repeat of what PSTs already learned in elementary schools when they were students is not a solution. Changes can be made to help PSTs to learn and improve their performance in solving and posing problems, not only those one-step problems with one correct solution but also those complex and open-ended problems. PSTs' beliefs and conceptions of mathematics and mathematics teaching can also be changed through carefully designed and systematic instruction and support over the years. It would be unrealistic to expect many PSTs to figure out by themselves what is important in mathematics and to realize that what they know is not enough. It is the job of mathematics teacher educators to reflect on what is offered in each course of their preparation program and possible connections across these courses, what changes may be needed, and how to make those needed changes. To facilitate such a process, it is important for us to learn from the literature what common approaches are used. This is the purpose of the remaining discussion.

2.3 Common Approaches Developed and Used in Providing Mathematical Training for Elementary Teachers

The need of improving mathematical training for elementary teachers, especially in the United States, has long been recognized (e.g., AMTE, 2017; CBMS, 2001, 2012). There are ongoing efforts to make changes and improvements in teacher preparation or professional development programs (Han & Li, 2015; Li, Silver, & Li, 2014; Wu, 2011a). As discussed by Han and Li (2015), there have been ongoing movements in changing from the typical professional development in the USA, with a patchwork of courses, due to its questionable effectiveness (Darling-Hammond, Wei, Andree, Richardson, & Orphanos, 2009; Wilson & Berne, 1999), to innovative approaches by closely connecting teachers' learning with their classroom practice (Ball & Cohen, 1999; Borko, 2004; Franke, Kazemi, Shih, Biagetti, & Battey, 2005; Garet, Porter, Desimone, Birman, & Yoon, 2001). Some notable approaches include studying children's mathematical understanding, attending summer institutes to boost mathematics knowledge, engaging in lesson study activities, creating school-based learning opportunities, and using online technology to improve teachers' knowledge for teaching mathematics.

The question about what teachers need to know has been a topic of interest to many educators, researchers, and professional organizations for the past several decades (Mewborn, 2001). In general, different approaches for improving mathematical education for teachers can be classified into two categories: one with a focus on knowledge improvement, and the other with an emphasis on the development of thinking and problem-solving. Although many approaches in these two categories cannot simply be separated in absolute terms, the classification reflects different emphases. In the following sections, we discuss further these two categories of approaches.

2.3.1 Knowledge-Oriented Approaches

It is commonly acknowledged that knowing mathematics is the foundation for effective mathematics teaching. In a way, this conception is self-explanatory, as we can question how one can teach mathematics effectively without knowing the subject. Knowing mathematics is thus a "must" condition for effective teaching of mathematics. Placing an emphasis on PSTs' learning of mathematics content is common in the United States (AMTE, 2017; CBMS, 2001, 2012; Mewborn, 2001; NCTM, 1991) and other educational systems (Li, Ma, & Pang, 2008; Li, Zhao, et al., 2008). For example, Mewborn (2001) summarized efforts in the United States on emphasizing mathematics knowledge preparation for elementary PSTs. Many policy documents were developed with an assumption that teachers' content knowledge will have a significant impact on students' learning. She also cited a report by the American Council on Education (ACE) as an example, "A thorough grounding in college-level subject matter and professional competence in professional practice are necessary for good teaching. The data are unequivocal: students learn more mathematics when their teachers report having taken more mathematics" (ACE, 1999, p. 6). Likewise, we can easily find from many documents and reports created by other professional and national organizations to emphasize and specify different mathematics content that PSTs would need to learn (e.g., CBMS, 2001, 2012; National Commission on Teaching & America's Future, 1996; NCTM, 1991). At the state level, mathematics exams for teacher certification also place clear emphasis on the acquisition of specific content topics and skills.

2.3.1.1 Considerations About the Nature of Mathematics That PSTs Need to Learn

Considerations about what mathematics would be important to PSTs in fact have been ongoing for many years. For example, Begle (1972, 1979) conducted studies to investigate the relationship between teachers' content learning and their students' mathematics achievement. But these studies did not find any statistically significant correlations between them, a result that was discouraging and counterintuitive. Although these studies had certain limitations, they were carefully planned and the result had been supported by a growing body of evidence (e.g., Ball, 1990a; Eisenberg, 1977). At the same time, it should be pointed out that teachers' content learning in Begle's studies includes general measures such as number of years of experience, major in mathematics, and college mathematics grade point average (GPA), in addition to mathematics tests. Further investigations beyond such general measures had been called for, but little progress had been made in the 1960s, 1970s, and 1980s (Mewborn, 2001).

Efforts to outline mathematics knowledge needed for teaching have become a big topic area in mathematics teacher education, with some well-known work by Shulman (1986), Ball and her colleagues (2008), to specify the structure of mathematics knowledge for teaching (MKT). Consistent with expertise research in psychology that highlights the importance of knowledge in expertise acquisition and development (Li & Kaiser, 2011), the importance of knowledge in teachers' expertise also goes beyond a simply count of courses and GPA measures to include knowledge structure with certain depth. Such a knowledge-based characterization of teachers' expertise has been used in both mathematics education studies (e.g., Hill, 2007; Hill, Rowan, & Ball, 2005) and large-scale international comparative studies, e.g., the Teacher Education and Development Study in Mathematics (TEDS-M, see Blömeke et al., 2014; Döhrmann, Kaiser, & Blömeke, 2018). In particular, Hill et al. (2005) revealed that gains in students' mathematics achievement is significantly related to teachers' mathematical knowledge when measured in terms of the specialized mathematical knowledge and skills used in teaching mathematics other than some general measures such as years of experiences, average time spent in mathematics instruction, and certification status. However, designing courses that reliably develop MKT is still a question.

Emphasizing PSTs' learning of mathematics knowledge with a close connection with what they will teach in the future has been clearly specified in many professional documents and reports over the past decade. For example, the Conference Board of Mathematical Sciences released an updated report about mathematical education for teachers (CBMS, 2012). The report provided specific guidance for what elementary PSTs would need to learn:

> Programs designed to prepare elementary teachers should include 12 semester hours focused on a careful study of mathematics associated with the CCSS (K–5 and related aspects of 6–8 domains) from a teacher's perspective. …… Number and operations, treated algebraically with attention to properties of operations, should occupy about 6 of those hours, with the remaining 6 hours devoted to additional ideas of algebra (e.g., expressions, equations, sequences, proportional relationships, and linear relationships), and to measurement and data, and to geometry. (CBMS, 2012, p. 31)

The report also indicated that studies of college-level mathematics should not be used to substitute the study of elementary mathematics from a teaching perspective, although it might be a possible addition. The emphasis on PSTs' learning of mathematics that they will teach is very much consistent with what educators have emphasized in some other education systems (e.g., Li, 2002; Li, Ma, & Pang, 2008).

Clarifying the nature of mathematics that PSTs need to learn does not specify how such knowledge should be organized and offered in different courses. In fact, it is often the case that PSTs are left to make possible connections by themselves after learning separate pieces of knowledge. In many teacher preparation programs, mathematics content courses are typically offered by a mathematics department, and pedagogy courses are delivered in an education department. These two departments may communicate little, if at all, about the content and instruction of these courses for the same group of PSTs. Checking whether PSTs are ready or not for their professional career in teaching often results in course counting rather than examining what is offered in these courses and how the various topics in them can and should be connected. There is much work remaining to improve PSTs' training in mathematics.

Mathematics educators at Michigan State University became aware of such issues many years ago. They designed a sequence of three mathematics courses, a methods course, and a curriculum seminar for elementary education majors. The sequence of these courses aimed to reorganize the mathematics curriculum around concept development and problem-solving. Schram, Wilcox, Lanier, and Lappan (1988) reported some success in helping preservice teachers expand their conceptual understanding through a mathematics content course, specifically a change in PSTs' conception of what mathematics is and a change in their notion of what it means to know mathematics and in their knowledge of how mathematics is learned. However, when they followed three teachers into their first year of teaching, they found that some of them struggled to replicate their own learning experiences in their classrooms. Others did not appear to attempt to replicate the type of learning environment they had experienced (Wilcox, Lanier, Schram, & Lappan, 1992). The researchers were led to conclude the following:

> Disciplinary study is necessary to develop in novice teachers a set of intellectual tools and a disposition to engage in mathematical inquiry themselves. But disciplinary study alone may be insufficient to overcome preservice teachers' deeply held beliefs about young children, what they can and should learn in the elementary mathematics classroom, and how they might learn that which is of most worth. Modeling new practices and nontraditional conceptions of mathematical pedagogy in the study of content may be insufficient to develop in beginning teachers the knowledge, skills, and beliefs to conceive of teaching as something other than telling or as more than a matter of technical competence. (Wilcox et al., 1992, p. 25)

These findings suggest the complexity of providing adequate training to elementary PSTs, in terms of not only PSTs' acquisition of mathematics knowledge at some levels by themselves but also how to teach mathematics conceptually to their students.

To build connections among different types of knowledge needed in mathematics teaching, Li and his collaborators put forward the conception of mathematics conceptual knowledge for teaching (MCKT) (Li & Howe, 2017; Li, Pang, et al., 2019). MCKT means the conceptual knowledge needed for understanding, explaining, as well as teaching specific mathematics content topics with connections. It can be specified as containing the following three knowledge components that can and should be offered in the same courses:

1. Knowing and being able to explain the meaning of a specific content topic.
2. Being able to connect and justify the main points of a content topic, and to place it in wider contexts.
3. Knowing and being able to use various representations for teaching the content topic, and being able to teach the relations between them.

Based on this conception, recent studies of sampled elementary PSTs from South Korea and China helped reveal that the structure of MCKT with three knowledge components provides an important and feasible lens for us to know the strength and weakness of teachers' knowledge (Li, Pang, et al., 2019). For example, for the case of Mainland China, the results suggest that prospective teachers likely gain much more on mathematics, somewhat less on mathematical pedagogy, but very limited on connections of mathematical ideas through their program studies (Li, Zhang, & Song, 2019). The findings lend further support to the need of developing elementary PSTs' mathematics knowledge beyond the traditional count of school mathematics.

2.3.1.2 Efforts to Improve PSTs' Learning of Mathematics in Other Education Systems

An emphasis on mathematical training is also evident in many other education systems, especially those high-achieving education systems in East Asia. For example, Li, Ma, and Pang (2008) studied the elementary teacher preparation programs in six selected education systems (i.e., Hong Kong, Japan, Mainland China, Singapore, South Korea, and Taiwan) in East Asia. In general, mathematics is seen as an important subject in elementary teacher education in all these six education systems. Specific survey results obtained from sampled PSTs in Mainland China and South Korea supported the perception that elementary PSTs in East Asia have a strong preparation in mathematics content knowledge. However, it seems that teacher educators sampled in China and South Korea still seek to provide even more training in mathematics, especially in school mathematics, to elementary PSTs.

Specifically, in the case of mainland China (Li, Zhao, et al., 2008), there are great variations in content course requirements across three different program structures in elementary teacher preparation (i.e., integrated preparation program, focus-area specified program, and middle-ground type of preparation program), with the integrated preparation type having the least amount of mathematics requirements and the focus-area specified type having the most. One sample program of the integrated preparation type requires about 12 credit hours in mathematics-related courses. One sample program of the focus-area specified type for those to focus on mathematics would require as many as 44 credit hours (a total of nine courses) in mathematics, plus a minimum of 13 elective credit hours (out of 22) in mathematics. Based on the availability of different courses in mathematics, elementary PSTs in the focus-area specified type are typically expected to study mathematics systematically and in depth. In reference to CBMS's recommendation of having 12 credit hours in mathematical training for elementary PSTs (CBMS, 2012), it is clear that all teacher preparation programs in China provide an adequate amount of credit hours in mathematical training for elementary PSTs.

Taking South Korea as another example, although South Korea has a different system for elementary teacher preparation as generalists, elementary PSTs are required to take the same courses with only the exception of their focus areas. With regard to requirements common in mathematics, elementary teacher education programs consist of the liberal arts courses and the major courses. Among the liberal art courses, prospective teachers take a compulsory course dealing with the foundations or basics of mathematics. Among the major courses, prospective teachers take two to four compulsory courses in mathematics. In the study (Li, Ma, & Pang, 2008), most of the teacher educators surveyed in South Korea thought that all PSTs should have more opportunities to learn mathematics content and method, partly because mathematics is one of the most important subjects in elementary schools, and it is taught more hours per week than other subjects such as social studies or music. Some teacher educators worried about PSTs' lack of profound understanding of basic but important mathematical concepts. Others thought that PSTs would have confidence in elementary mathematics content, but not in collegiate mathematics (i.e., the compulsory course in the category of liberal arts courses). But in fact, the survey of their PSTs' showed strong performance in mathematics. At the same time, one interesting aspect is that many teacher educators underlined the strengths of current teacher preparation in that elementary PSTs learn how to teach every subject. In particular, the comprehensive programs help PSTs make good connections among multiple subjects for developing an interdisciplinary approach, and have good understanding of children and general pedagogical knowledge.

2.3.1.3 Can an Elementary Mathematics Specialist Program Be the Solution?

The lack of sufficient mathematics training for elementary teachers in the United States led to calls of establishing the elementary mathematics specialist program (e.g., Fennell, 2006). Li (2008) argued to the contrary that the preparation of mathematics specialists may not be the solution for the long-standing concern about US teachers' knowledge in mathematics and their performance in mathematics teaching as documented in many studies (e.g., Li, 2007; Ma, 1999; Stigler & Hiebert, 1999).

First of all, elementary school teachers as generalists are not necessarily poor in knowing and understanding school mathematics. In fact, Li, Ma, and Pang (2008) indicated that mainland China is the only one out of six education systems in East Asia (i.e., Hong Kong, Japan, Mainland China, Singapore, South Korea, and Taiwan) that have elementary school teachers concentrated on teaching a main school content subject, such as mathematics. Therefore, the practice of having elementary teachers as generalists is common and does not necessarily damage teachers' knowledge and teaching in mathematics.

Moreover, as discussed above, the results from examining some US PSTs' performance in mathematics and their self-confidence suggested that although PSTs did not have high performance in mathematics except for numerical computations, they tended to have high confidence in their mathematics preparation (Li & Kulm, 2008). In fact, many PSTs are not clear that they do not know what they should know, when in fact cross-national studies also revealed a preparation gap between the United States and other high-achieving education systems such as Chinese Taipei and South Korea (Schmidt et al., 2007). Thus, it is necessary for US teacher educators to identify what specific knowledge in mathematics and pedagogy PSTs need to learn, and how their learning can be facilitated with a sequence of well-designed courses. These considerations and efforts are likely to be more important than simply changing elementary teacher preparation program from preparing generalists to content specialists.

2.3.2 Thinking-Oriented Approaches

An emphasis placed on the acquisition of mathematics knowledge for elementary PSTs will not tell the whole story. As discussed above, a shift has taken place to change from the acquisition of college mathematics to school mathematics that PSTs will teach in the future. Moreover, the learning of school mathematics in teacher preparation programs is not simply to repeat what PSTs learned in elementary schools before, but focuses on conceptual understanding of school mathematics in a way consistent with the notion of "elementary mathematics from an advanced point" used by Klein (1932) in his work with high school teachers. Moreover, changes in PSTs' beliefs and conceptions of mathematics and mathematics teaching are important and necessary.

The emphasis placed on conceptual understanding and conception changes requires the development of mathematical thinking. As noted through our review above, PSTs have long been accustomed to memorize and apply procedures and algorithms, but struggled with learning how to think and reason mathematically. Possible changes in having habits of thinking mind and beliefs would also take long time and persistent effort. Thus, approaches to take thinking development as a focus would be different from those approaches that focus on the coverage of mathematics knowledge and different knowledge types. In the following sections, we review and discuss several approaches in the literature with a focus on PSTs' thinking development.

2.3.2.1 Approaches Focusing on Mathematical Thinking

The importance of mathematical thinking has been emphasized in many professional writings and documents produced by professional organizations (e.g., CBMS, 2001, 2012; NCTM, 2000). In CBMS's report (2012), it highlights the importance for PSTs to be comfortable with mathematical practices, and to acquire them through carefully designed experiences of doing mathematics. It specifically recommends to provide time in coursework for PSTs to engage in reasoning, explaining, and making sense of the mathematics they will teach in the future. It further highlights the needs of developing PSTs' habits of mind as follows:

> Recommendation 4. All courses and professional development experiences for mathematics teachers should develop the habits of mind of a mathematical thinker and problem-solver, such as reasoning and explaining, modeling, seeing structure, and generalizing. Courses should also use the flexible, interactive styles of teaching that will enable teachers to develop these habits of mind in their students. (CBMS, 2012, p. 19)

It should be pointed out that an emphasis on the development of mathematical thinking does not mean to abandon the learning of mathematical knowledge, but to shift from the acquisition of knowledge as fixed or given products to sensemaking knowledge and learning how to approach various problems as making mathematical sense (Li & Schoenfeld, 2019). In particular, research concerning PSTs' mathematical thinking in specific content areas would inform instruction. As stated in *The Mathematical Education of Teachers*, "The key to turning even poorly prepared prospective elementary teachers into mathematical thinkers is to work from what they do know." (CBMS, 2001, p. 17)

As society keeps changing, it is difficult to predict what knowledge would be essential for the workforce in the future and thus for school education. A simple focus on knowledge acquisition will not be adequate for students, and so also for PSTs. But helping students and PSTs to develop mathematical ways of thinking would continue to be important for them as lifetime learners. It is in this perspective that Cuoco, Goldenberg, and Mark (1996) discussed extensively the importance of developing students' habits of mind as an organizing principle for high school mathematics curricula. Through such mathematics curricula, Cuoco et al. envisioned that students should develop a repertoire of useful habits through mathematics learning as listed below and understand when to use what: (1) as pattern sniffers, (2) as experimenters, (3) as describers, (4) as tinkerers, (5) as inventors, (6) as visualizers, (7) as conjecturers, and (8) as guessers.

As a famous mathematician and mathematics educator, Pólya (1959) advocated the importance of plausible reasoning with mathematics in the making. Going beyond mathematics as a finished science heavily equipped with definitions and demonstrative reasoning, he argued about the need for students to experience the process of mathematical practices through asking questions and developing reasonable and plausible arguments in problem-solving. Although he provided specific examples that are pertinent to high school students, the ideas are well feasible for students at all K–12 grade levels.

Likewise, Jacobbe and Millman (2009) extended the argument for the importance of developing elementary PSTs' habits of mind in connection with Pólya's problem-solving principles. They indicated that limited research attention has been placed on the development of mathematical thinking in K–12 mathematics education. To them, mathematical habits of mind can be developed through different practices in and through problem-solving activities, such as (1) exploring mathematical ideas, (2) formulating questions, (3) constructing examples, (4) identifying and generalizing problem-solving approaches useful to many other problems, and (5) checking and reflecting on problem solution as the "looking back" step. The development of mathematical habits of mind does not require a deep mathematical background as a pre-condition, but an inquiring mind to think in a mathematically creative way. And the learning and use of Pólya's problem-solving principles can help provide a technique for PSTs when thinking how to approach and solve mathematically rich problems.

2.3.2.2 Approaches Focusing on Students' Mathematical Thinking

Advocating the development of mathematical thinking as discussed above is mainly derived from views about mathematics itself as a scientific discipline. However, mathematics teaching and learning is not concerned about mathematics discipline only, but involves other aspects. One distinguished and important aspect is about students as learners. Without considering students, it is difficult if not impossible to make fundamental and effective changes in teaching practice in mathematics. For example, Sztajn (2003) studied two teachers who held similar beliefs about mathematics. But they taught in very different contexts and differed in their teaching practices. She found that attending to the teachers' beliefs about mathematics itself was insufficient for explaining the teachers' practices. In order to account for the differences in the teachers' practices, she needed to consider the teachers' beliefs about children, society, and education.

The findings from Sztajn's study (2003) remind us that an emphasis on the development of mathematical thinking should not be considered in a vacuum but needs to have direct connections with the learners. In fact, the success of the CGI program presented a compelling case for the need of knowing and building upon what students know.

The CGI program was started as a research and professional development program that helps primary teachers to learn and use research-based knowledge about children's thinking to guide their classroom instruction (Fennema et al., 1992). The program is based on the cognitive research that children enter school with well-developed informal or intuitive systems of mathematical knowledge that can be used as a basis for developing their understanding of mathematical concepts and procedures. Teachers can use CGI to design and carry out classroom instruction based on their own teaching styles, knowledge bases, beliefs, and the needs of their students. Instead of asking children to master isolated and abstract mathematical concepts and procedures, students typically spent most of their time solving problems. These problems are often well-designed word problems. For example, the following problems can be solved by kindergarteners or first graders through modeling or the use of concrete strategies (Fennema et al., 1992, p. 6).

1. Amy had 12 marbles. She gave 8 of them to John. How many marbles did Amy have left?
2. Amy had 8 marbles. John gave her some more and then she had 12. How many marbles did John give her?
3. Amy had some marbles. John gave her 8 more, then she had 12 marbles. How many marbles did Amy have before John gave her any?

As Fennema et al. (1992) indicated, children can come to recognize that it is not necessary to make a set of 8 objects to solve the second problem. They can find the answer just by counting from 8 to 12 and keeping track of the number of counts. Fennema et al. further argued that knowledge of place value and computational algorithms can also be developed through problem-solving. For example, children can learn to use symbols through inventing procedures for solving two- and three-digit problems without counters (p. 7)

Mary had 27 stickers. She bought 34 more stickers. How many stickers did she have then?

Julie: Mmmm 27, 37, 47, 57. Now I need 4 more. Well. 58, 59, 60, 61. (Julie mentally separated 34 into 30 and 4, counted 30 by 10s, and then counted 4 more.)

Jason: Well 20 and 30, that's 50; and the 4 and 7, that's 11. So it's 61. (Jason combined the 10s to get 50, combined the ones to get 11, and then combined the two sums.)

Many studies have been conducted over the years to explore the effectiveness of the CGI program. There are some interesting longitudinal studies that documented its success with primary teachers in changing their beliefs and instructional practices (e.g., Fennema et al., 1996; Fennema, Franke, Carpenter, & Carey, 1993). Efforts to help elementary PSTs' learning and use of children's mathematical thinking present aspects with varied success (e.g., Vacc & Bright, 1999).

Vacc and Bright (1999) conducted a study with 34 elementary PSTs with the introduction of CGI as part of a mathematics methods course in the first semester of the senior year. PSTs were introduced problem types for the basic operations and children's solution strategies (Carpenter, Fennema, & Franke, 1993) in the first four sessions and knowledge of children's geometrical thinking in the fifth session. For the first four sessions, PSTs were typically asked to solve a mathematical story problem and share different solution strategies, and then watched videotaped examples of children's solutions to the same problem and discussed how children's solutions may differ from their own. Further discussions were also followed about possible problems to give to children next. PSTs then were asked to use knowledge of children's thinking to plan instruction. Vacc and Bright used the CGI Belief Scale (Peterson et al., 1989) to assess possible changes in these PSTs' beliefs about teaching and learning of mathematics. They found significant changes in PSTs' beliefs about mathematics instruction occurred across the 2-year (i.e., junior and senior years) sequence of professional course work and student teaching during their undergraduate program. However, by following two selected PSTs as cases for in-depth study in their senior-year field experiences, they reported that PSTs' use of knowledge of children's mathematical thinking in instructional planning and teaching was limited. Specifically, PSTs may acknowledge the tenets of CGI and yet be unable to use them in their teaching. Vacc and Bright hypothesized several reasons that might contribute to this situation:

1. Unlike in-service teachers in previous CGI studies, it may take more time and efforts for PSTs to gain more experiences in mathematics instruction before they would be able to develop instructional strategies based on the CGI tenets.
2. PST's own levels of mathematical understanding may have impact on their use of CGI principles in mathematics instruction.
3. PSTs may need to have consistent support for learning and using the CGI principles in mathematics instruction in both teacher preparation program and field experience.

2.3.2.3 Approaches Focusing on Pedagogy in Contextualizing Mathematics

The CGI program discussed above shows not only the importance of learning and using the knowledge of children's mathematical thinking but also the expectation for teachers to develop the capacity of developing their own instruction based on such knowledge. It is not surprising to learn about possible challenges that teachers may face, or that long-term efforts would be needed to bring possible changes in teachers' beliefs and instructional practices.

There are other approaches in mathematics education that focus directly on curriculum and instruction for developing students' mathematical thinking. And teachers are expected to learn and implement those approaches in their classrooms. For example, curriculum reforms in many education systems over the years are sample approaches in this category. Specifically, Common Core State Standards (CCSS) (CCSSI, 2010) have been developed and adopted by many states in the United States. CCSS as developed by a group of professionals provides a guidance for schools and teachers to use for making changes in mathematics curriculum and instruction, including guidance on developing student thinking, especially through "mathematical practice standards."

Another well-known example, Realistic Mathematics Education (RME, Van den Heuvel-Panhuizen & Drijvers, 2014) initiated in the Netherlands in 1968 bears important features of developing students' mathematical thinking through instruction. RME was started as an approach to respond to the dominant approach of mechanistic teaching formal mathematics, instead of adopting "New Mathematics" in the 1960s in the Netherlands. As one leading scholar for this approach, Freudenthal (1968) stated: "What humans have to learn is not mathematics as a closed system, but rather as an activity, the process of mathematizing reality and if possible even that of mathematizing mathematics" (p. 7). RME promotes mathematics teaching as relevant to students and values that students are being offered opportunities for guided re-invention of mathematics to move from informal to formal representations. It takes doing mathematics as a primary way of learning mathematics, and one essential aspect is to provide opportunities for students to solve everyday life problems ("real-world" situations) and to make sense of mathematics that they need to learn.

RME has been adopted and used in many educational systems over the years, and it continues being implemented in the Netherlands (Van den Heuvel-Panhuizen, 2020). Its effectiveness in developing students' problem-solving and beliefs has been documented (e.g., Van den Heuvel-Panhuizen, 2020; Yuanita, Zulnaidi, & Zakaria, 2018). At the same time, however, there have been concerns about its effectiveness of developing students' computational competence (Colignatus, 2018) and the lack of mathematical rigor (see Van den Heuvel-Panhuizen, 2010).

2.4 Toward a Thinking-Oriented Training in Mathematics for Elementary PSTs

Our review of these two categories of approaches (i.e., knowledge-oriented approaches and thinking-oriented approaches) leads us to several observations.

1. Knowledge-oriented approaches do not entail simply the learning of school mathematics as students learned in elementary schools; they include also different types of knowledge that would be needed and important for teachers to teach elementary school mathematics. The knowledge acquisition and capacity development should be applied to all elementary PSTs who will teach mathematics in classrooms, not just for a small group of mathematics content specialists. The learning of advanced college mathematics might be nice to have, but should not be thought of as replacement.

2. Thinking-oriented approaches are not restricted to the development of mathematical thinking from a disciplinary perspective, but can and should include the importance of learning how to build on what students know and think.
3. Designing and providing specific curriculum and instructional principles for teachers can be productive, but their effectiveness would depend on the development of teachers' mathematical proficiency, their skills in assessing their own students' knowledge and mathematical thinking, and what is valued in a specific education system.

At the same time, however, each of the approaches discussed above tends to emphasize only one aspect of teachers' mathematical proficiency. Since each of these approaches has its own strength, thinking about possible integration may allow us to combine the strengths of different approaches.

On the basis of our literature review in Sect. 2.2, we would propose a thinking-oriented training in mathematics for elementary PSTs. From the literature discussed in Sect. 2.3, we suggest this perspective should encompass the following three features.

2.4.1 Connectedness of Knowledge Learning

Knowledge is important and should provide a foundation for thinking mathematically. Instead of acquiring separated pieces of knowledge, however, PSTs need to organize their mathematics conceptual understanding for teaching (MCKT) into topic-based knowledge packages that connect their component parts to make a coherent whole. At the same time, PSTs' development of mathematical thinking can and should be based on the development of conceptual understanding of mathematical knowledge and connections, reasoning, communicating, explanation, and justification. For example, elementary PSTs need to learn and develop confidence with the structure of our pervasive base ten system of arithmetic, in terms of "place value parts" (Beckmann, 2018, p. 6). The form in which we write numbers employs multiple conventions to encode substantial algebraic structure, which lies behind the power of the system. In previously published papers (Howe, 2015; Howe & Epp, 2008), five stages in the interpretation of place value notation are distinguished:

$$\begin{aligned}356 &= 300 + 50 + 6\\ &= 3 \times 100 + 5 \times 10 + 6 \times 1\\ &= 3 \times (10 \times 10) + 5 \times 10 + 6 \times 1\\ &= 3 \times 10^2 + 5 \times 10^1 + 6 \times 10^0\end{aligned}$$

The first equation shows that a base ten number is a sum of pieces of a special kind, as displayed in the second expression, which is often called expanded form. The expanded form is shown to students, perhaps in second grade, but it does not seem to receive much emphasis.

Instead of letting PSTs to learn the five stages as separate pieces of knowledge, it is important for PSTs to learn and understand their connections and to fully understand place value notation as involving knowledge development over the full 8 years of elementary and middle school. It would take a remarkably coherent curriculum to ensure that students master this structure over the years, and in fact, studies (e.g., Thanheiser, 2009) have revealed that only a small minority of PSTs at a prominent institution of teacher preparation even think in terms of the third stage. Moreover, it is important for PSTs to learn and understand why and how the place value structure can be used to explain the main algorithms for computing with base ten numbers, both in detail, and from a global viewpoint. For instance, we can expect that PSTs will understand why subtraction is harder than addition, and will be able to help their students by examining key examples.

Given the limited time in teacher preparation program, PSTs' learning of mathematics knowledge must by necessity focus on few key topics selected for depth and the promotion of PSTs' thinking habit rather than attempting to cover every single topic in elementary school mathematics. As discussed by Li and Schoenfeld (2019), a shift should take place from learning mathematical knowledge as *fixed products* to thinking about *ideas* in mathematics. Both sensemaking and making sense are important to have for PSTs to experience in and through knowledge learning. Li and Schoenfeld (2019) shared the learning and understanding of long division as an example. Specifically, Brownell (1945) discussed how arithmetic can and should be taught and learned not only as procedures but also as a meaningful system of thinking. He shared many examples like the following division,

$$8\overline{)576}$$

Brownell suggested to ask questions: what does the 5 of 576 mean? Why must 57 be the first partial dividend? Do you actually divide 8 into 57, or into 57 . . . 's? etc., instead of simply letting students memorize how to carry out the procedure. What Brownell advocated has been commonly accepted and emphasized in mathematics education nowadays as sensemaking (e.g., Schoenfeld, 1992).

One can certainly ask different questions for sensemaking of the computation. For example, one can ask first, how many 8s can be in 576? Can there be 100? No, because $8 \times 100 = 800$, which is more than 576. So, the quotient is less than 100, so at most a two-digit number. Can there be 10? $8 \times 10 = 80$, which is less than 576, so there can be several 80s in 576. How many? And continue like this, next justifying that the 6 is irrelevant for determining the number of 80s in 576.

PSTs can take a step further to think about why the above division procedure is performed from left to right (i.e., high-value to low-value places), which is different from the other operations (addition, subtraction, and multiplication) that are all performed from right to left (i.e., low-value to high-value places). PSTs can be encouraged to explore if the division can also be performed from right to left (i.e., starting from one's place). One can ask first, can 6 be divided by 8? No, because 6 is smaller than 8, 6 is not enough. So, can 76 be divided by 8? Yes, how many 8s can be in 76? 10? $8 \times 10 = 80$ which is more than 76. 9? $8 \times 9 = 72$, and there are $76 - 72 = 4$ left (and recording 9 at side). Now, you need to go a place value up further, and do the division of 500 by 8. But you need to get the 4 left from the division procedure and add it to 500 first, and then do the division of 504 by 8. You then need to figure out how many 8s can be in 504? Now, once you get to this step, it is very similar to what we just did above for dividing 576 by 8 directly as starting from hundreds place. In other words, the division process starting from the low-value place would require repeated processes of going from a high-value place to the low-value place; as a result, it is inefficient. PSTs can eventually discover that it can be done from right to left but not efficiently. Going through such a process would allow PSTs to experience the process of making mathematics make sense (McCallum, 2018), a process going beyond sensemaking of the division procedure as existing.

2.4.2 Mathematical Practices Dominated with Problem-Solving and Posing

Mathematical thinking is a broad concept that includes problem-solving and posing, as well as some other aspects such as modeling and abstraction (Drijvers, Kodde-Buitenhuis, & Doorman, 2019). As discussed by Schoenfeld (2020), descriptions about mathematical thinking present itself a long history in mathematics, starting from Plato's Meno dialog to work in the twentieth century such as Pólya's problem-solving (Pólya, 1945, 1959). However, it has been argued that most of what has been done in school mathematics education is about content rather than about practices (Li & Schoenfeld, 2019; Schoenfeld, 2020). We believe that it is not possible, or at least, not productive for learning mathematics, to separate content from practice. Much more work and attention need to be given to mathematical practices that can provide ample opportunities for students' development of mathematical thinking.

Although mathematical practices can take different forms such as mathematical modeling, problem-solving and posing are fundamental activities of mathematical practices not only for knowledge application but also for knowledge development. It is thus important for PSTs to be comfortable by themselves with problem-solving and posing, and also learn how to develop and use problem-solving and posing as mathematical practices for their students. Toward this end, we discuss more about solving and posing word problems below as an example.

As students, PSTs need to receive systematic training for developing their capacity and skills of working with word problems. Word problems are a notably weak point in US mathematics instruction. We posit that weakness in dealing with word problems contributes to overall difficulties with mathematics because word problems are the glue that connects mathematics to the real world. When word problems are not seen as sensible opportunities for mathematical practices, the mathematics itself tends to be regarded as a formal exercise. We need to provide training to PSTs so that they can work with their classes to understand how to make sense out of word problems, and to develop in their students the habit of doing so. The thinking involved in this process will be useful in dealing with much more than mathematics.

One important instrument for helping PSTs become comfortable with word problems is the taxonomies of one-step word problems in the appendix to the CCSSM (CCSSI, 2010, pp. 88–89). There is one taxonomy for addition/subtraction problems and one for multiplication/division problems. They give, in all, 14 types of addition/subtraction word problems, and 9 types of multiplication/division problems. These are far from arbitrary lists. They contain a lot of internal structure that places each type of problem in a sensible hierarchy; but on the other hand, they do serve to emphasize the variety of word problems that students need to deal with, even at the most basic level. A valuable corollary is that these lists serve as a reminder for teachers to make sure that their students are indeed exposed to a good sampling of each problem type, to ensure that they develop a robust sense of the meaning and applications of the operations. Moreover, study of the taxonomy makes clear that common rules of thumb, such as the "key word" guide to solving such problems, is inadequate. In fact, it is exactly the problem types where the key word rule fails (aka, the types with "inconsistent language") that highlight the inverse relationship between addition and subtraction, and these types should be used as occasions to help students understand and master this relationship.

The value of the taxonomies, furthermore, extends well beyond the arena of one-step problems. One-step problems can be combined to create two-step and three-step problems. If one posits that any one of the 23 types (14 types of addition/subtraction and 9 types of multiplication/division) can be combined with others at any step, there would be $(23)^3$, which is over 12,000, types of three-step arithmetic word problems involving the four operations. Even if some triples of types cannot be realized in actual problems, this number makes evident the size and complexity of the total corpus of arithmetic word problems. Learning the taxonomies, followed by studying examples of how the one-step problems combine to make multistep problems, and practice in analyzing the structure of multistep problems, can give PSTs tools for dealing effectively with this large literature, and confidence to help their students learn to analyze and solve the many possible problems they encounter. Further, it can give PSTs the ability to construct their own culturally responsive problems, to increase the engagement of their students in learning mathematics. As discussed in previous sections, some instructional programs, such as Cognitively Guided Instruction (e.g., Carpenter et al., 1997, 1998), have focused on similar topics and illustrated the great value of developing in-depth conceptual understanding of these topics.

Beyond arithmetic word problems lie algebra word problems. We can distinguish between the two by the nature of the solution process. If a word problem can be decomposed into a linear sequence of one-step problems, we consider it an arithmetic problem, but if steps must be combined in more complicated ways, we consider it an algebra problem. An example of a problem that may at first reading seem to be a two-step addition and subtraction problem, but is actually subtler is:

There are 36 chocolate bars in Ms. Smith's candy box. There are 8 more almond chocolate bars than dark chocolate bars. How many dark chocolate bars are in the class?

This problem involves a part-part-whole situation: (number of almond chocolate bars) plus (number of dark chocolate bars) equals (number of chocolate bars). It also involves a comparison, more situation: (number of almond chocolate bars) is eight more chocolate bars than (number of dark chocolate bars). However, neither the number of almond chocolate bars nor the number of dark chocolate bars is given; both must be determined. There is no first step and no second step. The two equations that express the given relationships both involve both unknowns. A problem that seems to be made of two addition/subtraction situations actually requires division for its solution.

In the United States, the typical reaction to this problem is to introduce variables, perhaps one or perhaps two, and to derive an equation. For example, if D stands for the number of dark chocolate bars in the candy box, the information in the problem can be summarized in the equation: $D + (D + 8) = 36$, which can then be solved by the standard methods of elementary algebra to find $D = 14$. However, one of the steps in the solution is dividing by 2. This step involving an operation that did not seem to be involved in the single-step situations that make up the problem is what invites the use of algebra for the solution.

However, the problem can also be approached in an "arithmetic" spirit, without introducing variables. We can say, if there are eight more almond chocolate bars than dark chocolate bars, let us imagine that one day, eight of the almond chocolate bars are absent. Then there will be only $28 = 36 - 8$ chocolate bars in the box, and there will be equal numbers of almond chocolate bars and dark chocolate bars, so each will be half of 28, or 14. Since all the dark chocolate bars are present, this is the number of dark chocolate bars in the box.

It is important for PSTs to experience and understand different ways of approaching similar problems, and in particular, to be able to use arithmetic approaches to solve a large variety of problems that, in the United States, would normally be treated with algebra. It will help PSTs become more capable of helping their students make sense of and solving this sort of problems. Combined with the increased capacity for dealing with variables as described above, PSTs can also be well equipped to prepare their students for algebra, which is currently one of great topics of debate in US mathematics education.

Consistent with an emphasis on mathematical practices, it is important for PSTs to have chances to think like teachers in practice. They need to develop a sense of knowing and building on children's mathematical thinking. For example, integrating problem-solving with knowledge learning for their students as discussed by Fennema et al. (1992) in their CGI study can help PSTs become sensible about various opportunities for learning and thinking development. And learning about common mistakes students make in solving word problems can help PSTs learn and develop student-centered instruction.

2.4.3 *Mathematical Pedagogy That Engages Students and Builds Upon Their Knowledge and Thinking*

Although PSTs will have opportunities of learning and developing mathematical pedagogy as teachers-in-service in the future, they will also be very busy, and may have little time to reflect and analyze. There is relatively limited support in the United States for in-service teacher learning and improvement outside of Massachusetts (Howe, 2020), unless the professional development can show impact on students' learning (Yoon, Duncan, Lee, Scarloss, & Shapley, 2007). It is important for PSTs to have opportunities to experience and learn how to translate what they learned as students into instructional practices.

Although much of existing work in mathematics education have focused on content rather than practices and thinking, Schoenfeld (2020) points out the existence of some work that encompass more than simply content such as RME (Van den Heuvel-Panhuizen, 2003; Van den Heuvel-Panhuizen & Drijvers, 2014). As we discussed above, RME has been developed and used as offering students opportunities of doing mathematics as a primary way of learning mathematics. Solving everyday life problems is an essential aspect, and students are being offered opportunities for guided re-invention of mathematics. It was developed as a pedagogical approach to move away from mechanistic teaching formal mathematics in the 1960s in the Netherlands.

We agree with Schoenfeld (2020) that RME presents more than simply mathematical content with its emphasis on mathematical practices and students' thinking development. Thus, RME can be adapted and used as a possible approach for PSTs to learn and experience how to design and develop thinking-oriented instruction. Placing mathematics in a context can certainly help engage students, make mathematical ideas more accessible and meaningful to students, and integrate problem-solving and posing activities well in instruction. At the same time, the demands in time and efforts for the use of RME need to be considered. The extensive use of RME in classroom instruction would likely make the intended goal of students' guided re-invention of mathematics through RME questionable. It is more important for PSTs to learn and develop their own proficiency in mathematics and pedagogy through different activities and experiences, rather than simply acquiring RME as the taken-for-granted instructional approach.

2.5 Coda: The Need of Developing New Scholarship to Support a Thinking-Oriented Approach

Developing and using a thinking-oriented approach for PSTs' mathematics training would require research support. However, existing work in mathematics education has focused much more on the teaching and learning of mathematics content knowledge, with some supplements needed by teachers (see Blömeke et al., 2014;

Mewborn, 2001), or on seeking to provide in-depth and rigorous training in school mathematics and its structure from a disciplinary perspective (e.g., Wu, 2011b). In comparison, there has been limited research on the development of mathematical thinking, especially with teachers (Jacobbe & Millman, 2009).

There are also many other types of in-service teacher professional development activities with diverse foci and formats. For example, lesson study in Japan is a popular approach to engage teachers in a systematic and collaborative inquiry into teaching practice to improve students' learning (Fernandez, 2002; NRC, 2002). And multiple approaches commonly used in China, such as intensive textbook studies (Ding, Li, Li, & Gu, 2013; Pu, Sun, & Li, 2018), teaching research group (Yang & Zhang, 2018), and master teacher workstations (Li, Tang, & Gong, 2011). But these approaches typically focus on the improvement of teachers' understanding of curriculum and textbook content, instructional design development, and the improvement of teaching practice. A focus on mathematical thinking, especially teachers' own mathematical thinking, is typically absent in teacher education.

In fact, existing research on expertise in mathematics instruction typically takes a focus on knowledge and its structure, rather than on mathematical practices and thinking (see Li & Kaiser, 2011). Much more research would be needed to develop needed scholarship for a thinking-oriented approach in mathematical training for PSTs.

Taking an analytic look at each of the three aspects discussed in Sect. 2.4, we can find studies available in the literature that help us gain much insight for each aspect. However, there are no systematic studies available that connect these three aspects in efforts to train PSTs with mathematical thinking as a constant focus throughout. As argued by Hiebert and Morris (2009), it is important to build a useful knowledge base if a coherent system of mathematical training for PSTs is to be established. Toward that end, we are at a starting point to develop new research and scholarship to support mathematical training for elementary PSTs.

References

American Council on Education (ACE). (1999). *To touch the future: Transforming the way teachers are taught. An action agenda for college and university presidents*. Washington, DC: American Council on Education (ACE). Retrieved from http://www.acenet.edu/resources/presnet/report.cfm.

Anhalt, C. O., & Cortez, R. (2016). Developing understanding of mathematical modeling in secondary teacher preparation. *Journal of Mathematics Teacher Education, 19*, 523–545.

Association of Mathematics Teacher Educators (AMTE). (2017). *Standards for preparing teachers of mathematics*. Raleigh, NC: Author. Retrieved from amte.net/standards.

Ball, D. L. (1990a). The mathematical understandings that prospective teachers bring with them to teacher education. *Elementary School Journal, 90*(4), 449–466.

Ball, D. L. (1990b). Prospective elementary and secondary teachers' understandings of division. *Journal for Research in Mathematics Education, 21*(2), 132–144.

Ball, D. L. (1991). Research on teaching mathematics: Making subject matter knowledge part of the equation. In J. Brophy (Ed.), *Teachers' subject matter knowledge and classroom instruction* (Advances in research on teaching) (Vol. 2). Greenwich, CT: JAI Press.

Ball, D. L., & Cohen, D. K. (1999). Developing practice, developing practitioners: Toward a practice-based theory of professional education. In L. Darling-Hammond & G. Sykes (Eds.), *Teaching as the learning profession: Handbook of policy and practice* (pp. 3–32). San Francisco, CA: Jossey-Bass.

Ball, D. L., Thames, M. H., & Phelps, G. (2008). Content knowledge for teaching: What makes it special? *Journal of Teacher Education, 59*(5), 389–407.

Banilower, E. R., Smith, P. S., Weiss, I. R., Malzahn, K. A., Campbell, K. M., et al. (2013). *Report of the 2012 national survey of science and mathematics education*. Chapel Hill, NC: Horizon Research. Retrieved from http://www.nnstoy.org/download/stem/2012%20NSSME%20Full%20Report.pdf.

Beckmann, S. (2018). *Mathematics for elementary teachers with activities* (5th ed.). New York, NY: Pearson.

Begle, E. G. (1972). *Teacher knowledge and student achievement in algebra*. School Mathematics Study Group Reports, No. 9. Retrieved from https://files.eric.ed.gov/fulltext/ED064175.pdf.

Begle, E. G. (1979). *Critical variables in mathematics education: Findings from a survey of empirical literature*. Washington, DC: Mathematics Association of America and the National Council of Teachers of Mathematics.

Blömeke, S., Hsieh, F.-J., Kaiser, G., & Schmidt, W. H. (Eds.). (2014). *International perspectives on teacher knowledge, beliefs and opportunities to learn: TEDS-M results*. Dordrecht: Springer.

Borko, H. (2004). Professional development and teacher learning: Mapping the terrain. *Educational Researcher, 33*(8), 3–15.

Brownell, W. A. (1945). When is arithmetic meaningful? *The Journal of Educational Research, 38*(7), 481–498.

Browning, C., Edson, A. J., Kimani, P., & Aslan-Tutak, F. (2014). Mathematical content knowledge for teaching elementary mathematics: A focus on geometry and measurement. *The Mathematics Enthusiast, 11*(2), 7.

Browning, C., Thanheiser, E., Edson, A. J., Kimani, P., Olanoff, D., Tobias, J., & Whitacre, I. (2014). Prospective elementary teacher mathematics content knowledge: An introduction. *The Mathematics Enthusiast, 11*(2), 3.

Carpenter, T. P., Fennema, E., Peterson, P. L., & Carey, D. A. (1988). Teachers' pedagogical content knowledge of students' problem solving in elementary arithmetic. *Journal for Research in Mathematics Education, 19*(5), 385–401.

Carpenter, T. P., Fennema, E., & Franke, M. L. (1993). *Cognitively guided instruction: Children's thinking about whole numbers*. Madison, WI: Wisconsin Center for Education Research.

Carpenter, T., Fennema, E., & Franke, M. (1997). Cognitively Guided Instruction: A knowledge base for reform in primary mathematics instruction. *Elementary School Journal, 97*, 3–20.

Carpenter, T., Franke, M., Jacobs, V. R., & Fennema, E. (1998). A longitudinal study of invention and understanding in children's multidigit addition and subtraction. *Journal for Research in Mathematics Education, 29*(1), 3–20.

Carpenter, T. P., Fennema, E., Franke, M. L., Levi, L., & Empson, S. (1999). *Children's mathematics: Cognitively guided instruction*. Portsmouth, NH: Heinemann.

Colignatus, T. (2018). *The Math War between traditional and "realistic" mathematics education and its research. An analysis in institutional economics on research on education in arithmetic and algebra, with a focus on long term memory of pupils and using a causal model for valid testing on competence*. MPRA Paper No. 88810. Retrieved from https://mpra.ub.uni-muenchen.de/88810/.

Common Core State Standards Initiative (CCSSI). (2010). *Common core state standards for mathematics*. Retrieved from http://www.corestandards.org/Math/Practice.

Conference Board of the Mathematical Sciences (CBMS). (2001). *The mathematical education of teachers*. Retrieved from https://www.cbmsweb.org/archive/MET_Document/index.htm.

Conference Board of the Mathematical Sciences (CBMS). (2012). *The mathematical education of teachers II*. Retrieved from https://www.cbmsweb.org/archive/MET2/met2.pdf.

Cooney, T. J. (1988). The issue of reform: What have we learned from yesteryear? *The Mathematics Teacher, 81*(5), 352–363.

Cooney, T. J. (1999). Conceptualizing teachers' ways of knowing. *Educational Studies in Mathematics, 38*, 163–187.
Copur-Gencturk, Y., & Papakonstantinou, A. (2016). Sustainable changes in teacher practices: A longitudinal analysis of the classroom practices of high school mathematics teachers. *Journal of Mathematics Teacher Education, 19*, 575–594.
Crespo, S. (2003). Learning to pose mathematical problems: Exploring changes in preservice teachers' practices. *Educational Studies in Mathematics, 52*(3), 243–270.
Crespo, S., & Sinclair, N. (2008). What makes a problem mathematically interesting? Inviting prospective teachers to pose better problems. *Journal of Mathematics Teacher Education, 11*(5), 395–415.
Cummins, D. D., Kintsch, W., Reusser, K., & Weimer, R. (1988). The role of understanding in solving word problems. *Cognitive Psychology, 20*, 405–438.
Cuoco, A., Goldenberg, E. P., & Mark, J. (1996). Habits of mind: An organizing principle for mathematics curricula. *The Journal of Mathematical Behavior, 15*, 375–402.
Darling-Hammond, L., Wei, R., Andree, A., Richardson, N., & Orphanos, S. (2009). *Professional learning in the learning profession: A status report on teacher development in the U.S. and abroad*. Oxford, OH: National Staff Development Council.
Ding, M., Li, Y., Li, X., & Gu, J. (2013). Knowing and understanding instructional mathematics content through intensive studies of textbooks. In Y. Li & R. Huang (Eds.), *How Chinese teach mathematics and improve teaching* (pp. 66–82). New York, NY: Routledge.
Döhrmann, M., Kaiser, G., & Blömeke, S. (2018). The conception of mathematics knowledge for teaching from an international perspective: The case of the TEDS-M study. In Y. Li & R. Huang (Eds.), *How Chinese acquire and improve mathematics knowledge for teaching* (pp. 57–81). Leiden, The Netherlands: Brill|Sense.
Drijvers, P., Kodde-Buitenhuis, H., & Doorman, M. (2019). Assessing mathematical thinking as part of curriculum reform in the Netherlands. *Educational Studies in Mathematics, 102*, 435–456.
Eisenberg, T. A. (1977). Begle revisited: Teacher knowledge and student achievement in algebra. *Journal for Research in Mathematics Education, 8*(3), 216–222.
Ellerton, N. F. (2013). Engaging pre-service middle-school teacher-education students in mathematical problem posing: Development of an active learning framework. *Educational Studies in Mathematics, 83*, 87–101.
Fennell, F. (Skip). (2006). *We need elementary school mathematics specialists NOW*. Retrieved November 28, 2007, from http://www.nctm.org/uploadedFiles/About_NCTM/President/2006_11pres.pdf.
Fennema, E., Carpenter, T., & Franke, M. (1992). Cognitively guided instruction. *NCRMSE Research Review: The Teaching and Learning of Mathematics, 1*(2), 5–9.
Fennema, E., Carpenter, T. P., Franke, M. L., Levi, L., Jacobs, V. R., & Empson, S. B. (1996). A longitudinal study of learning to use children's thinking in mathematics instruction. *Journal for Research in Mathematics Education, 27*, 403–434.
Fennema, E., Franke, M. L., Carpenter, T. P., & Carey, D. A. (1993). Using children's mathematical knowledge in instruction. *American Educational Research Journal, 30*, 555–583.
Fernandez, C. (2002). Learning from Japanese approaches to professional development. *Journal of Teacher Education, 53*(5), 393–405.
Franke, M., Kazemi, E., Shih, J., Biagetti, S., & Battey, D. (2005). Changing teachers' professional work in mathematics: One school's journey. In T. A. Romberg, T. P. Carpenter, & F. Dremock (Eds.), *Understanding mathematics and science matters* (pp. 209–229). Mahwah, NJ: Erlbaum.
Freudenthal, H. (1968). Why to teach mathematics so as to be useful. *Educational Studies in Mathematics, 1*, 3–8.
Garet, M. S., Porter, A. C., Desimone, L., Birman, B. F., & Yoon, K. S. (2001). What makes professional development effective? Results from a national sample of teachers. *American Educational Research Journal, 38*(4), 915–945.

Goldstein, D. (2019, December 3). 'It just isn't working': PISA test scores cast doubt on U.S. Education Effort. *The New York Times*. Retrieved August 12, 2020, from https://www.nytimes.com/2019/12/03/us/us-students-international-test-scores.html.

Gregg, J. (1995). The tensions and contradictions of the school mathematics tradition. *Journal for Research in Mathematics Education, 26*, 442–466.

Han, X., & Li, Y. (2015). Selected approaches to professional development of mathematics teachers in the United States. In S. Li & Y. Li (Eds.), *Curriculum • Teacher • Classroom Instruction—Mathematics curriculum reform in China and the United States* (pp. 116–130). Beijing: Beijing Normal University Press.

Heyd-Metzuyanim, E., Tabach, M., & Nachlieli, T. (2016). Opportunities for learning given to prospective mathematics teachers: Between ritual and explorative instruction. *Journal of Mathematics Teacher Education, 19*, 547–574.

Hiebert, J., & Morris, A. K. (2009). Building a knowledge base for teacher education: An experience in K-8 mathematics teacher preparation. *The Elementary School Journal, 109*(5), 475–490.

Hill, H. C. (2007). Mathematical knowledge of middle school teachers: Implications for the no child left behind policy initiative. *Educational Evaluation and Policy Analysis, 29*(2), 95–114.

Hill, H. C., Rowan, B., & Ball, D. L. (2005). Effects of teachers' mathematical knowledge for teaching on student achievement. *American Educational Research Journal, 42*(2), 371–406.

Howe, R. E. (2015). The most important thing for your child to learn about arithmetic. In X. H. Sun, B. Kaur, & J. Novotna (Eds.), *Proceeding of ICMI STUDY 23: Primary mathematics study on whole number* (pp. 107–114). Macao, China: University of Macau. Retrieved from http://www.umac.mo/fed/ICMI23/proceedings.html.

Howe, R. E. (2020). Knowing and teaching elementary mathematics—How are we doing? *Notices of the AMS, 67*(6), 842–845.

Howe, R. E., & Epp, S. S. (2008). *Taking place value seriously: Arithmetic, estimation, and algebra*. Retrieved from http://www.maa.org/sites/default/files/pdf/pmet/resources/PVHoweEpp-Nov2008.pdf.

Jacobbe, T., & Millman, R. S. (2009). Mathematical habits of the mind for preservice teachers. *School Science and Mathematics, 109*(5), 298–302.

Kastberg, S., & Morton, C. (2014). Mathematical content knowledge for teaching elementary mathematics: A focus on decimals. *The Mathematics Enthusiast, 11*(2), Article 6.

Klein, F. (1932). *Elementary mathematics from an advanced standpoint: Arithmetic, algebra, analysis*. London: Macmillan.

Knapp, N. F., & Peterson, P. L. (1995). Teachers' interpretations of "CGI" after four years: Meanings and practices. *Journal for Research in Mathematics Education, 26*, 40–65.

Larson, M. (2017). *Why is change so hard?* NCTM President's messages. Retrieved from https://www.nctm.org/News-and-Calendar/Messages-from-the-President/Archive/Matt-Larson/Why-Is-Change-So-Hard_/.

Leavy, A., & Hourigan, M. (2020). Posing mathematically worthwhile problems: Developing the problem-posing skills of prospective teachers. *Journal of Mathematics Teacher Education, 23*, 341–361.

Li, Y. (2002). Knowing, understanding and exploring the content and formation of curriculum materials: A Chinese approach to empower prospective elementary school teachers pedagogically. *International Journal of Educational Research, 37*, 179–193.

Li, Y. (2007). Examining and understanding mathematics classroom instruction and possible contributing factors through a cross-national lens. *School Science and Mathematics, 107*, 308–310.

Li, Y. (2008). Mathematical preparation of elementary school teachers: Generalists versus content specialists. *School Science and Mathematics, 108*, 169–172.

Li, Y., & Howe, R. E. (2017). Developing pre-service teachers' mathematics conceptual knowledge for teaching. In B. Kaur, W. K. Ho, T. L. Toh, & B. H. Choy (Eds.), *Proceedings of the 41st Conference of the IG-PME* (Vol. 1, pp. 84–87). Singapore: PME.

Li, Y., & Kaiser, G. (Eds.). (2011). *Expertise in mathematics instruction*. New York, NY: Springer.

Li, Y., & Kulm, G. (2008). Knowledge and confidence of pre-service mathematics teacher: The case of fraction division. *ZDM Mathematics Education, 40*, 833–843.

Li, Y., Ma, Y., & Pang, J. (2008). Mathematical preparation of prospective elementary teachers—Practices in selected education systems in East Asia. In P. Sullivan & T. Wood (Eds.), *International handbook of mathematics teacher education: Knowledge and beliefs in mathematics teaching and teaching development* (Vol. 1, pp. 37–62). Rotterdam, The Netherlands: Sense.

Li, Y., Pang, J. S., Zhang, H., & Song, N. (2019). Mathematics conceptual knowledge for teaching—Helping prospective teachers know mathematics well enough for teaching. In D. Potari & O. Chapman (Eds.), *International handbook of mathematics teacher education: Knowledge, beliefs and identity in mathematics teaching and teaching development* (Vol. 1, 2nd ed., pp. 77–104). Leiden, The Netherlands: Brill|Sense.

Li, Y., & Schoenfeld, A. H. (2019). Problematizing teaching and learning mathematics as "given" in STEM education. *International Journal of STEM Education, 6*, 44.

Li, Y., Silver, E. A., & Li, S. (Eds.). (2014). *Transforming mathematics instruction: Multiple approaches and practices*. Cham, Switzerland: Springer.

Li, Y., & Smith, D. (2007). Prospective middle school teachers' knowledge in mathematics and pedagogy for teaching—The case of fraction division. In J. H. Woo, H. C. Lew, K. S. Park, & D. Y. Seo (Eds.), *Proceedings of the 31st Conference of the International Group for the Psychology of Mathematics Education* (Vol. 3, pp. 185–192). Seoul, The Republic of Korea: PME.

Li, Y., Tang, C., & Gong, Z. (2011). Improving teacher expertise through master teacher workstations: A case study. *ZDM: International Journal on Mathematics Education, 43*, 763–776.

Li, Y., Zhang, H., & Song, N. (2019). Mathematical preparation and confidence of prospective elementary teachers in China. In M. Graven, H. Venkat, A. Essien, & P. Vale (Eds.), *Proceedings of the 43rd Conference of the International Group for the Psychology of Mathematics Education* (Vol. 3, pp. 33–40). Pretoria, South Africa: PME.

Li, Y., Zhao, D., Huang, R., & Ma, Y. (2008). Mathematical preparation of elementary teachers in China: Changes and issues. *Journal of Mathematics Teacher Education, 11*, 417–430.

Lo, J.-J., Grant, T. J., & Flowers, J. (2008). Challenges in deeping prospective teachers' understanding of multiplication through justification. *Journal of Mathematics Teacher Education, 11*, 5–22.

Ma, L. (1999). *Knowing and teaching elementary mathematics: Teachers' understanding of fundamental mathematics in China and the United States*. Mahwah, NJ: Erlbaum.

McCallum, W. (2018). *Sense-making and making sense*. Retrieved from https://blogs.ams.org/matheducation/2018/12/05/sense-making-and-making-sense/.

Mewborn, D. A. (2001). Teachers content knowledge, teacher education, and their effects on the preparation of elementary teachers in the United States. *Mathematics Education Research Journal, 3*, 28–36.

National Association for the Education of Young Children & National Council of Teachers of Mathematics (NAEYC & NCTM). (2010). *Position statement. Early childhood mathematics: Promoting good beginnings*. Retrieved November 2012, from http://www.naeyc.org/positionstatements/mathematics.

National Commission on Teaching & America's Future (NCTAF). (1996). *What matters most: Teaching for America's future*. New York, NY: Author.

National Council of Teachers of Mathematics (NCTM). (1991). *Professional standards for teaching mathematics*. Reston, VA: Author.

National Council of Teachers of Mathematics (NCTM). (2000). *Principles and standards for school mathematics*. Reston, VA: NCTM.

National Mathematics Advisory Panel (NMAP). (2008). *Foundations for success: The final report of the national mathematics advisory panel*. Washington, DC: U.S. Department of Education.

National Research Council (NRC). (2002). *Studying classroom teaching as a medium for professional development: Proceedings of a U.S.-Japan Workshop*. Washington, DC: The National Academies Press. https://doi.org/10.17226/10289

National Research Council (NRC). (2011). *Successful K-12 STEM education: Identifying effective approaches in science, technology, engineering, and mathematics*. Committee on Highly Successful Science Programs for K-12 Science Education. Board on Science Education and Board on Testing and Assessment, Division of Behavioral and Social Sciences and Education. Washington, DC: The National Academies Press.

Olanoff, D., Lo, J.-J., & Tobias, J. (2014). Mathematical content knowledge for teaching elementary mathematics: A focus on fractions. *The Mathematics Enthusiast, 11*(2), Article 5.

Peterson, P. L., Fennema, E., Carpenter, T. P., & Loef, M. (1989). Teachers' pedagogical content beliefs in mathematics. *Cognition and Instruction, 6*, 1–40.

Philipp, R. A. (2007). Mathematics teachers' beliefs and affect. In F. K. Lester (Ed.), *Second handbook of research on mathematics teaching and learning* (pp. 257–315). Charlotte, NC: Information Age Publishing.

Pólya, G. (1945). *How to solve it*. Princeton, NJ: Princeton University Press.

Pólya, G. (1959). Mathematics as a subject for learning plausible reasoning. *The Mathematics Teacher, 52*, 7–9. (Originally published in German in Gymnasium Helveticum, 1956).

Pu, S., Sun, X., & Li, Y. (2018). How do Chinese teachers acquire and improve their knowledge through intensive textbook studies? In Y. Li & R. Huang (Eds.), *How Chinese acquire and improve mathematics knowledge for teaching* (pp. 165–184). Leiden, The Netherlands: Brill.

Rattan, A., Good, C., & Dweck, C. S. (2012). "It's ok—Not everyone can be good at math": Instructors with an entity theory comfort (and demotivate) students. *Journal of Experimental Social Psychology*. https://doi.org/10.1016/j.jesp.2011.12.012

Reys, R., Reys, B., McIntosh, M., Emanuelsson, G., Johansson, B., & Yang, D. C. (1999). Assessing number sense of students in Australia, Sweden, Taiwan, and the United States. *School Science and Mathematics, 99*, 61–70.

Schmidt, W., Tatto, M. T., Bankov, K., Blomeke, S., Cedillo, T., Cogan, L., et al. (2007). *The preparation gap: Teacher education for middle school mathematics in six countries*. East Lansing, MI: Michigan State University.

Schoenfeld, A. H. (1992). Learning to think mathematically: Problem solving, metacognition, and sense-making in mathematics. In D. Grouws (Ed.), *Handbook for research on mathematics teaching and learning* (pp. 334–370). New York, NY: MacMillan.

Schoenfeld, A. H. (2020). Mathematical practices, in theory and practice. *ZDM: The International Journal on Mathematics Education, 52*, 1163–1175.

Schram, P., Wilcox, S., Lanier, P., & Lappan, G. (1988). *Changing mathematical conceptions of preservice teachers: A content and pedagogical intervention* (Research Report 88-4). East Lansing: Michigan State University, National Center for Research on Teacher Education.

Shulman, L. S. (1986). Those who understand: Knowledge growth in teaching. *Educational Researcher, 15*(2), 4–14.

Simon, M. (1993). Prospective elementary teachers' knowledge of division. *Journal for Research in Mathematics Education, 24*(3), 233–254.

Stigler, J. W., & Hiebert, J. (1999). *The teaching gap—Best ideas from the world's teachers for improving education in the classroom*. New York, NY: The Free Press.

Strand, K., & Mills, B. (2014). Mathematical content knowledge for teaching elementary mathematics: A focus on algebra. *The Mathematics Enthusiast, 11*(2), Article 8.

Sun, K. L. (2018). The role of mathematics teaching in fostering student growth mindset. *Journal for Research in Mathematics Education, 49*(3), 330–355.

Sztajn, P. (2003). Adapting reform ideas in different mathematics classrooms: Beliefs beyond mathematics. *Journal of Mathematics Teacher Education, 6*, 53–75.

Thanheiser, E. (2009). Preservice elementary school teachers' conceptions of multidigit whole numbers. *Journal for Research in Mathematics Education, 40*(3), 251–281.

Thanheiser, E., Browning, C., Edson, A. J., Lo, J.-J., Whitacre, I., Olanoff, D., & Morton, C. (2014). Prospective elementary mathematics teacher content knowledge: What do we know, what do we not know, and where do we go? *The Mathematics Enthusiast, 11*(2), Article 9.

Thanheiser, E., Whitacre, I., & Roy, G. (2014). Mathematical content knowledge for teaching elementary mathematics: A focus on whole-number concepts and operations. *The Mathematics Enthusiast, 11*(2), Article 4.

Thompson, A. G. (1984). The relationship of teachers' conceptions of mathematics and mathematics teaching to instructional practice. *Educational Studies in Mathematics, 15*, 105–127.

Thompson, A. G. (1992). Teachers' beliefs and conceptions: A synthesis of the research. In D. A. Grouws (Ed.), *Handbook of research on mathematics teaching and learning* (pp. 127–146). New York, NY: Macmillan.

Tirosh, D. (2000). Enhancing prospective teachers' knowledge of children's conceptions: The case of division of fractions. *Journal for Research in Mathematics Education, 31*, 5–25.

Vacc, N. N., & Bright, G. W. (1999). Elementary preservice teachers' changing beliefs and instructional use of children's mathematical thinking. *Journal for Research in Mathematics Education, 30*, 89–110.

Van den Heuvel-Panhuizen, M. (2003). The didactical use of models in realistic mathematics education: An example from a longitudinal trajectory on percentage. *Educational Studies in Mathematics, 54*, 9–35.

Van den Heuvel-Panhuizen, M. (2010). Reform under attack: Forty years of working on better mathematics education thrown on the scrapheap? No way! In L. Sparrow, B. Kissane, & C. Hurst (Eds.), *Proceedings of MERGA 33 Conference: Shaping the Future of Mathematics Education* (Vol. 1, pp. 1–25). Fremantle, Western Australia: MERGA.

Van den Heuvel-Panhuizen, M. (Ed.) (2020). *International reflections on the Netherlands didactics of mathematics: Visions on and experiences with realistic mathematics education.* ICME-13 Monograph. Cham, Switzerland: Springer.

Van den Heuvel-Panhuizen, M., & Drijvers, P. (2014). Realistic mathematics education. In S. Lerman (Ed.), *Encyclopedia of mathematics education* (pp. 521–525). Dordrecht: Springer.

Wilcox, S. K., Lanier, P., Schram, P., & Lappan, G. (1992). *Influencing beginning teachers' practice in mathematics education: Confronting constraints of knowledge, beliefs, and context.* (Research Report No 99-1). East Lansing, MI: NCRTL, Michigan State University.

Wilson, S. M., & Berne, J. (1999). Teacher learning and the acquisition of professional knowledge: An examination of research on contemporary professional development. In A. Iran-Nejad & P. D. Pearson (Eds.), *Review of research in education* (Vol. 24, pp. 173–209).

Wu, H.-H. (2011a). The mis-education of mathematics teacher. *Notices of the AMS, 58*(3), 372–384.

Wu, H.-H. (2011b). *Understanding numbers in elementary school mathematics.* Providence, RI: American Mathematical Society.

Yang, D. C. (2007). Investigating the strategies used by pre-service teachers in Taiwan when responding to number sense questions. *School Science and Mathematics, 107*(7), 293–301.

Yang, D. C., Reys, R. E., & Reys, B. J. (2009). Number sense strategies used by pre-service teachers in Taiwan. *International Journal of Science and Mathematics Education, 7*, 383–403.

Yang, Y., & Zhang, B. (2018). In-service mathematics teachers' professional learning in teaching research group: A case study from China. In Y. Li & R. Huang (Eds.), *How Chinese acquire and improve mathematics knowledge for teaching* (pp. 209–223). Leiden, The Netherlands: Brill.

Yoon, K. S., Duncan, T., Lee, S. W.-Y., Scarloss, B., & Shapley, K. (2007). *Reviewing the evidence on how teacher professional development affects student achievement* (Issues & Answers Report, REL 2007–No. 033). Washington, DC: U.S. Department of Education, Institute of Education Sciences, National Center for Education Evaluation and Regional Assistance, Regional Educational Laboratory Southwest. Retrieved from http://ies.ed.gov/ncee/edlabs.

Yuanita, P., Zulnaidi, H., & Zakaria, E. (2018). The effectiveness of realistic mathematics education approach: The role of mathematical representation as mediator between mathematical belief and problem solving. *PLoS One, 13*(9), e0204847.

Chapter 3
The Missing Element for Teachers: Learning What Mathematics *Is*

E. Paul Goldenberg, Albert A. Cuoco, and Cynthia J. Carter

In a set of recommendations for a book with this title, what you would likely expect us to say is that most elementary teachers do not really know the content; whatever content they "know," they do not really understand; and often they do not realize that there *is* anything to understand. The obvious recommendation would be more mathematics courses, especially about the content they will teach.

While some of that is true, we think it also misses something important. There is no kindergarten teacher anywhere who does not know how to count and add and subtract, which is most of what her children will encounter during the year. And if the teacher is not sure of the name of some particular geometric shape, that is way down in the noise of what will matter for teaching. In the earliest grades, a lack of mathematical information—even a lack of understanding of why particular algorithms work—is not the biggest roadblock to good mathematical teaching.

In fact, as we see it, a major part of what hurts elementary mathematics teaching is some of the ways in which teachers know too much "math" without a tempering sense of what it is to do mathematics or to think mathematically and, therefore, without a sense of what *not* to teach. Our view is that the mathematical preparation of elementary school teachers needs to be less about "mathematics"—explaining how things work and what they are called—and more about *doing* mathematics, immersion in mathematics using what they already know to explore a territory and figure out "new" things. And only then formalizing those "new" things to add to content knowledge. To be clear, understanding "how things work and what they are called" is also important, but concise summaries, formalizations, and nomenclature are clarifications and codifications of results, not introductions or declarations.

E. P. Goldenberg (✉) · A. A. Cuoco
Education Development Center, Waltham, MA, USA
e-mail: pgoldenberg@edc.org; acuoco@edc.org

C. J. Carter
The Rashi School, Dedham, MA, USA

Y. Li et al. (eds.), *Developing Mathematical Proficiency for Elementary Instruction*, Advances in STEM Education, https://doi.org/10.1007/978-3-030-68956-8_3

Doing mathematics means (1) starting with what you know, (2) using it to derive something new, and (3) only *then* formulating the new result clearly. Both doing and learning require what we call "experience before formality."

There is of course plenty of mathematical learning for teachers-in-preparation (or in service) to do, but we believe that by treating the goal as "more math to know," we are actually exacerbating the very problems we see in classrooms. To become effective for more than a few students, mathematics education needs a paradigm shift and teacher education is the only place that can happen on a large scale. That, of course, will require a change not only in how we prepare elementary (and secondary) teachers, but a change in how their preparers are prepared. For undergraduate education to change, attention must be given to the education and support of the faculty.

For mathematicians, mathematics is not about what we know but about what we figure out—it is more like solving a puzzle than like grinding out a routine. Making sure that teachers and students, at the level of *their* knowledge, experience mathematics as mathematicians do would give them a more faithful picture of the discipline, knowing the joy and surprise as well as the content. Properly done, teachers would consequently be better prepared for *doing* mathematics and also more effective at *learning* the kinds of facts and procedures that states and schools test. Such teachers, if they have the support of their schools—which remains another huge issue to solve—would be in a better position to help their students do mathematics and be more effective learners.

3.1 Experience Before Formality: Stories from Classrooms

Over a span of years and in several schools, we have watched many fourth-grade teachers as they present $7 \times (\square + \bigcirc) = 56$, which their curriculum uses as a 2-min warm-up puzzle in advance of a unit that introduces the distributive property for students to apply to multi-digit multiplication. In each case, as soon as the children saw the puzzle, many blurted out that $(\square + \bigcirc)$ had to be 8, justifiably pleased that they knew the multiplication fact 7×8. Having seen equations like $\square + \bigcirc = 8$ before, they (like most children) also treated the $(\square + \bigcirc)$ as a call to suggest several possibilities for $\square$ and $\bigcirc$. Such spontaneous comments and observations are generally lively and a sufficient entry point for exploration of the distributive property.

At this point, what we hoped we would see is teachers acknowledging what students have figured out on their own and then letting them have the satisfaction of abstracting a common theme from independent work on a short, crafted sequence of similar problems that their curriculum provided. And then, *after* that, helping them pull it all together.

Instead, most teachers skip to the end in one way or another, explaining the "point" of the lesson. They teach that $7 \times (\square + \bigcirc) = 7 \times \square + 7 \times \bigcirc$, or have students test their suggested values of □ and ◯ to verify that 7 × □ + 7 × ◯ gave 56. There is lots of writing, lots of symbols, lots of words, and a group pace that is dreary to some and distracting to others. The teachers' inclination to give the punchline first steals the surprise—a motivator of interest—and gives students insufficient time to make sense of the ideas. And the independent work, originally designed to support the sensemaking and abstraction, no longer does that; it is now merely repetitive, a grind and not a puzzle.

One teacher wrote out $7\overline{)56}$ as a "justification" for the 8 that the children already shouted out. One teacher put up a table to show how values for □ and ◯ "should" be recorded. None of that contributed at all to the lesson at hand. Many introduced irrelevant vocabulary and other formalities, like declaring that □ and ◯ were called "variables." Several teachers asked questions like "What do you think the word 'variable' means?" More than once, after such a question elicited only a frozen silence from the children, we heard teachers define a variable as "a letter, like *x*, that stands for a number" leading to blank stares. What children saw and responded to when shown 7 × (□ + ◯) = 56 did not use (or need) the word *variable*, nor did it contain "a letter, like *x*," but teachers recognized things they knew and auto-piloted into teaching them, probably unaware of how much of their own knowledge—knowledge the children lacked—was required to make any sense of what they said. What had begun as lively, engaged responding and learning—children "getting" the puzzle and chirping a bunch of correct values for □ and ◯—ended with many looking as though they once again felt that they just did not understand math.

In one classroom, after children had offered a few possibilities for □ and ◯, the teacher took extra time to say that 4 + 4—which no child had even suggested—"would be wrong because then the equation would need to have two squares or two circles."

The teacher's statement was wrong, but there is a more pervasive and, we believe, bigger problem than teaching a wrong "fact." Most of our observations on this puzzle and extensive observation of other teaching suggest that teachers often feel that whatever they know about the mathematics, especially vocabulary, is on the table for teaching *now*.[1] Often, as in the examples just given, we would see teachers

[1] For one other example, when we were researching ideas and images for curriculum materials, first-grade teachers in Texas loved images like  because they supported a rich variety of grade-appropriate ways for children to enumerate: counting, adding 4 + 4, adding 2 + 2 + 2 + 2, skip counting, and more. They also liked similar rectangular arrays of stars or teddy bears. But mathematics coordinators rejected 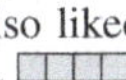. When we looked mystified, not understanding what the difference could possibly be, one state coordinator explained in a way that came to make sense to us. While arrays of circles, bears or squares would all feel the same to children, teachers *knew* more. They would recognize arrays of squares as the array model of multiplication (a fourth-grade standard) and feel they were therefore supposed to teach 4 × 2 and 2 × 4 in first grade. When we asked the teachers themselves about 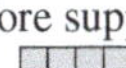, they openly wondered if that was the intent. Routinely, we see teachers seeming to believe "If you *see* something, *teach* something."

"improve" an exploratory puzzle or problem, one that children could handle on their own, by adding formal details or terms that children did not need, shifting students' attention from the core ideas and blurring the focus. Lively contribution ends, engagement fades.

Timing matters. Rules, methods, tricks, vocabulary, and shortcuts, if provided before students have built a foundation of experience and investigation, appear to work against students developing their mathematical minds. Shortcuts can be efficient in the short run, speeding the process of giving students knowledge of vocabulary or procedure. That can be useful for imminent tests—and that must be taken into account, at times—but the knowledge tends to be fragile and neither generalizes to new situations nor lasts intact, contributing to the common teacher lament that students' knowledge seems to evaporate in a few months. What might at first appear to be a time-saver ends up costing time in the form of greater review. It also costs interest: to students, as to teachers, review feels like a lack of progress. But *after* students have built understanding and experience, "shortcuts" (of the right kind) can be consolidating, simplifying, and even lasting.

Teacher preparation must help teachers know what (and when) *not* to teach.

3.2 Vocabulary, Convention, and Mathematics

Using vocabulary and symbols correctly is essential in mathematics—conventions for clear and correct expression are even more critical in mathematics than they are in text writing—but none of this *is* the mathematics. It is only the language with which we express and record mathematical ideas to help us process and communicate them.

3.2.1 Vocabulary and Convention Are Not Mathematics

The particular example we gave of what not to teach included an intrusion of vocabulary (*variable*), definition (letters standing for numbers), and conventions (in our example, a mistaken claim about the values variables can hold). Yet, intrusion of vocabulary—in fact, *starting* a mathematics lesson with vocabulary and notation—seems nearly universal, even among teachers who know from their language arts instructional methods that children learn vocabulary best when used in context.[2]

[2]In fact, that's how they learned *all* their language before school, which, for many, accounts for nearly half of the daily vocabulary they will ever use.

In a way, teachers' tendency to focus on vocabulary should not be surprising. "Word walls"—records of new vocabulary encountered in social studies, science, or books students read as a class—are common practice outside of mathematics. But even in mathematics, textbooks for teacher preparation are generally organized by topics, all of which have names: factors and multiples, commutative property, integers, algorithm, repeating decimals, irrational numbers . . . . And teachers, now more than ever, must teach to lists of these same names. Standard writers certainly do not believe that the names *are* the mathematics, but it can be no surprise if teachers and their supervisors feel that teaching names is necessary for, perhaps the essence of, teaching concepts, and this often leads to thinking that vocabulary (like a title for a chapter) must be the first thing to teach.[3] In fact, teachers are often required to write these on the board.

Mathematics teachers at all levels (including teacher educators) must be prepared to distinguish conventions and vocabulary from what can be reasoned out or understood. Children did not need to be asked "What do you think $7 \times (\square + \bigcirc) = 56$ might mean?" to blurt out that $(\square + \bigcirc)$ had to be 8. But when teachers did ask that question, it seemed to engender lively thought and participation even though it is only about a notation; the form was familiar enough for students to make sensible comments. By contrast, the question "What do you think *variable* means?" silenced the class. Children who knew *vary* might make up something reasonable to say (relevant or not) but, in this context, to these fourth graders, *variable* could mean anything.

Defining variable at all, let alone as "a letter, like x, that stands for a number" was not needed at this stage and certainly did not clarify anything. The harm is not just wasted time but blurred focus and an image of mathematics as about form, not reason.

Preparing teachers to know what is *not* relevant is hard because they cannot just follow a rule like "stick to your lesson." We really do want teachers to sense when a slight nudge will extend students' learning beyond what the curriculum provides or what the district's "list" mandates. To decide what is plausibly successful and to reject what is not—to figure out what to teach and what to leave out—takes mathematical knowledge (recognizing worthy opportunities) and sophisticated pedagogical judgment. The latter requires understanding children's thinking well and having enough teaching experience to make sense of it. Preservice teachers typically start with none of these backgrounds.

[3] A teacher we recently observed taught the words *monomial*, *binomial*, *trinomial*, and *polynomial* at the outset of a unit on polynomials, and had students write the "definitions" in their notebooks. Dismayingly, the teacher defined *polynomial* not inclusively but only as the leftovers—expressions with "four or more terms." Far, far worse, though, was the students' experience of meaninglessness and arbitrariness. Even correct definitions would depend on understanding what a *term* is, which the students did not, as this was their introductory lesson. The teacher was teaching what *he* "knew" (ignore the error) without considering what the students knew, or whether this was useful information or a useful time to impart it.

Just as learning mathematics requires some doing, so does learning to teach. In both cases some of the doing can be engineered into classes.

For starters, teacher preparation courses must model this kind of learning. We must give teachers the experience of doing mathematics without starting with or focusing on names and notations. To achieve this, the teacher preparers, themselves, need to have had mathematical experience before formality, and need mathematically clean texts that support such an approach with their teachers-in-preparation.

3.2.2 The Role of Vocabulary in Mathematics

The role and utility of names and notations—and, importantly, the understanding of when fancy terminology and symbols are *not* appropriate—might even be teachable in an explicit way. Giving teachers the experience of needing more precise (or concise) language and *then* filling that genuine need[4] can help them appreciate the difference between memorizing definitions and learning terms or symbols that make communication easier and clearer. This would also have greater fidelity to the work and thinking of mathematicians, who create special names and notations for expressions like $7 \times 7 \times 7 \times 7 \times 7$ or $5 \times 4 \times 3 \times 2 \times 1$ not to be obscure or fancy but to be clear, and because *these* particular expressions are common enough and useful enough to warrant easy notations and special names. Some of these notations even give us other benefits. For example, exponential notation can be used to show a parallel between addition and multiplication, and the extension of admissible values for exponents to rational and later real values can be based on already known "laws of exponents." Good notation highlights key features while suppressing camouflaging detail: 7^5 conveys more easily and more reliably the exact intent of a computation than having to count (or write) all the 7s. Such a notation illustrates ideal mathematical communication: precision with concision. Factorial notation does not suggest a parallel with earlier operations or notation,[5] but it also

[4]Referring to this as the "necessity principle," Guershon Harel writes "For students to learn the mathematics we intend to teach them, they must see a need for it, where 'need' means intellectual need, not social or cultural need" (Harel, 2008, p. 900). His context is teaching and learning of higher-level mathematics, but the same principle appears, unsurprisingly, to apply to elementary school teachers and children as well. For children, "intellectual need" often means deep curiosity about things that surprise or puzzle them, one of our reasons for putting experience first and cautioning against spoiling the punchline.

[5]Triangular numbers occur in many contexts but do not typically appear in a learner's experience prior to algebra and are so easily represented with $n(n + 1)/2$ that they need no special notation of their own.

serves concision and clarity. Factorial shows up in many calculations, especially those concerning counting and enumerating, even by middle school. Good notation leads to functional equations that can be used to help students make sense of the values of 1! and 0! instead of seeing them as arbitrary "rules" and can give students a glimpse of surprises to come, even before calculus provides the tools to use properties of factorial to extend its domain further (Goldenberg & Carter, 2017).

Mathematicians also create special names when casual language becomes ambiguous. Casually referring to the sides and corners of a square creates no confusion, but those terms become ambiguous when referring to features of a cube. Our room looks roughly like a cube and has four sides we can go to and four corners we can stand in, but it is natural to refer to a die (a cube) as having six "sides" and eight "corners." So, in the context of a cube, "sides" and "corners" become ambiguous. Mathematics needs to avoid such ambiguity, so it invents terms like *face*, *edge*, and *vertex* for clarity, not academic pomposity or just some "rule" of mathematics. Teachers (and later, their students) need to know that the justification for vocabulary, in mathematics as everywhere else, is *clarity* of communication. They also need to know that terms that *reduce* clarity or do not serve a mathematical need at the time, even if they are mathematically correct, are *not* appropriate.[6] This is not an anti-terminology screed; it is a call for understanding *what* terminology and *when*. As Ecclesiastes might have said it, there is a time for everything, a time for technical talk and a time for casual talk.

---

[6]Many of the geometric terms that are no longer in most elementary school standards still survive in classrooms. Terms like *parallelogram* or *isosceles triangle* gain their importance when membership in that class of figure tells us features that we can use for further reasoning. In elementary school, where class membership *does* remain part of many school standards (quadrilateral, rectangle, square, rhombus, all with class inclusion) but the reasoning based on it does not, those terms are distractions, and some of them routinely raise questions for teachers. All one can do with the terms is show one knows them. The ubiquity of pattern blocks in schools gives teachers a reason to name trapezoids (which appears in kindergarten standards), and it has the same utility for kindergarteners as star-shaped or moon-shaped or boat-shaped. But no *category* is implied—they are gestalts; even teachers would be unlikely to recognize  and  as trapezoids—nor are the features that define trapezoids age-appropriate (though they are often mentioned, with the expectable distraction or confusion or flat-out error). Those pattern blocks also contain *rhombuses*, which teachers often name right away (because it is useful to be able to refer to them) but again as gestalts. *Rhombus* appears in third-grade standards, where class membership becomes the purpose. Rather than using new and unfamiliar terminology to serve that genuinely useful mathematical purpose, might it not be better served by, for example, living thing ⊃ animal ⊃ dog ⊃ poodle?

In a similar way, when we are first teaching the meaning and operations on fractions are we communicating more clearly and unambiguously when we refer to *numerator* and *denominator* which beginners may continue to confuse for a while, or to the top and bottom, which all students can easily track? And is the significance of *denominator* as a kind (denomination) of unit and *numerator* as a count (enumeration) of them actually ever understood well enough to justify their use, and, especially, their continued use with rational expressions in algebra?

This recommendation is another facet of experience before formality. Teaching (and hence teacher preparation and the texts written for that preparation) should structure content so that definitions and vocabulary are capstones to experience, not foundations for it.

3.3 Preparing Teachers to Make the Best Use of Any Curriculum

Of course, curriculum materials influence what happens in class. Some provide well-crafted problem sequences and mathematical explorations building experience from which students can abstract ideas that can then be formalized into knowledge with understanding and can be honed over time into skill. Others present rules for action followed by repetition to build skill and leave understanding to emerge from application and experience over time as students' ideas grow and they have more knowledge to interconnect. And, of course, there are other styles, including ones in which districts give teachers only a list of objectives and a miscellany of web resources. These all make different demands on the teacher.

Good preparation for these varied curricular theories of learning is, we believe, largely the same. Preparation must help teachers learn to see the "core" of the mathematics-du-jour, so they can decide what is essential, what that core is built on and where it leads, and what is extraneous. That helps teachers make sensible decisions—based on listening closely to students' ideas, insights, and questions—about what to add and what *not* to add in order to support understanding (or even raw skill). Further, teacher preparation must give teachers mathematical experiences that show how investigation and exploration, surprise, and sensemaking can be inserted productively even into fact-centric curricula that are so common in schools. And it can help them see what a difference an approach centered on understanding makes. It is common in teacher preparation workshops we have given to hear teachers say that they wish they had learned mathematics this way as a child because it makes so much sense.

For teachers this is, of course, mathematical learning, but not so much *more* mathematics as a vastly different experience of the mathematics they already learned in K–12.

3.3.1 A View from Secondary School: Another Lens for Looking at Elementary Instruction

Though this chapter is about elementary school, it is instructive to look at a story from high school that shows the effect of secondary curricula and teacher preparation on middle and high school.

The class is working on graphing linear equations, using the "slope-intercept" method, one among several lessons, each about different—and disjoint—"methods" for graphing lines. For readers outside the culture of secondary school, this generally means that you transform the equation into the form $y = mx + b$, and, from this, you can produce the graph. As an example, consider the equation $2x + 3y = 9$. There is, of course, a sensible and simple method for graphing the equation as is, but on this day, the teacher's goal was to use the slope-intercept method.

So, students transformed the equation into $y = \frac{-2}{3}x + 3$. Then, to produce the graph, a three-step procedure is used: (a) go up 3 units on the y-axis and put a point; (b) from here, go to the right 3 units and down 2, and put a point; (c) connect the two points.

One of us (AC) walked around the room, and, sure enough, most students followed the procedure and produced the correct graph. Almost as an afterthought, I asked one student if the point $(1, 2.5)$ was on the graph. The student looked baffled, and, after using his pencil to plot it, said that it looked as if $(1, 2.5)$ was indeed on the graph. I asked if $(300, -595)$ was on the graph. He said there was no way to tell—it was off the paper.

This happened several times with different students in that class, and it is a phenomenon that has been seen in too many classes (Greenes, Chang, & Ben-Chaim, 2007). For many students, the curriculum treated $y = \frac{-2}{3}x + 3$ as a kind of code. From it, one obtained three numbers (-2, 3, and 3), and, from that, could produce a picture. Completely missing was the idea that one can test a point to see if it is on a graph by testing to see if its coordinates satisfy the graph's equation. And the fact that it was missing flew under the radar of the class assessments; after all, given an equation, students could transform it to slope-intercept form and produce a correct graph. The goal of this lesson was skill-aimed, all about procedure, so the gaping hole in students' understanding remained hidden.

This example might seem just plain weird to many readers outside the culture of secondary school teaching—demanding that students use a particular form just to practice switching forms—but this kind of low-level exercise is quite common in school mathematics. And there are enforced special-purpose methods as well. There is the "box method" for setting up equations to model word problems, another box method for factoring quadratics, the "switch x and y and solve for y" method for inverting functions, and a host of other "tools" that have no existence or purpose outside of school.

Note all the parallels to the fourth-grade example. The high school example is from a classroom that used a curriculum structured around explanatory exposition followed by worked out examples, followed by exercises. The fourth-grade examples were from classrooms with a curriculum that provides well-crafted problem sequences and explorations. Yet in both, the teaching emphasis was on *form*—a particular *way* of writing and doing—which students then gravitated to, becoming less curious, less exploratory, and less able to use what they know to solve new problems.

The "good preparation" that we have in mind (page 56) can go a long way to enhance good curricula in elementary and high school and ameliorate the effects of those that promote questionable mathematical habits.

Courses (in both mathematics and education) should distinguish convention from substance. They can distinguish core mathematical ideas from technical details. They can unflatten the "everything is as important as everything else" syndrome and, as a result, keep the focus squarely on doing and understanding the mathematics.

Some may say that it is a stretch to trace the roots of such stories back to teacher preparation. We do not think so. Yes, there are other forces at play—curricula, pressures from high-stakes exams, oppressive working conditions, school lore, administrators who know only lists and have not studied teaching mathematics. But a preparation that focuses on the *doing* as well as the learning of mathematics would give prospective teachers some tools to overcome the schoolish nonsense common in commercial curricula, to prepare students for state tests while immersing them in real mathematics, and to downplay the clutter in district syllabi so that there is time to concentrate on what is really important. That does not have to mean ignoring the district syllabi. Often a teacher cannot. Instead, one might find some mathematical context, topic or activity of genuine intellectual worth as a vehicle for presenting the otherwise lightweight clutter.[7]

This suggests an important role for a close and mutually respectful collaboration of mathematicians and teacher educators, looking seriously at school curricula, thinking hard about what part is mathematics and what part is school-only clutter, and preparing teachers to distinguish between the two and to concentrate on matters of mathematical substance.[8] Most teachers know that what *they* value and devalue,

---

[7]Clutter reduction, for its own sake and in service of leaving space for a coherent mathematical core, was part of the standards effort. But absence from a list does not guarantee absence from teaching, even when meticulously aligned to the standards. If "subtraction" is on the list, a teacher, supervisor, or curriculum writer may erroneously assume that teaching *minuend* and *subtrahend* is part of meeting that standard.

[8]One kind is idiosyncratic or curriculum-specific terms like "number buddies." Another is idiosyncratic uses of terms, like "friendly numbers," that *have* a mathematical definition, but appear in school with completely unrelated meanings. Children and teachers *will* hear these in school, and they may sometimes even be useful in school, but they will never be used outside of school. It is impossible to catch all of these, and probably counterproductive to try playing language police, but

and love and fear is communicated to their students. When teachers come to understand the heart of mathematics, they communicate this focus to students even when a particular day's lesson is all about the etudes or "stuff you have to know for the test."

3.4 Prepare Teachers to Make the Best Use of Any Curriculum

We propose several suggestions for teacher preparation programs, based on what we have discussed so far.

3.4.1 Learn from Expert Practitioners

A third-grade class was learning to use fractions to describe (name) the shaded portion of figures like these 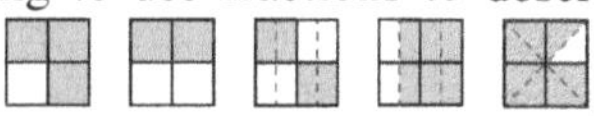, using fractions with denominators 2, 4, and 8. One boy looked at the last figure and wrote $\frac{3\frac{1}{2}}{4}$. The teacher came over to one of us (PG) and asked "Is that right?" The question clearly showed that she knew it *could* be, that it made sense, that she understood the *real* mathematics, but she was not sure whether it was "legal." As is common enough to be taken seriously in the mathematical preparation of elementary school teachers, this shows not a lack of content understanding but a misperception that vocabulary, convention, and notation might override logic and sense.

A fourth-grade teacher in the same school often presented puzzles in which students were given a long list of clues to the identity of some secret four- or five-digit number, clues like "I'm a multiple of five" or "my thousands digit is prime" or "my tens digit is twice my hundreds digit." The children loved the puzzles and the teacher would wander by and casually ask questions like "Do you need all the clues?" "If you use the clues in a different order, can you find the number with fewer clues?" "What's the fewest clues you can use?" That style of question and his general disposition to puzzle things through logically in his teaching made us imagine that he might have had a mathematics or computer science background. When the topic was equivalent fractions and children were generating them using the common "rule" of multiplying the numerator and denominator by the same number, one child listed $\frac{2}{4}$, $\frac{3}{6}$, and $\frac{4}{8}$ as equivalent to $\frac{1}{2}$, but was not sure whether that meant they were also all equivalent to each other. He asked the teacher whether $\frac{2}{4}$ was equivalent to $\frac{3}{6}$. The child's question was much like the question asked by

teachers need to know that some of the language they feel *must* be taught in order to be "correct" is simply lore and not law.

the third-grade teacher described above: asking the question was, itself, evidence that the child at least suspected transitivity—if both were equivalent to $\frac{1}{2}$, they should be equivalent to each other—but was not sure how the definition of the term "equivalent" worked. It *could* refer only to numbers generated by the rule the children were using. The teacher answered incorrectly but, always citing logic not authority, gave a reason based on what he thought the definition was. He said "no," explaining "there is no number that you can multiply 2 and 4 by to give 3 and 6." But the child persisted because the question did not feel settled. He said "but they're both equivalent to $\frac{1}{2}$, so…." Again, the teacher stuck with logic: "Oh! Right! They should be! So now I'm not sure. I'll have to check how *equivalent* is defined." He did, and he clarified in a very simple way for the whole class. This and other periodic lapses of fact showed that the teacher's background was not mathematics, but he was thinking mathematically. That approach, even in the absence of many mathematical facts, made for a very effective mathematics teacher. As can be seen from the continued lively engagement of his student, his approach produced a great environment for growing genuine curiosity around, and understanding of, underlying ideas. It also allowed the student along with the teacher to develop a firm understanding of important concepts and even explore the notion of definition.

In both cases, the teachers were implicitly wondering if mathematics was as logical as their students were. *That* is a message for teacher preparation, quite different from "more mathematics courses."

3.4.2 Learn from Other Disciplines

Mathematics is the least authoritarian subject there is. None of us have any way to know that George Washington was the first president of the United States of America other than to trust an authority—a person, book, document, website. But if we share a common language and know the meanings of "three," "four," "seven," "plus," and "equals," we do not need an authority to tell us that three plus four equals seven. High school geometry typically is (well, *was*) all about that kind of reasoning based on common understanding of terms and agreement on a (very) small number of facts and principles. Most of elementary school mathematics could be treated that way, too. And yet, as we have shown, many teachers expect mathematics to be even more authoritarian than other subjects, with rules that do not adhere strictly to what seems most sensible. To correct that misperception it may be helpful to draw some important ideas about the teaching of mathematics (not so much about mathematics itself), from other subject areas in which elementary school teachers are often more comfortable. Here is one example.

Preschool writing is often a letter-salad. A child might proudly fill a space with whatever letters she knows and then "read it" as a story. At this stage, the child treats writing as if it means making marks like these (see the figure on next page); reading is looking at marks like these (see the figure on next page) and telling a story.

Children later refine the idea, treating marks on paper not just as representing speech but as representing *specific sounds* of speech. Modern teachers understand that invented spellings like "I LIC SRC" (I like sharks) or "YE ET SAPR AT 6 OCIK" (we eat supper at 6 o'clock) are evidence that children have taken an important cognitive step. Despite the many inconsistencies of English spellings, letters do reflect the sounds we say; we want children to use *that*—the sounds they make, even if they are the wrong sounds (e.g., "fwee" for "three")—to guide their writing. The reason we don't immediately squash these invented spellings for the conventional ones is that we know, and want the children to know, that they now have the *right* idea. Allowing students to solidify this step provides a firm foundation on which to build the next one in which children treat conventional spelling as refinement, not a rejection of the principle of seeking and using an underlying logic.

It is for that reason—not to coddle little children's self-esteem—that good teachers carefully decide when to correct invented spellings and when not to. Teaching is about learning, not just about getting right answers. The first advance beyond preschool letter salads is learning that letters represent sounds, that *particular* letters represent (at least sometimes) *particular* sounds, and that we can use the way we hear our own speech as a clue to those letters. Then, as that begins to solidify, teachers add more knowledge: some letters can represent more than one sound, some sounds can be written in more than one way or require combinations of letters, and so on.

Over time, there has been considerable change in how teachers think about teaching reading and writing. In the bad old days, even in primary grades, teachers red-penciled every error in capitalization, spelling, punctuation, and even the neatness of the printing. Teachers now realize that, though the mechanics matter, good teaching involves choosing the right time: focusing too early or too much on conventions can get in the way of young children learning to write at all (or even to recognize that they have something to communicate). Focusing too late can also handicap them. There is a time to every purpose.... Deciding on the right time takes education, thought, and experience.

Another thing that elementary school teachers now understand about language arts is that language is a *mental* product; speech is one (ephemeral) expression of that language; writing is a record of that language; and reading and writing growth depend heavily on the growth of ideas and language. That chain of understanding could inform mathematics teaching, too.

Finally, teachers really understand that becoming immersed in a book—even a very simple picture book that a little child loves so much that he wants it read over and over and over—aids reading powerfully. Research shows that after children learn to decode—that is, after they acquire this most basic idea that particular letters and combinations represent particular sounds and the few details that match letters to sounds—reading skill improves more from the act of reading than from instruction in reading. We might apply this understanding to mathematics teaching: after counting and cardinality, mathematics growth depends heavily on doing mathematics not just receiving instruction about it.

For little children, writing is hard. The physical effort and even discomfort gets in the way of 7- or 8-year-olds immersing themselves in writing. But as with reading, the act of writing and receiving appropriate feedback—questions or comments that respond more to the content of the communication than to the mechanics—engenders more growth than being taught rules for writing. Conventions and rules remain important, but because they serve the goal of communication, increasing ease and fidelity in the communication; they have no *separate* importance. The intended message, the coherence and clarity of thought and expression, and the conventions do not all have equal weight.

Teachers can learn a lot from their knowledge about teaching, reading, and writing, but also need to recognize how prose writing and mathematical notation differ. One idea worth addressing explicitly is the level of redundancy.

In the multi-thousands of years of the development of mathematics, the notation we use today—the way we write numerals, expressions, and equations—is quite young, barely 500 years old. This modern notation, by expressing certain mathematical ideas more precisely and concisely than prose can, vastly revolutionized both what we can express and what we can do mathematically.[9] But those very virtues—precision and concision—make it difficult for new learners. If a child writes a short story or a sentence describing a picture and misspells many words or even omits some, we can still generally understand what the child was trying to say. The meaning remains clear enough. But if one single character in a mathematical expression or equation is wrong—omitted, substituted, or extraneous—the meaning changes, often without sufficient clues to detect which "misspelling" caused the change.[10] The notation is wonderful once we are skilled with it, but unforgiving

[9]And who knows whether another revolution in notation might take place in the next 400 years?

[10]Word problems are also an unforgiving dialect. Storybook prose has enough redundancy to help a child understand an unfamiliar word or at least get the gist without understanding that word. Word problems rarely do. Except when they are deliberately trying to be confusing (a perversion of good communication), they present just the necessary information. And word problems should not be confused with mathematics: *life* does not throw us word problems; it throws us situations to

to the unskilled, and perhaps, as Scott Kim, a puzzle and game designer, put it, "not a good user interface" for people who are just beginning to learn mathematics (personal communication).

Nevertheless, mathematics is typically treated as if it were entirely a written discipline. Manipulatives might be used as "aids" to get an answer,[11] and arithmetic facts might be memorized, but a ubiquitous view is that you have not "done" the mathematics until you have a proper written expression. Kindergarteners learn to write (prose) *after* they are fluent in the language they are writing; mathematics even in kindergarten tends to *start* with written forms and give them more weight than the ideas that are behind them. The writing of mathematical sentences often begins before the children understand what they are recording, and long before there is any perceived need to record it.

Mathematics, like language, is a *mental* product, and written mathematics is a record of that product. Kindergarteners who know that four plus four is eight will also yell out confidently and proudly that four hundred plus four hundred is eight hundred, and forty plus forty is eighty. (See also chapter 2 in Goldenberg, et al., 2015.) They would not necessarily know how to read or write those latter statements and might not even know how to write "$4 + 4 = 8$" correctly. But should the generalization wait for the writing? The children's generalization is, of course, linguistic not mathematical, but there is value not to be lost. Helping a second grader use $28 - 8$ as an "easy" starting place for a computation like $28 - 9$, a starting place that does not require a rule or memorized fact, involves *mathematical* thinking (if subtracting 8 gives 20 then subtracting 9 must give the next lower number), but making sense of $28 - 8$ in the first place—it rarely starts out clear—might also take advantage of linguistic rather than mathematical ideas. Twenty Eight and Jane Doe are names, and removing Eight or Doe leaves Twenty or Jane. Only one of those is a mathematical claim, but numbers were not born with names; we gave them the names to reflect a truth.[12] Even in cases like "five hundred plus five hundred equals

mathematize. Word problems are just an artifact of the attempts to bring "real-life" problems into the classroom. If identical problems arose not from an artificially constructed stream of words but from actual projects children were engaged in, the useful redundancy—the contextually supplied meaning—would be available to support understanding.

[11]In our view, manipulatives and visualizations should be used for getting the *idea*, for building a mental model, and not for getting the *answer*. Using a number line to organize one's thinking about computations like $^{-}127 - 45$ and $^{-}127 - {}^{-}45$ shows quickly—if one mentally places -127 and 45 and $^{-}45$ on the number line and thinks of subtraction as finding how far apart the numbers are (ignoring the sign of that difference, for the moment)—which distance is greater. No "rules" about when to "add" and when to "subtract." For a short time, while little children are getting the idea, *counting* on the number line might help build the idea of what they are doing, but the number line should then not be used for getting the answer. Alas, many times in elementary school, one sees manipulatives used as if they were calculators, to *get* the answer, bypassing the conceptual. This is also worth addressing in teacher preparation.

[12]Some number names—like eight, twelve, hundred—give no clue, except in their etymology, about the number they name. But most do give very explicit clues. This is why it is relatively easy for a 7-year-old to jump quickly from knowing half of eight and six to stating half of eighty or

ten hundred," linguistic generalization gives a mathematically correct result despite the unconventional name.

Teachers need to understand that mathematics is not an alien thought form; it is a refinement and extension of logic we already have—and if we do not already have the logic (or are not encouraged to use it) we cannot refine or extend it and mathematics just becomes a set of rules, truly alien and not a thought form. When a fourth grader responds to the *written* expression $\frac{5}{8} + \frac{5}{8}$ by writing $\frac{10}{16}$, we know that the child is seeing the + sign as something like "add everything in sight," but, more importantly, is not using the logic that she would have used in kindergarten if she had been asked *verbally* "What is five-eighths plus five-eighths?" Kindergarteners, when asked that question, always get it right. My (Goldenberg, Miller, Carter, & Reed, 2017) favorite response was "Ten ayfs! What is an ayf?" That comical answer is proof that *that* child had mastered an important idea. We can build on it in fourth grade only if we appeal to the logic first and *then* show how it is notated.

Placing primary emphasis on written forms or vocabulary as if they *were* the mathematics elevates form over substance. It runs throughout schooling: Proof in geometry often starts with a two-column structure, the statement and reason in T-form. It is not a bad structure but is merely a way a proof is presented, not conceived.

Precision in elementary school may require a different approach from the familiar starting with definitions and reasoning from them—precise vocabulary is often the capstone to deliberations, not their foundations. In elementary school, vocabulary and conventions should not be the way we introduce new ideas. Teachers who understand the difference can help their students use the drive for precision to help them build new ideas. We conjecture that understanding that difference is in part, perhaps large part, the result of having had such experiences for themselves in some domain.

Elementary school teachers also need to know when a push for language can get in the way of mathematical learning. Though all of us at any age can find it very hard to articulate what we are thinking, that difficulty can hobble young children. Just listen to any 7-year-old explain a movie she saw or a book she just read: the child starts, stops, rephrases, gesticulates, breathes in the middle of words. The whole enterprise is wonderful, and the child really learns a lot from trying to tell that story, but it clearly demands a lot of focus and cognitive effort to get through.

Of course, vocabulary and writing are not *only* a capstone. As we know, again from prose writing, setting ideas down even in a preliminary way can help us rethink and refine them. But that is true only if we can fluently and accurately record what we are thinking and fluently reread and review it. For young children, all of writing and reading is less fluent; mathematical writing is, because of its concision (and non-ubiquity), all the harder. Trying to learn a subject through a modality that is still a challenge may be too much cognitive load.

---

sixty or even eighty-six, but half of twenty-two or eighteen can remain totally mysterious to the same child. Our names reveal (somewhat imperfectly and inconsistently), the base ten structure we use to label numbers. The individual words are arbitrary, but they get used as parts of a systematic process.

Later, once one is comfortable enough with mathematical notation, it is more than a record of thought. It also provides the mechanism for crafting intermediate formulations that make further investigations possible. In Cuoco, Goldenberg, and Mark (1996) we talked about how mathematicians use "chants" when trying to shoehorn a nascent insight into a statement that is open to investigation. This often involves reformulating the insight, writing down the different half-baked formulations (often over and over), and using language and the written records to unfold the idea, look at it from different angles, and push it to a form where you can understand (or at least say to yourself) what you feel in your gut. This is also true for young students. All of us have seen students across the grades struggle to say what they know and to use scribbles and chants to get there. One high school student was struggling to get something down and started squeezing his head in an attempt to get the idea out and precise.

Trying to explain one's thinking in a mathematical argument can help a child clarify and refine and solidify that thinking. But the cognitive challenges can also get in the way. By the time we are adult mathematicians, the expressive part—the "chants" and writing down half-baked formulations over and over—is no longer the challenge it once was, and we can use it to help us clean up our mathematical idea. We have often seen 7-year-olds land "ballistically" on a correct answer, then explain their idea inarticulately and then reject their correct answer (and sometimes the reasoning that led to it) because the words they hear themselves saying overpower the thought that presumably led to those words. With young children it can be safer to wait until the child has a chance to *show* how their idea is correct before pushing for spoken (or especially written) formulation.

Again, to everything there is a season.[13] Students must eventually acquire fluency in reading and writing correct conventional mathematical notation—to keep track of where they are in multipart calculations, to record intermediate results while seeking a pattern, to seek structure in some process, to communicate with others, to pursue mathematical reasoning for which casual language is just plain clumsy—but much of the early *doing* and *learning* of mathematics can be done without notation. Curricula, by their nature, convey their messages in print (on screens or paper), but teacher preparation can make clear through experience that that's the *medium* not the message, not the mathematics.[14]

> Emphasizing mathematical thought and exploration over writing is essential to young students' understanding mathematics as logic and ideas. It serves to grow mathematical doers and thinkers.

[13]Ecclesiastes sometimes seems depressive but gets a lot right.

[14]Computer programming is a different kind of medium. It allows children to translate their own thinking into working computational models. More on that later.

3.4.3 Learn from Other Professions

Great musicians invested effort into developing *études*—practice exercises that were genuine music, not just random drill—to help their students become skilled players. We can do this in mathematics, suiting activities to each stage of learning that are true to the nature of mathematics, not just to its facts and routines. But that is not what we generally see in school. As a single example, fact practice, instead of being random (depending solely on recall), can be built around arithmetic properties. Second graders who can halve 4, 6, and 8 are delighted to discover (in a playful verbal, not written, exercise) that they can, with no instruction at all, halve four hundred, six hundred, eight hundred, four million, six thousand, and eighty. And with barely more than a comment, then (still verbally, without writing) halve 408, 604, even 412, and then 48. They have not yet built a robust version of the distributive property—their success is still largely a linguistic "trick," halving each part and then saying the numbers they have in their heads, not yet perceiving it as adding. But the next step, when students are eager for one, is not hard. When they are learning to add 10 to anything—a place value goal—that, too, can be aural/oral rather than a digit-based action on paper, and then the "number fact" goal of adding 9 can be a composition of functions ($+10 - 1$) that serves the learning of single-digit facts but also lets children add 9 (or later 19 or 99) mentally to anything. If they are already fluent with pairs that make 10, second graders often (again, speaking not writing) extend without instruction to pairs that make 20 or 30.[15]

The vignette of the high school classroom illustrated how mathematics is often presented as a collection of special-purpose methods that allow one to perform specific tasks. These are the calisthenics—the finger exercises—of mathematics, but not crafted to be "music." Unfortunately, rather than strategically using skill practice to reinforce what students have already learned and helping them to build automaticity informed by logic and understanding, skill practice is often used too early with the effect of teaching students a rote skill unconnected to other mathematical ideas. Exercise as an end in itself produces muscle-bound results that actually impede performance. Knowing how to transform an equation to some canonical form is a valuable skill, developed best through orchestrated exercise. But knowing *when* to use a particular form—realizing that some forms reveal features that other forms hide—is much closer to what mathematics is about. Again, it is the *doing* of mathematics that gives people an understanding of the discipline. *Learning* mathematics is important too but, by itself, leads to a view of the subject that places emphasis on getting to $y = mx + b$ rather than understanding the connection between an equation and its graph. And, by itself, it leads to a view of the subject as about how much you know rather than about how much you can figure out.

[15]See, e.g., https://www.youtube.com/watch?v=p7w0A2T6Eyk. at time-marker 1:55.

3.4.4 Learn to Trim and Sequence

The problem that our stories illustrate is not just what is missing, but what is *there*–a view of mathematics incongruent with that of most mathematics professionals. Wu (2015) has written about "textbook school mathematics" as a dialect of the discipline that lives in precollege curricula. Wu's main criticism is the lack of precision, the sloppy (or missing) definitions, the absence of logical sequencing, and the missing distinctions between assumptions (again, ill-formulated) and the results that follow. These are valid observations. But the classroom stories illustrate a ubiquitous problem that is deeper and more fundamental than imprecision and the "flatness" that raises convention to the same level of importance as matters of mathematical substance. We see a view of mathematics that is totally devoid of the perspective that learning mathematics means developing a collection of practices that help you *figure out* what to do when you do not *know* what to do—developing the habits of mind that underlie flexible proficiency in the discipline.

We also see a mad rush to get every possible nuance of a topic read into the record of instruction, right from the day it is introduced. Joan Ferrini-Mundy and colleagues introduced the notion of "trimming."

> In school algebra, as well as other areas of the curriculum, we propose that teachers have to trim mathematical or contextual content in a way that is mathematically acceptable but leaves intact the content to be learned. This trimming entails retaining important mathematical features while reducing complexity in ways that make the content accessible to students. (McCrory, Ferrini-Mundy, Reckase, & Senk, 2012)

The geometry of quadrilaterals is full of subtle delights, but there is no need to frontload these when the idea is first introduced. Developed over time, they are far more compelling. (See footnote 6 for another pitfall of premature geometric vocabulary.)

And finally, early or excessive focus on convention in mathematics gets in the way of young children doing mathematics just as early and excessive focus on the conventions in writing gets in the way of their writing. Young children who proclaim themselves to be "bad at mathematics" often have not been given the opportunity to use their own creativity on figuring out mathematics. They have not been allowed the chance to see how good they actually are at the logic and reasoning. Mathematical processes and ideas can be figured out, but how results "should" be named or notated is convention, which *cannot* be "figured out." Emphasizing the memorization of vocabulary and formulas at the beginning of every lesson gives students the false and often destructive idea that those are the center of mathematical understanding.[16]

---

[16]Summaries at the end of every lesson have a more subtle but similar effect. Many things that are important to learn take time to develop and percolate. The end of a 45-min class cannot be expected *always* to be the right time to deliver the punchline. Conversely, if we teach in a way that *can* always be summarized at the end of 45 min, we are dicing mathematics too fine. That is not how it works.

Both understanding and proficiency are goals. They're different. Proficiency requires skill, not just knowing what "should" be done. And building skill—in mathematics, soccer, piano, writing, cooking, anything—takes practice. Teacher preparation needs to help classroom professionals sort out when skill practice is likely to further the aims of students becoming lively mathematical thinkers and doers, and when it is more likely to shut that process down.

3.5 Prepare Teachers to "Teach on the Fly"

The old saying, "kids say the darndest things" is never so apparent as it is in mathematics class. But young people (unlike many adults) do not say something unless they have ideas. Picking at the germs of insights in what seems like a half-baked statement by kids is one of the hardest and most important skills that teachers can develop. Much of this skill is learned on the job, but teacher preparation can help prospective teachers build the habits of careful listening and constructive reply.[17]

We need to help teachers learn how to build on student insights and productive ideas.

3.5.1 Errors Are Not Sinful; Closure Is Not Sacred

These days, teachers receive lots of propaganda about "celebrating" or "embracing" errors—cringe-worthy terms not just from overuse but because of their combination of passion and emptiness. Other than to say "we learn from our errors"—a salve, and still fairly empty—there is little said about why we should be so celebratory. The language of "celebrating" and "embracing" errors addresses a real problem—the dread of errors—but just as a mantra. Nobody hopes for the most errors just so that they can have a party. The propaganda aims to take the sting away but does not help teachers *or children* stop feeling that the goal is still to reduce errors as fast as possible. No error left uncorrected seems to be a common theme K–12, not just elementary.

[17]Careful listening is one of the key ideas in Deborah Schifter's work (e.g., Schifter, Bastable, & Russell, 2017, 2018, 2019).

A better understanding of slips and errors would be that they are neither to mourn nor to celebrate but are just normal when learning is going on.

Some courses give exercises about trying to infer from students' errors what wrong fact or thought might have led to those errors. But teacher preparation (and in-service learning) appears rarely to give the message that some errors can be evidence of exactly the *right* kind of thinking. Recall those invented spellings!

Of course, some errors *are* evidence of wrong logic. Sometimes wrong logic can be addressed without waiting; sometimes it cannot be, and no disaster will occur if it is not. (Teachers often lament that children cannot remember right ideas from one day to the next and yet still worry that children will remember wrong ideas lifelong if we do not correct them immediately. How logical is *that*?!)

And some wrong answers are random slips worth ignoring totally—not evidence of either right thinking or wrong thinking but perhaps just of momentary distraction.

In any case, thinking that produces errors—whether from good (but interim) logic or genuinely wrong thinking—must eventually be qualified or refined. There is no rule that says we must wait for children to mend their ways spontaneously; we are allowed to intervene. We might craft situations that reveal contradictions in children's logic—the term "creating cognitive dissonance" is adopted from a different context to describe this kind of teaching—and let them work out the resolution. We might craft situations that call upon more robust logic that students already have: five dogs plus five dogs, five cats plus five cats, five-eighths plus five-eighths. We might even *teach* the resolution, when appropriate. But, just as with invented spellings, we choose the right time thoughtfully.

Preservice education needs to make clear why we must choose the time to correct and the time to stand down. It is fine to *stretch* students' logic, but if we defy it—if we present a case that students cannot yet make logical sense of and assimilate—we risk teaching that mathematics is about authority, not logic.

Possibly related to the fear that no error should be left uncorrected are the ideas that no question should be left unresolved, and no lesson unsummarized. Language growth is often understood as a work in progress, but the common lore on mathematics is that lessons require summaries and closure at every step. This is a truly wrong idea; it is just not how learning works. There is value in consolidating ideas, pulling experiences together, summarizing, naming concepts, and so on. But not before there are ideas to consolidate, experiences to formalize. Worthy intellectual content does not always come in 45-min chunks. Worse, attempts to chunkify mathematics reduce it to small and discrete skills, matters of notation, or convention. Ruminating on a new idea or complex skill a bit longer, getting more experience, sleeping on it, may make more sense—that is, may help *students* make more sense of the idea—than trying for small goals that allow instant closure.

Learning to multiply, for example, may take a few days of experience, comfort with the imagery and ideas, before a "closing summary" makes sense. Premature closure may be a real spoiler for conceptual development.

Preservice education must help teachers treat closure not as a tic but as a conscious choice.

A wonderful feature of mathematics is that many errors can be repaired, and the repair of an idea often takes much less work than it takes to come up with the idea in the first place. "If a number divides a product, it has to divide one of the factors." No—look at 6, a factor of 4×3. But there is a modification of the statement that is true and appears with a few numerical examples. Wiles' original proof of Fermat's Last Theorem had a flaw. It took a great deal of work to repair it, but that was nothing compared to the work it took Wiles to come up with the ideas in the first place. That example was for you, not for elementary teacher preparation, but even in elementary school, mistakes should not be thrown away; they should be salvaged in a way that recovers the germ of insight that led to them in the first place (see, e.g., Goldenberg et al., 2017). Proper preparation can help teachers approach their own errors this way as a model for learning how to help their students do the same. In PROMYS for Teachers (PfT, 2017), there is a problem type nicknamed "PODASIP": "Prove or Disprove and Salvage if Possible." This recommendation is very different from the "embracing mistakes" mantra—it is a call for emphasis on checking and refining conjectures with concrete examples.

3.5.2 Immersion in Doing Mathematics: Supporting Students and Teachers in Doing Mathematics Worthy of Thinking About and Communicating

How do we help teachers learn how to *do* and then teach their students to *do* mathematics, to use language and writing to refine, express, and *record* mathematical thought and action rather than treat it as *being* the mathematics and the goal of mathematical instruction, and to see the growth and potential for further growth in partially correct answers?

In fact, we might wonder why teachers are not teaching this way already. Teachers could treat the early learning in mathematics the same way they treat early learning in language arts but, in general, they do not. We know of no research-based explanation of why not, but direct and repeated observation offers a plausible hypothesis. While most elementary teachers have some background in mathematical facts and computation, they have little experience with the doing of mathematics.[18] It is unheard of for an elementary teacher to not be able to read and write or to have never picked up a book. But it is largely the case that elementary teachers have never done a piece of mathematics—investigating and solving a problem rather than rehearsing a learned technique—even at an elementary school level. And it is largely the case that elementary teachers have never even been exposed to serious mathematics (even at elementary level). It is very common in school to hear teachers describe mathematics as being exclusively about learning facts and formulas, vocabulary, and proper ways to write and perform computations on paper.

To help students learn by doing mathematics, teachers need experience *doing* mathematics.

3.6 Four Models for Immersion: *Doing* Mathematics in Teacher Preparation

It is hard to shift perspectives from *acquiring* to *doing* when one has not, oneself, experienced doing.[19] So, for teachers, just as for students, the goal is to provide the experience of doing mathematics. That experience need not be exactly the mathematics the teachers expect to teach—after all, the teachers are adults—but to see how they might apply what they learn, teachers must be able to recognize a connection between what they expect to do in class and the content and the style of work they did in their preparation. They are getting a course in applied mathematics: mathematics "applied" to their intended profession. (See Cuoco, 2018.) Elementary school teacher preparation can provide such experiences in several forms. Just as we cannot legislate what happens in elementary school classrooms, we cannot do so for college classrooms preparing the teachers. But here are four models that we see as contributing, in some combination, to a sensible course.

[18]And they have generally been taught ways that things "must" be done, rather than the logic that lets them choose how to do things. An example that Roger Howe contributed is that teachers have generally learned, and teach, that in adding, you must always start with the ones.

[19]More generally, teachers-in-preparation are given many exhortations that (like the very mathematics teaching that we lament) amount to authority rather than reason. Even data-supported claims like "encourage growth mindset, not fixed mindset" become, essentially, propaganda if the teachers do not *have* growth mindsets themselves. People can *mouth* a change in belief but rarely *change* beliefs—for example, that mathematical ability is built in and immutable—just because someone says they should or that the belief is harmful to students. *Experience* is a more powerful change agent. Experience *does*, of course, include what we learn from data and trusted authorities—a kind of vicarious experience—but often needs support from direct experience.

3.6.1 *Adaptations of the Approach in PROMYS for Teachers and PCMI*

An immersion model could be built by adapting the Program in Mathematics for Young Scientists (PROMYS, 2017) or PROMYS for Teachers (PfT, 2017) to suit preparation of elementary school teachers. PROMYS is an immersion program for mathematically strong and interested high school students. PfT is aimed primarily at in-service secondary teachers. We know of no program similar to PfT at the elementary level. Creating one for both preservice and in-service use could be a productive effort of mathematicians working closely and respectfully with teachers. Why *mathematicians*, in particular? Partly because mathematicians have had the experience of doing—starting with something they know and puzzling through something they do not know. The frontier for teachers-in-preparation will not be the frontiers of mathematics, but the spirit can be the same. Another role for mathematicians, in particular, is the long-range view: It is easier to find an intriguing problem to solve with "elementary" techniques when one can see the surprising results that can result from research using nothing more than credibly grade-level-appropriate content. Why "working closely and respectfully with teachers"? Perhaps obvious at this point. We have shown how elementary school teachers often do not recognize how much their own knowledge (knowledge their students lack) helps them understand a "simple" explanation that remains opaque to their students. Mathematicians building courses for teachers need to avoid a similar trap.

Whatever else may be part of a teacher preparation course, we believe this element to be essential. It may tolerate some commentary (e.g., "this, not that" or to bring explicit attention to some issues of mathematics learning, like the "necessity principle"), and even some methods (e.g., "this experience was for *you*, adult learners, and here is how you might give that kind of experience to an 8-year-old"), but the core must be the doing of mathematics. The remaining three models below suggest ways that this core might need supplements.

3.6.2 *Using Case Studies: Making Sense of Student Thinking Requires Doing Mathematics*

Learning to teach mathematics involves *doing* in both the mathematics and the teaching, and also involves learning how to get *students to do* the mathematics. An immersion experience in mathematics is partial support for the first of those—as we said, we do not think the job can be done without that—but immersion in teaching, especially immersion in *mathematics* teaching, goes beyond what any student-teaching practicum (especially with all of the counterproductive pressures[20] it often has) now offers. This aspect of learning to teach mathematics needs more thought and design at the preparation level but almost certainly cannot be completed there; it requires continued in-service learning. To help teachers learn to listen closely to students' thinking and to *use* that thinking to support decision-making about what and how to teach, Schifter, Bastable, and Russell (2017, 2018, 2019) designed an approach *for practicing teachers*—in-service, not preservice—based on close analysis of case studies documented by practicing teachers. For a clientele that has generally had little to no experience in classrooms except as students, this would require some rethinking, but the idea of using real case studies might add depth that student-teaching contacts cannot do. This approach focuses on an essential skill of teaching—attending closely to students and trying to understand what they are saying/showing—that is not mathematics but involves teachers and potentially pre-teachers in dissecting and understanding student's mathematical thinking, which also requires teachers to *do* mathematical thinking. Understanding what elements of students' thinking are correct—especially when what they have said/done is, as a finished product, not correct—involves flexible and deep understanding of the mathematics. Case study analysis of student thinking is not, of course, a substitute for building one's own experience doing mathematics—in fact, analyzing the mathematical ideas students present requires a fair amount of mathematical knowledge and versatility on the part of the teacher, but listening to students and being able to make sense of what they are saying is an essential element of good teaching.

3.6.3 Programming in an Accessible, Mathematically Focused Environment

Because "coding" has become the hot topic du jour, a "new literacy" or a "civil right" that schools are scrambling to insert somewhere into an already crowded day, this first paragraph is intended to make clear that, for the mathematical preparation of elementary school teachers, we believe that coding for coding's sake is a distraction, not a help. But there are ways that programming can serve elementary school mathematics if the activities and contexts are thoughtfully mathematical, and the language and approach are optimized for teachers and children. End of preamble.

Most teachers have had the experience of being immersed in reading. Few have had the same in mathematics. In 1971, as one of the early position statements on Logo, co-invented by Wally Feurzeig, Seymour Papert, and Cynthia Solomon, Papert wrote a paper titled "Teaching Children to be Mathematicians Versus Teaching About Mathematics" (Papert, 1972). He proposed a particular use of

[20]Such as feeling that one must "look professional" and be much like the teacher one had as a child or the teacher one is currently working with, complete with organized, "efficient" lesson plans.

computer programming as a context for immersing children in mathematics. Though the context (programming) was new, the theme was mathematics not technology: *doing* mathematics, not just acquiring mathematical facts or procedures. Papert's particular use of programming in mathematics education for children faithfully mirrors our call for immersion in *doing* mathematics for secondary students and for teachers.

But Papert's images of what to do with children did not fit school mathematical contexts then and they fit even worse now. They were conceived for informal settings and ignored many school realities. And what *was* done in school, even if mathematical, lacked "face validity" as mathematics.

Subsequent work *in mathematics classes in elementary schools*, described below, has shown ways that programming *can* serve school's mathematical goals and can be implemented by schools. We know of no model designed for teacher preparation using programming as a context specifically for mathematical exploration, expression, and learning, but the work with students—e.g., ScratchMaths (UCLIE, n.d.), or the *Math+C* (EDC, 2020) student materials with teacher guides currently being developed at EDC—might serve as a suitable basis for designing such a preparation for teachers.

The ScratchMaths team found that programming in Scratch supported by a mathematics-focused curriculum that they designed made the doing of mathematics accessible to young children. The ScratchMaths project shows how programming, suitably used, can be *not* just one more job skill added to an already crowded elementary school regimen, but a venue and vehicle for developing (better) the mathematics one is already trying to teach. They describe their idea this way:

> Maths is difficult in part because of the language in which it is expressed. Can we find a different language—and set of ideas and approaches—that is more open, more accessible and more learnable. And can we find it without sacrificing what makes mathematics work? Our tentative answer is "yes"—the language of programming might, if we design it right, be just such a language. (UCLIE, n.d.)

Again, note that they are not *replacing* mathematical language and conventions with computer code. Rather they are arguing that coding might provide early access to the *doing* of mathematics and the learning by doing that results, which can then, when appropriate, be expressed in more conventional mathematical expressions.

How might programming (suitably designed and used) be a better initial "user interface" than conventional written mathematical notation? For one thing, it is a *live* notation. Jottings on paper just sit there, correct or incorrect. To check their veracity, a child must reread (hard) and mentally reinterpret and reenact them (also hard) or, more commonly, rely on external authority. When a student writes "to make an even number you times by 2" on paper, the notation records the idea, but does not help perform the next calculation. And it does not help you make your expression more precise. Computer programs do all that and more: they let you then build on your idea to generate the next. Jottings in the form of computer commands and functions can be reenacted by the machine. That is the sense in which programs are a "live notation."

A programming language is a *formal* language for expressing certain kinds of ideas—process and algorithm in particular—in a precise way. Learning to use such a language can augment students' expressive power in mathematics and make key ideas more accessible. Programming, despite its logical grounding, is not, of course, a magic pill. What one does with it—the ideas and approaches one uses—must also be intellectually worthy, mathematically and developmentally appropriate, and appealing to students.

More directly to the point, a programming language *is a language*, one well suited to computational thinking and, potentially, very supportive of mathematical learning. It gives students a language with which to express their thinking precisely and have that thinking enacted by the computer, making it easier for students to assess, refine, and extend that thinking. Learning to express *mathematical* ideas in a precise computational way—in a program—and to enact those ideas, get feedback on the enactment, and assess its fidelity to the original idea is a large part of what CT is. Expressing mathematical ideas as runnable code rather than static writing—building and manipulating mathematical objects on a computer—lets students experiment with mathematics and gather results from the experiments much as they would manipulate physical objects to explore their properties and behavior. The same, designed appropriately, could serve teachers as well.

For our EDC work on the *Math+C* project, a project designed for use *in regular mathematics classes in school* with children from age 7 through 11, our specific goals led us to choose Snap*!* (2020) instead of Scratch (2020) as the programming language.[21] While Scratch has been enormously successful with young children in informal settings and to introduce "coding" in school settings, it is optimized for those purposes and some of the language and environment choices that make it so successful in those settings limit it when the goal is to express and explore mathematics.

Snap*!* is a high-level, high-powered language that *looks* like Scratch but is optimized for broad programming use, specifically including mathematics and computer science, and is used in college-level CS courses (BJC, 2020). Though not designed for young children, we have found that, for our mathematical purposes, Snap*!* serves us better than Scratch, even in elementary school. It can, for example, let children and their teachers build programs that behave as composable functions, the mathematical process behind the nearly ubiquitous input–output tables children see. Perhaps these can even use the "function-machine" imagery sometimes used as the elementary school visualization of functions (Fig. 3.1).

Programming could be the "invented spelling" of conventional mathematical notation: fully expressive, accessible to beginners. With training, teachers can use this to help students *do* the early expressing, doing, exploring, and learning of

[21] Scratch projects can be executed in Snap*!*, so a teacher preparation program using Snap*!* can take full advantage of the curricular coherence of ScratchMaths (and whatever prior experience, if any, that teachers have had with Scratch).

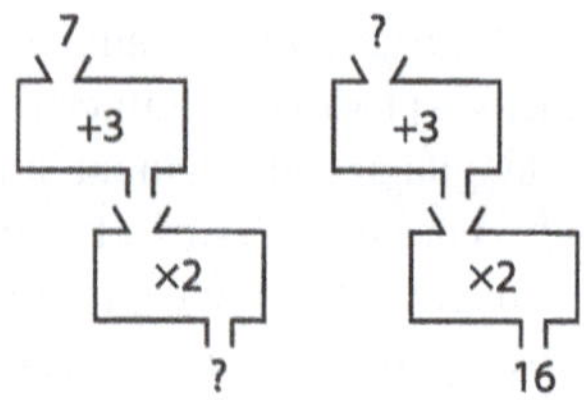

Fig. 3.1 Two function machines illustrating $x \rightarrow (x + 3) \times 2$, with input or output shown

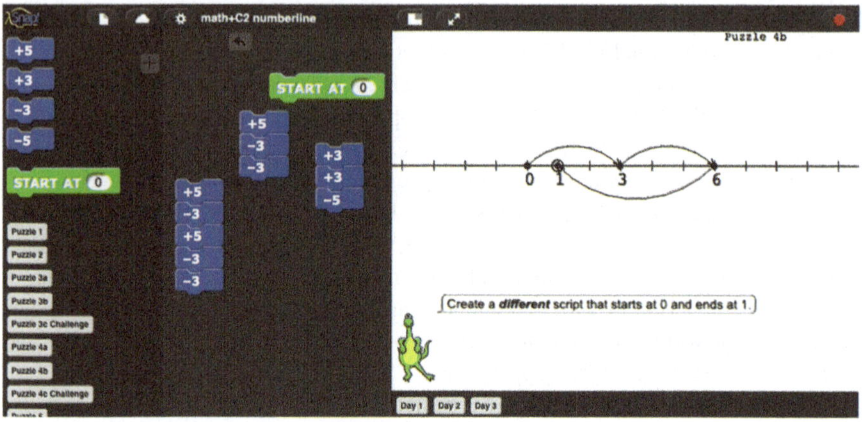

Fig. 3.2 A 7-year-old's work in a mathematical environment focused on the number line

mathematics before they have to encounter the potentially distracting extra demands of conventional notation.

The *context* of teacher preparation that uses Snap*!* (or Scratch) may be programming, but the *focus* must remain on what its details (any bit of content, programming, pedagogy) means for *mathematics*.

Here is one example (of many) that is designed for second graders who have had no prior programming experience in which we have observed teachers, themselves, expressing surprise and new learning. The environment (Fig. 3.2) shows a number line with tick marks; only the 0 (not at the leftmost tick mark) is labeled. A circle surrounds the dot at 0. After the teacher asks the children what languages *they* speak, they are told they will learn one language that the computer knows. At first, they will see only four of its "words"—the blue blocks—and use those words to make "sentences" which we call *scripts*. As children sit on the rug and look at the display, the teacher shows how to drag out blue blocks and click on them. That moves the circle, draws an arc, and labels a new number. Several children help demonstrate with other moves. They use the green block to clear the number line and start over at 0 (or elsewhere). They learn how to select new puzzles. That is it—generally no more than 5 min—and then they are on their own.

The puzzles focus on mathematical ideas of second grade but have a lot of richness. Though no puzzle refers to or requires movements to the left of zero, children naturally and excitedly wander there, and invariably know how to "get back" and where they will land, giving a casual experiential grounding for formality that will not become part of their official study for years. They also come up with observations on "doing and undoing" and various strategies for labeling all the numbers.[22] The initial programming is minimal—as suits a first experience—but introduces ideas that will later be needed for more advanced mathematical explorations. This same environment, in later puzzles, uses linear combinations of 300 and 500 instead of 3 and 5, and in later grades, linear combinations of three- and five-eighths +5/8, −3/8, −3/8, and −5/8 with a number line marked in eighths (Fig. 3.3) and a new START AT 3/4 block, showing that (deferring notational differences) addition and subtraction of fractions works like addition and subtraction of integers, and depends on *unit* fractions.

The teachers, in just this first number line environment for second grade, commented on how much thinking and predicting the children were doing, how engaged they were in the computation. Teachers also commented on how much *they*, themselves, were seeing that they had not noticed or fully understood before. Later, linear combinations of, say, 6 and 15, let children explore common factors.

In a different environment, designed to give second graders experience using ± 1, ± 10, and ± 100 to construct new numbers to add or subtract, our in-service teachers similarly commented on their own learning as well as that of the students. Children were amused (which we interpret as engagement of their logic) that they can create 291 in such very different ways as shown in Fig. 3.4. The left-side construction resembles what children see when they manipulate base 10 blocks; the middle is "expanded notation" in a form that children spontaneously create, and appear to find clearer than the conventional written form; and the third shows how mental addition of 291 to some other number might be performed most easily, why rounding or approximation is useful, not a consolation prize for the "correct" answer.

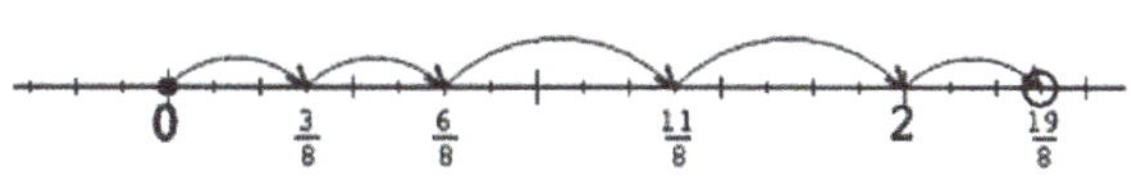

Fig. 3.3 A fraction number line for adding eighths

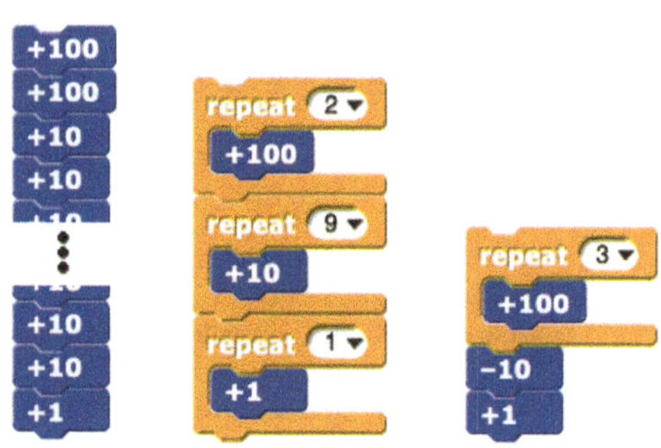

Fig. 3.4 Three (of many) ways to create 291 using ± 100, ± 10, and ± 1

[22]One 7-year-old said that she found a simple way to get to all the numbers. Just use $+3 + 3 - 5$ to get from 0 to 1, then "click it 15 times" (her words, which clearly exaggerated the reality).

Again, teachers noticed connections with other mathematics they find hard to teach, connections they did not consciously recognize themselves. More advanced programming environments let older children do their mathematics functionally, creating and composing functions, and testing and comparing results. All of this, we are beginning to see, is quite educative to teachers who already know the mathematics.

Exploring mathematics, even the mathematics of elementary school at a cognitively more adult level, would be a creditable *doing* experience for teachers. In the absence of seeing children in action, such experiences might not as readily convince preservice teachers of what children can do, but their own experiences with exploring might help fill that gap.

3.6.4 Building on Math Olympiad Problems for Elementary and Middle School

A fourth immersion model could be built around Math Olympiad problems for elementary and middle school (MOEMS, 2020). Like ScratchMaths and PROMYS, these are intended for children, not teachers. Unlike ScratchMaths and PROMYS, both of which are designed as curricula with structure and coherence, MOEMS problem sets contain random problems (see Figs. 3.5 and 3.6 for examples), not with the development of any particular topic in mind; and they are designed as tests, contests for the high-flyers in school. On the surface, these might seem a nonsensical basis for even a part of a preservice course. But the problems, themselves, ask for thought, not just fact-and-procedure memory. Each one requires the solver to *do* a piece of mathematics, rather than just store or demonstrate a piece of mathematical knowledge. With coherence and thought, the approaches that are recommended for coaching children for these contests could be adapted to provide the same mathematical-doing experience for teachers in preparation. In turn, teachers can use these with their students to give some logical grounding to any curriculum. The two examples below are both geometric, but Olympiad-like problems tap all of K–8 topical content and typically draw on insights from more than one topic

A jeweler bent a single strand of silver wire to make this earring consisting of five squares. The length from top (T) to bottom (B) is 5 cm. Find the length of the strand of wire used to make this earring.

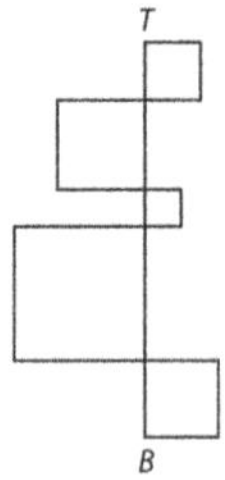

Fig. 3.5 An Olympiad-like problem for elementary school

A jeweler designed this pendant.

It's made entirely of silver wire bent into perfect circles, all linked together.

Here you see the pendant spread out on black velvet in its box in the display case.

The box is exactly 10 cm long.

Roughly how many centimeters of silver wire were used for the pendant?

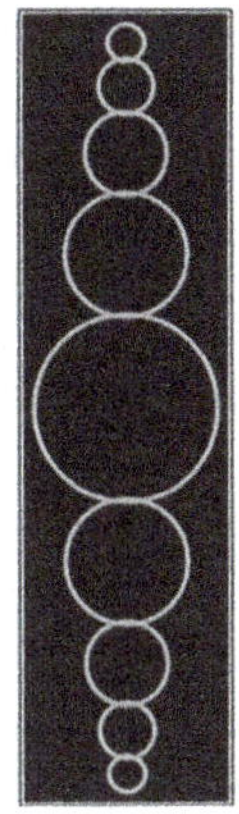

Fig. 3.6 Essentially the same problem but requiring middle school mathematical knowledge

Fig. 3.7 A "fancy" input–output table

□+△	7	10			29				20	26
□	4	8	9	10	20		16			
△	3	2	4	5		7		20		
□−△	1	6				0	8	40	6	8

area, fostering a unified rather than siloed view of the discipline and "reviewing" creatively.

Though the Olympiad problem sets, themselves, have no topical focus, books (e.g., Kalman, 2008) for team coaches do have a focus: helping teachers prepare children to be mathematical problem-solvers. This could be a useful tool for the professor who really understands that the goal is not to show how smart or knowledgeable the professor is, but how much the teachers-in-preparation already know (after all, they have all been to elementary, middle, and high school and even succeeded in being admitted to college) and how smart *they* can be when they learn that it is permissible—in fact, the central idea—in mathematics to figure things out. Many teachers have had experiences that tell them thinking and figuring out is *not* valued ("You should not have to think; you should *know* this by now!") and not needed (since methods are taught without understanding).

A problem from second grade in *Math Workshop* (Wirtz, Botel, Beberman, & Sawyer, 1964) illustrates how even pure practice in age-appropriate addition and subtraction—everyday stuff, not for Olympiads, and, despite its vintage, fully consistent with today's Common Core content and practice standards—can have that extra element of puzzle that is so rare in school curricula and teacher preparation (Fig. 3.7). Note especially the last two columns.

Exposing teachers to problems like these can potentially change their perceptions of what mathematics is. And problems like these can be an essential resource for teachers using a facts-first curriculum, giving them ideas for how to insert into their lessons targeted experience with investigation and surprise that can become a basis in understanding for mastering facts.

Presenting Olympiad-like problems as *puzzles to develop mathematical thinking*, not as *tests of mathematical prowess*, and discussing them in the ways that these coaching books suggest might well help new teachers see both how real problems differ from practice exercises, and also how "challenging" does not just mean bigger numbers. As in R. P. Burns's (1997) *A Pathway Into Number Theory*, problems can be the *starter*, not the culmination or exercise, of an idea. And the problems can also be openings for dealing with unknown or incompletely understood content. The challenging part of the two jewelry problems above, for example, is that they appear not to provide enough information to solve them, but the knowledge needed for them is only that the four sides of squares have equal length and that the circumference of a circle is determined by its diameter. Playing (not being tested!) with these problems can uncover gaps (e.g., having forgotten how to determine a circumference from a diameter) that can be filled unsystematically as needed or can lead to more systematic discussion.

None of the models above, used as is, would constitute a preservice course in itself, but all four have potential for use in the thoughtful creation of preservice courses that focus on giving teachers the experience of *doing* mathematics, a kind of course that, to the best of our knowledge, does not exist or, at least, is not at all common. We can imagine creating a PROMYS-for-Teachers-like experience for elementary teachers; a DMI-like structured analysis of student thinking that works with students who have not yet taught; a programming experience that leads with the mathematics, not the coding; and a Math-Olympiad-like experience in problem solving and coaching children to be problem-solvers.

3.7 Prepare Teachers to Take a Developmental Perspective

Nicolina Malara and Giancarlo Navarra compare the development of mathematical language—in particular, the writing of mathematical expressions—to the development of natural language and coined the term "algebraic babbling." They posit that in both languages, the babble is a natural, even important, stage en route to the mature forms.

> The control of the syntactical aspects of a new language is obtained through its semantic control. On acquiring a natural language, the child gradually realizes its meanings and the rules that support them, which gradually develop up to school age, when he learns to read and reflect on the *structural* aspects of the language. Similarly, it is believed that the mental models of algebraic thinking should be organized right from the early years of primary school, constructing algebraic thinking, in a tight intermingling with

> arithmetic, by starting from its *meanings*. It is therefore necessary to build an environment promoting the autonomous development of formal encodings of sentences in the natural language and their collective comparison. This allows the experimental acquisition of the new language [mathematical symbolism], whose rules gradually ripen within a didactic contract which *tolerates* initial, syntactically shaky moments. We call this process of construction/interpretation/refinement of the 'rough' writings *algebraic babbling*. (Malara & Navarra, 2015)

This is analogous to what teachers already accept about invented spellings and mispronounced words. They are works in progress. It is worthwhile remembering that symbolic algebra with the use of variables is less than 500 years old, and much of the notation of arithmetic (for example, place value notation) is not much older. People have been solving algebra problems for 3000+ years, but without the notation.

The developmental perspective we should help teachers acquire must also include that mathematical growth is not monotonic. It is not that children know it one day and forget the next; it is that the whole process is messy. They learn things, learn new things, reassemble their learning, appear to backtrack, and only later does it click. Proof that they are not totally forgetting? Well, if they really *do* forget, then the next encounter is once again their first encounter. We are doomed. Learning takes some focused clock time; it also takes calendar time—sleeping on it, so to speak, and coming back to it later. "Spiral" can be overdone, and can be badly done, but the need for breaks and revisiting is the reason that "spiral curriculum" ever came into existence. It is the way we *all* learn, except when we are cramming. And then, as we all know, the cramming works wonders, but only for the duration of the exam. We need to revisit later in order to regain and reassemble that learning.

And finally, taking a developmental perspective involves seeing age-appropriate elementary school learning in the context of the long view toward mathematics.

3.8 Learning to Relax

When one of us (EPG) was teaching at a moderately prestigious graduate school, he read applications from students who had studied at (mostly other) moderately prestigious colleges. Many, not just a few, wrote that they wanted to focus on the primary grades "because I know I can do the math." This is an indictment of the entire K–16 education system. And it is not because these students do not know more than second grade math—they most certainly do. In fact, many had taken and succeeded in AP calculus in high school and then more (though only rarely more advanced) mathematics in college preparing for teaching. It is an indictment of their K–16 experience because they *feel* that all they can do is second-grade math. What they are lacking is a sense that they understand any of it. As for the earliest primary grade mathematics, they likely do not feel there is anything *to* understand—it is all either so obvious or so mechanical that it poses no cognitive problem for them.

But we can see teachers' residual anxiety in how they approach mathematics in class. Correcting every last thing may well reflect that anxiety. They feel that they, themselves, did not learn well and they want to do better by their students, so they leave no room for error.

We need to help preservice (and in-service) teachers relax so that they can, in their mathematics teaching, use the logic, common sense, and judgment that they clearly have and use elsewhere. Wine might help but is frowned upon in school.

Teachers would certainly feel more relaxed if they felt more confident that they really knew and understood their mathematics well: as we have said before, part of teacher preparation clearly does need to work on their mathematical understanding. But by the time they are applying to graduate school, pre-teachers have had a lot of experience; if it has not been the right experience, merely "more" is not the solution. Even in undergraduate school, an approach to mathematics that is primarily instruction on facts and how they work will not change the experience teachers have already had: they are still being shown, not being given the time and guidance to puzzle it out themselves. They are acquiring information and applying it in exercises, not being immersed in the doing of mathematics. At this point, their idea of what mathematics is may well be pretty fixed: it is stuff to know. Focusing (as we should) on how it works may easily be seen as just *more* stuff to know because that is what they believe the whole enterprise to be.

They need something new. Some of that could certainly be explicit in the "methods" courses: Relax. Slow down. Pick up the pace. (Wait! Didn't we just say "slow down"? The first is about being more patient. The second is about being less draggy and repetitive.) Let children self-correct, puzzle things out. Or, as Dan Meyer (2010) suggests, "Be less helpful" (see also Peck, 2018). After all, children trip on shoelaces and fall off bikes, and they somehow work it all out. And they might well have the principles and logic, but just lack uniformity—learning is not monotonic—so we can take it a bit slower about correction. But that is just propaganda until they have had those experiences themselves. Telling teachers how to feel and giving them "teacher moves" (like ease up on correcting) is not enough. If we are truly serious about children learning to *think* mathematically, not just follow rules blindly (even if the rules, followed correctly, work), then we must treat teachers the same way. Relax cannot mean "do not care." Slow down cannot mean "be monotonous." Let children puzzle things out cannot mean "just let them flounder." So exactly *how* unhelpful must I be? And what help really does *help* rather than replace the child's thinking?

Olympiad-like problems appear even on some licensure tests (the Massachusetts Tests for Educational Licensure (MTEL, 2020), for one example), giving them a kind of credibility that teachers-in-preparation might find comforting. But the experience needs to be playful, puzzles to solve, not hoops to jump through or test-prep cramming. We do puzzles for entertainment, for the sheer joy and challenge of solving them, not for the solution: the solution is not mission-critical to anything. It is just a puzzle—a *puzzle*, not a *problem*. The puzzles also have to be not the gotcha kind that depend on reinterpreting the information in some special way that is idiosyncratic to that puzzle: they have to be things that are totally solvable with normal thinking and course-relevant knowledge (in our case, the arithmetic, algebra, and geometry of elementary school) but, like the examples shown above, require thought or lead to some interesting idea. Or can be done in a brute-force way but can also be approached more elegantly. A lot of genuine mathematics has arisen as ways to solve problems in easier ways, and this is how we would like our teachers to see the discipline: its goal is not to make life harder, but to make it easier.

Given a "math problem" to solve, many people assume that they are supposed to see immediately how to solve it and if they do not know what to do they are somehow deficient. Puzzles are different. You know, even before you start, that a crossword puzzle's 100 or so clues will be mostly useless at the start, but you search for the 1 or 2 you think you know, and then check your tentative answer with other clues. Unlike the classroom problem, which you may experience as a test and which raises anxiety if you do not immediately see how to do it, you give yourself more time on a puzzle to search the clues, experiment, and then make a next move. Nothing depends on the solution of a particular puzzle—no bridge will collapse if you do not get the right answer—but doing puzzles makes you better at doing more of them, just like reading makes you better at reading and, in general, immersion in a language makes you better at that language. Puzzles—even puzzles like SolveMe, https://solveme.edc.org/, or various others that, unlike crosswords, use mathematical content and build some skill with the content and the logic behind it—are not a substitute for an orchestrated sequence of mathematical acts that build mathematical learning, but we would love that puzzle-solving demeanor to be the feeling students bring to and experience in mathematics class, so we need teachers to have that feeling and take that approach. And some puzzles (one example is Futoshiki, https://www.futoshiki.org/), used well, may build some of the skills and dispositions (search for starting places, logic, persistence) that are needed for proof. But the real key here is just seeing "problems" as "puzzles." Research shows that presenting a set of items as puzzles generates higher performance than presenting the exact same items as a test (Brown, Day, 2006). The particular context of the Brown and Day work was stereotype threat, but the observation is robust and makes sense. Test anxiety is common, and we know that a large portion of our elementary school teachers are math-anxious.

Reduce that anxiety. Let teachers *play* with mathematics so that they can take the same relaxed approach with their students. Whatever role exercise-dressed-up-as-game—e.g., a shoot-'em-up video game to drill math facts—might serve, exercise-dressed-up-as-game is not at all what we mean by useful play. The "game" in those is entirely extrinsic to the mathematics. We mean the kind of tinkering, trying things, seeing what happens, *playing* with ideas, sensemaking, and reasoning that mathematicians do when they approach a new problem, but doing all that at a mathematical level that is suited to the teachers we are preparing—seeing the unknown as a puzzle to tackle, rather than as a threat or a problem to overcome. That is where suitably designed immersion models can help: PROMYS-like experiences, accessible mathematical programming environments, and Olympiad-like puzzles.

One can also tap KenKen puzzles, which do "drill" math facts,[23] but are more about the logic than about arbitrary random drill. Solvers of such puzzles are *doing* mathematics; it is super-limited and super-elementary, but they are *doing* the mathematics, not being shown how to do it.

One of us just asked his son how studying for the bar differed from the law school studies he had just completed. The answer had a fascinating ring for mathematics education: "In law school we were learning how to think through situations and make decisions when there is not already a law that covers that case; but we didn't learn many actual laws. That's what I'm studying now when I study for the bar." The law school study was about "figuring out what to do when you don't know what to do," the element of teacher preparation that, at the very beginning of this paper, we claimed is missing.

References

BJC. (2020). Retrieved from https://bjc.edc.org/ and https://bjc.berkeley.edu/.

Brown, R. P., & Day, E. A. (2006). The Difference isn't black and white: Stereotype threat and the race gap on Raven's Advanced Progressive Matrices. *J. Appl. Psychiatry, 91*, 979–985.

Burns, R. P. (1997). *A pathway into number theory*. New York: Cambridge University Press.

Cuoco, A. (2018). Some applications of abstract algebra to high school teaching. In N. Wasserman (Ed.), *Connecting abstract algebra to secondary mathematics, for secondary mathematics teachers*. New York: Springer.

Cuoco, A., Goldenberg, E. P., & Mark, J. (1996). Habits of mind: An organizing principle for mathematics curriculum. *Journal of Mathematical Behavior, 15*(4), 375–402.

EDC. (2020). *Elementary math at EDC*. Retrieved May 1, 2020 from https://elementarymath.edc.org/programming.

Goldenberg, E. P., & Carter, C. (2017). A student asks about (−5)! *Mathematics Teacher, 111*(2), 104–110.

Goldenberg, E. P., Mark, J., Kang, J., Fries, M., Carter, C., & Cordner, T. (2015). *Making sense of algebra: Developing students' mathematical habits of mind*. Portsmouth, NH: Heinemann.

Goldenberg, E. P., Miller, S., Carter, C., & Reed, K. (2017). *Mathematical structure and error in kindergarten* (Young children) (Vol. 72(2)). Washington, DC: NAEYC.

Greenes, C., Chang, K. Y., & Ben-Chaim, D. (2007). International survey of high school students' understanding of key concepts of linearity. In *Proceedings of the 31st Annual Meeting of the International Group for the Psychology of Mathematics Education* (Vol. 2, pp. 273–280). Seoul, Korea.

Harel, G. (2008). DNR perspective on mathematics curriculum and instruction, Part II: With reference to teachers' knowledge base. *Zentralblatt für Didaktik der Mathematik, 40*, 893–907. https://doi.org/10.1007/s11858-008-0146-4.

Kalman, R. (Ed.). (2008). *Math Olympiad contest problems* (Vol. 2). Bellmore, NY: Mathematical Olympiads for Elementary and Middle Schools.

[23]And variations on this theme, like the MysteryGrid, https://solveme.edc.org/mysterygrid/, puzzles that are part of the SolveMe suite, both expand the pure elementary-school fact practice and extend the logic to polynomial arithmetic.

Malara, N., & Navarra, G. (2015). *Generalization questions at early stages: The importance of the theory of mathematics education for teachers and pupils*. Retrieved July 28, 2017, from http://directorymathsed.net/montenegro/Malara.pdf.

McCrory, R., Ferrini-Mundy, J., Reckase, M., & Senk, S. (2012). Knowledge of algebra for teaching: A framework of knowledge and practices. *Journal for Research in Mathematics Education, 43*, 584–615. https://doi.org/10.5951/jresematheduc.43.5.0584

Meyer, D. (2010). Retrieved April 1, 2020, from https://www.ted.com/talks/dan_meyer_math_class_needs_a_makeover.

MOEMS. (2020). *Mathematical Olympiads for elementary and middle schools*. Retrieved April 1, 2020, from http://www.moems.org/.

MTEL. (2020). Retrieved April 1, 2020, from https://www.mtel.nesinc.com/.

Papert, S. (1972). Teaching children to be mathematicians versus teaching about mathematics. *International Journal of Mathematical Education in Science and Technology, 3*(3), 249–262. http://hdl.handle.net/1721.1/5837.

Peck, F. (2018). Learning to be less helpful. *AMS Blogs: On Teaching and Learning Mathematics*. Retrieved April 1, 2020, from https://blogs.ams.org/matheducation/2018/12/17/learning-to-be-less-helpful/.

PfT. (2017). *PROMYS for teachers*. Retrieved 29 July, 2017, from www.promys.org.

PROMYS. (2017). *Program in mathematics for young scholars*. Retrieved 29 July, 2017, from www.promys.org.

Schifter, D., Bastable, V., & Russell, S. J. (2017, 2018, 2019). *Developing mathematical ideas*. Casebooks, facilitators guides, and videotapes. Reston, VA: National Council of Teachers of Mathematics.

Scratch. (2020). Retrieved April 1, 2020, from https://scratch.mit.edu/.

Snap*!*. (2020). Retrieved April 1, 2020, from http://snap.berkeley.edu/.

University College London Institute of Education [UCLIE]. (n.d.). *ScratchMaths*. Retrieved March 22, 2017, from https://www.ucl.ac.uk/ioe/research/projects/scratchmaths.

Wirtz, R., Botel, M., Beberman, M., & Sawyer, W. W. (1964). *Math Workshop*. Retrieved from https://hdl.handle.net/2027/mdp.

Wu, H.-H. (2015). *Mathematical education of teachers, Part I: What is textbook school mathematics?* American Mathematical Society. Accessed March 5, 2017, http://blogs.ams.org/matheducation/2015/02/20/mathematical-education-of-teachers-part-i-what-is-textbook-school-mathematics/#sthash.3Gzvh1Hp.dpbs/.

Chapter 4
The Elementary Mathematics Project: Supporting Preservice Teachers' Content Knowledge for Teaching Mathematics

Suzanne H. Chapin, Lynsey K. Gibbons, Ziv Feldman, Laura Kyser Callis, and Alejandra Salinas

Mathematics teacher educators have always been concerned with how to prepare future elementary teachers to teach mathematics. Over the last 12 years, several policy reports have been compiled that aim to answer: What should future elementary teachers know about mathematics and be able to do when they graduate from a teacher preparation program (AMTE, 2017; CBMS, 2001, 2012; U.S. Department of Education, 2008; Greenberg & Walsh, 2008; Hill, Rowan, & Ball, 2005; Hill, Sleep, Lewis, & Ball, 2007)? The Elementary Preservice Teachers Mathematics Project, currently referred to as the Elementary Mathematics Project (EMP), has been designed to respond to the call to develop and deepen future elementary teachers' mathematics knowledge for teaching. Since 2009, the project has been developing and at the same time researching our instructional materials (curriculum) for use in mathematics content courses for prospective elementary school teachers. In this chapter, we draw on Clements' (2007) definition of the term, curriculum—

This material is based upon work supported by the National Science Foundation under Grant No. DUE-1625784, the Elementary Mathematics Project (2016–2021), Grant No. TUES-1323156 (2013–2015) and Grant No. CCLI-0837349 (2009–2011). Any opinions, findings, and conclusions or recommendations expressed in this material are those of the authors and do not necessarily reflect the views of the National Science Foundation.

S. H. Chapin (✉) · Z. Feldman · A. Salinas
Boston University, Boston, MA, USA
e-mail: schapin@bu.edu; zfeld@bu.edu; Salinas@bu.edu

L. K. Gibbons
University of Delaware, Newark, DE, USA
e-mail: lgibbons@udel.edu

L. K. Callis
Curry College, Milton, MA, USA
e-mail: laura.callis@curry.edu

Y. Li et al. (eds.), *Developing Mathematical Proficiency for Elementary Instruction*, Advances in STEM Education, https://doi.org/10.1007/978-3-030-68956-8_4

"curriculum is a written instructional blueprint and set of materials for guiding students' acquisition of certain culturally valued concepts, procedures, intellectual dispositions, and ways of reasoning (p. 36)"—to describe the focus of our work with preservice elementary teachers (PSTs). Our fundamental principles have not changed since the inception of the project, but the embodiment of our goals and guiding principles have been refined and improved.

This chapter shares information about the design, development, testing, and implementation of the EMP curriculum. The curriculum currently consists of seven units of study: Number Concepts, Addition and Subtraction, Multiplication and Division, Number Theory, Fractions, Geometry, and Geometric Measurement (see Appendix for a list of lessons in each unit). When used together, these units provide two or three courses (depending on credit hours) of mathematics content for preservice elementary teachers. Units vary in length based on the content, from 6 to 11 lessons in each. Lessons also vary in length from 60 to 120 min and are often taught over two class periods. The goals of the EMP project, focused on undergraduates' and instructors' learning, have guided the development and implementation of the curriculum. The curriculum is designed to

- Support preservice elementary teachers in deepening their knowledge of mathematics for teaching.
- Support post-secondary instructors in teaching mathematics content for teaching to preservice elementary teachers.
- Support preservice elementary teachers in developing their mathematical identities as mathematical thinkers and future mathematics teachers.
- Support both preservice elementary teachers and their instructors in using discussions to deepen their understanding of mathematics.

4.1 What Are the Current Recommendations for PSTs Experiences in Content Courses?

The Conference Board of the Mathematical Sciences (CBMS) in their reports titled *The Mathematical Education of Teachers I and II* (*MET I, MET II*) recommended the mathematical preparation of elementary teachers focus on developing a deep and comprehensive view of the mathematics in the elementary and middle school curriculum. The content they suggested for future elementary teachers aligns with the domains and practices detailed in the Common Core State Standards for Mathematics (CCSSM) at these grade levels (National Governors Association Center for Best Practices, Council of Chief State School Officers, 2010) and includes topics such as number and operations (whole numbers as well as rational numbers), algebra, data, measurement, and geometry. Subsequently, in 2017 the Association of Mathematics Teacher Educators released their report, *Standards for the Preparation of Teachers of Mathematics*. These standards concur with earlier content recommendations and further highlights the importance of elementary teachers understanding the mathematics they will teach. This knowledge of mathematics for teaching is often referred to as *content knowledge for teaching mathematics* (CKTM) (Wilhelm, 2014) and includes both common content knowledge (CCK) and specialized content knowledge (SCK) (Ball, Thames, & Phelps, 2008). Common content knowledge is the mathematical knowledge that an everyday citizen holds such as being able to calculate a percentage or add two fractions. Specialized content knowledge for teaching mathematics includes teachers' abilities to (1) analyze and interpret students' mathematical thinking and ideas, (2) use and connect multiple representations of mathematical concepts, (3) justify fundamental ideas in elementary mathematics, and (4) define terms in mathematically correct and accessible ways.

The authors of the *MET II* and the AMTE *Standards for Preparing Teachers of Mathematics*, along with other mathematics education researchers, also indicated how future teachers should learn mathematics. For example, "Effective mathematics teacher preparation programs ensure that candidates are immersed in mathematical practices and processes of reasoning, sense making and problem solving" (AMTE, 2017, p. 31). Furthermore, the *MET II* report acknowledges that "*engaging in mathematical practice takes time and opportunity*, so coursework... for teachers must be planned with that in mind. Time and opportunity to think about, discuss, and explain mathematical ideas are essential for learning to treat mathematics as a sense-making enterprise" (CBMS, 2012, p. 24, emphasis original). Other researchers highlight the need for future educators to be taught mathematics with understanding (Eisenhardt et al., 1993; Ma, 1999; Simon, 1993; Sowder, 2007) and for PSTs to develop mathematical proficiency, which includes a productive disposition as well as conceptual understanding, procedural fluency, strategic competence, and adaptive reasoning (National Research Council, 2001).

While there have been extensive recommendations about content courses for elementary PSTs, there are few studies of what is actually occurring in the field. Masingila, Olanoff, and Kwaka' (2012) study of 825 institutions that teach content courses found that at the time, the majority of these teacher preparation institutions were not meeting the course and content recommendations put forth by the CBMS in the *MET I* document (2001) or by the National Council on Teacher Quality (Greenberg & Walsh, 2008). In addition, Masingila's study found that instructors at these institutions were not extensively using methods that model the desired elementary classroom experiences: only 13% reported using primarily an activity-based format, 71% reported using a combination of lecture-based and activity-based formats, and 14% stated that the format was primarily lecture-based. Masingila's study also determined that the academic and professional background of instructors teaching mathematics content courses for elementary teachers varies considerably; roughly half of institutions report it is common for instructors to hold a masters or doctorate degree in mathematics, while the other half report that instructors hold a masters or doctorate degree in mathematics education. Regardless of background, the majority of instructors of these courses do not have elementary teaching experience. Studies of other mathematics-for-teachers courses have found that instructors of different academic backgrounds may not share the same understanding of the language typically used in the K–12 mathematics education community (Bleiler, 2014). This suggests that faculty members who teach these courses need support or training for teaching this population of students.

More recent research suggests that instruction in content courses for preservice elementary teachers (PSTs) is changing. As part of her dissertation study, Callis (2017) surveyed 396 college instructors to identify the instructional practices used in college-level mathematics content courses for elementary teachers. Instructors in her study indicated that their instruction was aligned with both content and pedagogical recommendations mentioned above. For example, PSTs in more than 75% of survey participants' classrooms spent less than 50% of their time listening to lectures and instead were engaged in more interactive instructional practices such as small group work, student presentations, and discussions.

4.2 What Content Should Be Learned?

One of our first considerations was what content to include in a curriculum for future elementary teachers. We decided to focus on CKTM as a way to support future teachers' understanding of the content described in the CCSS-M documents. Research on PSTs' learning of topics informed our choices around which mathematical topics to address (Browning, Edson, Kimani, & Aslan-Tutak, 2014; Kastberg & Morton, 2014; Newton, 2008; Olanoff, Lo, & Tobias, 2014; Thanheiser, Whitacre, & Roy, 2014). In addition, recommendations from the *MET I* and *MET II* reports (CBMS, 2001, 2012) identified the need for elementary teachers to develop a deeper understanding of elementary mathematics, with particular attention to understanding the properties of operations, making meaning for widely used whole number and fraction algorithms and measurement formulas.

An important consideration that informed our work is that PSTs have been exposed to elementary school content as younger students. They may not have had opportunities to deeply understand the content, but they do have some knowledge of it. Namely, we were not introducing content to them for the first time. With this in mind, we focused most of our work on developing problem-based lessons that helped future teachers deepen their understanding of concepts, procedures, and the underlying structure of our mathematical system. We did not include the reteaching of procedures per se, but instead focused our curriculum on more complex questions that require learners to explain, justify, create, conjecture, and connect ideas such as, "When subtracting, 74 − 29, what do we mean when we say we are 'borrowing from the 7'? How might a child interpret the word, borrow? What can we say that might better explain mathematically what is happening?" or "Describe the mathematical ideas and properties that make it possible for us to multiply two two-digit numbers together and get a correct product."

Yet, even when only focusing on CKTM and the content of the grades K–5 CCSS-Mathematics standards, the EMP team concluded that it was impossible to address all mathematics content and pedagogical implications suggested in policy documents in 8–9 credits of coursework for future K–5 teachers—there just is not enough time. Thus, we decided to focus the lessons on "high-leverage content"—the particular topics and practices that are vital for beginning K–5 teachers to understand and be able to teach. What content should be classified as high-leverage content? While there is some agreement among researchers as to what are "high-leverage teaching practices," there is scant information on high-leverage content. The Teaching Works website from the University of Michigan (2015) states, "Examples of high-leverage content in mathematics at the elementary level include place value, number concepts and operations, geometry, fractions, and representing and explaining mathematical ideas and relationships." We agreed that this list of topics was fundamental but felt that other content such as decimals, geometric measurement, and elementary number theory needed to also be addressed. Though the study of number theory is not supported in policy and standards documents to the same extent as the other content, researchers have found that prior to instruction, PSTs' knowledge of factors, multiples, prime factorization, and divisibility is uncoordinated and procedurally oriented (Feldman, 2012; Zazkis, 1998; Zazkis & Campbell, 1996a, 1996b). Our collective experience working with PSTs called attention to the importance of supporting their understanding of the multiplicative structure of integers and the connections between prime factorization and multiplication and division.

In order to reduce the number of topics, some elementary content is included but only peripherally—linear measurement is embedded in lessons on decimals and number lines but not in significant ways. We also made decisions to not include topics such as proportional reasoning, data, and ideas of algebra (e.g., expressions, equations, and linear relationships), topics that are the focus of instruction at the middle school level. It is never easy to leave out content that we know will provide future teachers with depth and breadth when teaching mathematics. Our choice of content provides just one approach to the dilemma of managing time and emphasis when preparing future elementary school teachers to teach mathematics.

4.3 How Should Content Be Learned?

Policy documents referenced earlier (AMTE, CBMS) suggest that content courses for elementary PSTs should be organized to support learners to engage in problem-solving and sensemaking. That teachers often teach the way they were taught (Feiman-Nemser, 2001; Lortie, 1975) highlights the critical need for PSTs to have experiences learning mathematics that are student-centered and focused on mathematical meaning. Studies have also shown that PSTs' achievement is higher in elementary content courses where instructors reported less lecturing and more student engagement (Castro Superfine, Li, & Martinez, 2013; McCrory, Zhang, Francis, & Young, 2009). In a study by Hart, Oesterle, and Swars (2013), undergraduates in mathematics courses for future elementary teachers described instructional practices that supported their learning such as working in groups, discussing ideas, being allowed to "do the thinking," and using manipulative materials. This was in contrast to practices that they described as not supporting their learning and also negatively affecting their views of mathematics—using lectures, note taking and Powerpoints where material was simply presented and expected to be understood. In general, the focus was not on understanding the concepts but instead being able to reproduce procedures at an abstract, theoretical level. While policy recommendations suggest that learning in these courses build connections to the elementary classroom, student subjects in this study felt that the non-engaging instructional approaches did little to address their future work as teachers.

In terms of the experiences we wanted PSTs to have with the EMP curriculum, the authors took a theoretical orientation based on sociocultural theory. Represented in the works of Forman (1996, 2003), Brown, Collins, and Duguid (1989), and Lave and Wenger (1991), sociocultural theories posit that learning occurs through coordinated social activity and is mediated by the use of language and the context in which the learning is situated. Learning mathematics involves participation in communities of practice. From this perspective, learning is conceptualized as coming to know how to participate in the discourse and practices of a particular community. Our premise is that what PSTs take as knowledge and how they think about and express ideas are the products of their interactions with the content and with each other in the classroom. They appropriate as their own the shared ideas, concepts, and conclusions in which this community engages through group work and discussions. Recurring cycles of engagement, similar to Simon's (1994) work in teacher education courses, support PSTs in becoming more skilled in using the concepts and forms of reasoning. Furthermore, as their dispositions shift to expecting that they will make sense of ideas, PSTs are further enculturated into the community, changing it through their new ways of thinking.

Thus, the EMP curriculum was written to engage PSTs in two communities of practice, a community of mathematical learners and a community of beginning mathematics teachers. Taking the perspective that PSTs belong to a community of mathematical learners, lessons support them in participating together in mathematical practices such as identifying patterns, making connections, justifying their thinking, and articulating generalizations. We wanted PSTs to have different experiences learning mathematics that engaged them and that helped them develop confidence in their understanding (Gibbons et al., 2018). In addition, we also wanted PSTs to be inducted into a community of beginning teachers. Our curriculum poses problems that help PSTs notice and reflect on what K–5 students understand as a way to develop their own mathematical understanding. Furthermore, questions and problems ask future teachers to articulate relationships that link the two communities (e.g., "Why can we decompose numbers in our place value system and how does this support children's computational strategies?" and "Create two word problems for $36 \div 9$ that elementary students might interpret differently. Explain your reasoning.").

The use of mathematical discussions is one pedagogical strategy we chose to emphasize because of its potential to support PSTs in engaging in both communities of practice: learners of mathematics and future teachers of mathematics (Carpenter & Lehrer, 1999; Chapin & O'Connor, 2007, 2012; Elliott & Garnett, 2008; Gibbons et al., 2018; Hiebert & Carpenter, 1992; O'Connor, Michaels, Chapin, & Harbaugh, 2017; Walshaw & Anthony, 2008). With this in mind, the EMP lessons were intentionally structured so that about 85% of every class period was allocated to group work followed by whole class mathematical discussions, leaving the rest of class time for other activities such as presentations. Yet using discussions as a mode of instruction can be challenging since it is unpredictable which ideas and questions will arise as PSTs respond to each other during small group and whole class discussions. We conjectured that instructors would need additional support in responding to the unexpected and in keeping the focus of the discourse on the mathematical meaning of important concepts and relationships. Depending on their background training as either a mathematician or a mathematics educator, instructors might not have had prior experiences with facilitating discussions, soliciting multiple responses, or pressing PSTs to connect approaches and meanings themselves. Hence, supports for instructors were created, including detailed Instructor Guides that explicate our methodologies, videos of lessons enacted at our university, and many examples of both PSTs' responses during discussions and instructors' in-the-moment decision-making.

4.4 The Development of the EMP Curriculum

Work on the EMP curriculum started in 2009 and continues as of the writing of this chapter. Much of our work has been funded by the National Science Foundation (Grant Nos. TUES-1323156, CCLI-0837349, and DUE-1625784), but there have been years when we have worked on refining materials without funding. Clements (2007) Curriculum Research Framework provided us with guidelines for categories and phases of curriculum development.

In Phase I (2009–2011), units were written and revised by a team of faculty members, graduate students, and teachers: whole numbers, fractions, number theory, geometry, geometric measurement, and ratio and proportion. First, a priori research on mathematics content and pedagogy was reviewed and used to make decisions about both the content and design of the units. This resulted in units consisting of 6–15 lessons for use with elementary preservice teachers. Next, these units were piloted, revised, and re-piloted in Boston University's mathematics content courses for preservice elementary teachers. Data were collected on PSTs achievement using pre- and post-unit project-generated assessments.

In Phase II (2013–2015), three of these units, fractions, geometric measurement, and number theory, were rewritten by a team to reflect our current thinking about these topics. We again examined the most recent research on the topics as well as refined our design principles regarding mathematics content and pedagogy. We solicited extensive feedback from experts in the field. Following three iterations of revisions, these three units were piloted at seven colleges with different student populations, revised again to reflect the feedback from pilot-testers, and field-tested nationally at 33 colleges and universities with 46 instructors and 1500 students. Using project-generated assessments, results showed statistically significant growth in undergraduate knowledge following instruction for all three units, with large effect sizes of 1.2–1.7 (Salinas, Feldman, Callis, & Chapin, 2016).

In addition, educative instructor guides, homework files, and answer keys for every lesson were created. Documents describing our pedagogical and content foci and suggestions on how to implement our materials also were written. The EMP curriculum for these units was made available to college instructors on a web-based platform (www.elementarymathproject.com). Finally, during Phase II, we videotaped every EMP lesson as it was taught in courses at Boston University. These videos are of undergraduates and graduate students discussing key mathematical ideas explored in the materials. Following editing, we uploaded two to five video clips (they range in length from 2 to 20 min) to the website for every lesson. Commentary accompanied each video, highlighting what the BU instructor was or was not doing to facilitate discussion, sensemaking, and community building.

Phase III of the project (2016–2021) has focused on the development of four new units. The original whole number unit was separated into three distinct units (Number Concepts, Addition and Subtraction, and Multiplication and Division), and all the lessons were rewritten, revised, or created from scratch. Likewise, our initial geometry unit underwent substantial edits. We refined our educative instructor guides to respond to users' suggestions and added additional homework problems to lessons based on instructor's requests. As before, at the start of the process, we examined a priori research, current reports and sought the advice of our advisory board. Our experiences from Phase II also informed us in terms of length of lessons and units, usability, and effectiveness of components. Pilot testing occurred at 11 institutions and information gleaned from these users informed us in terms of edits to the materials. In particular, we further shortened lessons, eliminated or clarified content, and included more information for instructors on how to support struggling students.

Professional development for instructors interested in using our materials has been provided through Zoom virtual communities, workshops, presentations at conferences, summer institutes held at Boston University, and with video clips of the enactment of all lessons. Similar to Phase II, we videotaped the lessons from these four units as they were taught at BU during 2017–2018, edited them, and uploaded selections to the website to support instructors. In addition, we created a video library consisting of hundreds of video clips of preservice elementary teachers engaged in learning mathematics from all seven units. There are clips of whole class discussions, problem-solving, student presentations, student debates, and review work. In all instances, the PSTs are working to make sense of the mathematics and explain their ideas. Faculty members can search the video library based on their questions about content or pedagogy.

During 2018 and 2019, we nationally field-tested the four new units. In this iteration of testing, we used standardized assessments (i.e., the Learning Mathematics for Teaching instruments from the University of Michigan) as well as collected survey data on undergraduates' beliefs and attitudes about mathematics. In particular, we are asking questions of PSTs about their confidence in making sense of elementary mathematics and their beliefs about the role of discussion in their learning following their engagement with the EMP curriculum. Finally, using surveys and interviews, we are collecting data from the field-testing instructors about the curriculum, Not only are we interested in implementation variables that instructors have dealt with, but we also are interested in learning whether the pedagogy inherent in the materials has influenced instructors and if so, how.

4.5 Design Principles

In this section, we describe our two design principles on which the instructional materials are based. We hold that PSTs' learning benefits from: (1) a focus on problems and discussion questions that address CKTM at the elementary level; and (2) engagement in recurring cycles of collaborative problem-solving and group discussion. Whereas the first design principle highlights the content of the materials, the second principle emphasizes the structure of the presentation and enactment of the lessons. Problems and discussion questions are cognitively demanding and focus on content and practices important to the learning and teaching of mathematics. As such, these two principles are not mutually exclusive; in fact, considerable overlap was built into the design of materials.

4.5.1 Design Principle I: A Focus on Problems That Address Content Knowledge for Teaching Mathematics (CKTM) at the Elementary Level

The EMP lessons were designed to include the following four types of problems: (1) connecting mathematical concepts and ideas; (2) understanding why formulas, rules, or algorithms work; (3) using mathematical structure; and (4) identifying the mathematics in elementary students' thinking. We argue that these four problem types not only help PSTs' deepen their content knowledge for teaching mathematics (CKTM), but they also support PSTs' engagement in mathematical practices such as explaining, justifying, making connections, conjecturing, and generalizing. It is important to recognize that the problem types are not mutually exclusive; in other words, a particular problem can often be classified as more than one type. For example, many of our problems focus on mathematical concepts and explaining why algorithms work. Likewise, problems that highlight elementary students' thinking often attend to underlying mathematical concepts, rules, or structures. While the first three problem types support PSTs in developing as mathematical learners, the fourth type of problem was written to support PSTs in joining a community of future teachers, in which they must notice and respond to elementary students' mathematical thinking. Below we briefly describe each problem type and provide an example. We then illustrate the problem types using excerpts of transcripts in which

PSTs at Boston University are exploring various computation strategies from one lesson, the Addition Strategies lesson.

4.5.1.1 Problem Type 1: Connecting Mathematical Concepts and Ideas

The first problem type focuses on helping PSTs deepen their understanding of mathematical concepts and ideas found in the elementary curriculum and then connecting these concepts and ideas. This can be done in a variety of ways, such as comparing and contrasting ideas or procedures, using physical or visual models to represent the same mathematical objects, or attending directly to mathematical properties and generalizing their usage. When individuals strengthen their knowledge of concepts, they can use their knowledge flexibly to identify and connect different representations (Carpenter & Lehrer, 1999; National Research Council, 2001). PSTs need to develop conceptual understanding of the mathematical ideas of the elementary curriculum, so they can support their students in making mathematical connections.

An example of this type of problem is shown in Fig. 4.1. Throughout the Addition Strategies lesson, PSTs are asked to analyze many different addition strategies that hypothetical students used to compute. Then at the conclusion of the lesson, PSTs compare and contrast the various addition strategies. These strategies are based on decomposing and then recomposing addends and/or compensation. In order to sort the strategies into groups, PSTs must recognize their similarities and the underlying mathematical concepts on which the strategies are based. As a result, the role of place value, properties of operations (e.g., commutative, associative, additive inverses, identity property of addition), and different ways to decompose/recompose addends become more salient and are connected. PSTs better understand these core components through classification and discussion.

4.5.1.2 Problem Type 2: Understanding Why a Formula/Rule/Algorithm Works

The second problem type attends to why a formula, rule, or algorithm works. This may include asking PSTs to explain why a strategy works or is mathematically valid, or to derive well-known formulas (e.g., the area of a trapezoid). This

> Examine the strategies used by Clarissa, Molly, Karl, Priyanka and Fahmil. Sort their strategies into groups where the strategies are similar. For each group, identify the common mathematical ideas.
>
> *Note: This question is at the end of the lesson after PSTs have worked on problems about the listed students' strategies. These strategies are shared with the reader in subsequent figures and examples.*

Fig. 4.1 *Connecting Mathematical Concepts and Ideas* problem from the Addition Strategies lesson

a) Fahmil found two ways to compute 37 + 98. Examine his work. Explain what Fahmil did and why it works.

37 + 98	37 + 98
= (37 + 3) + (98 − 3)	= (37 − 2) + (98 + 2)
= 40 + 95	= 35 + 100
= 135	= 135

b) Make up 3 new addition problems and use Fahmil's strategy to compute their sums.

Fig. 4.2 *Understanding Why Formulas/Rules/Algorithms Work* problem from the Addition Strategies lesson

problem type appears frequently throughout the EMP curriculum, as much of what elementary teachers do involves making sense of the strategies and algorithms that children work with in the elementary grades. If future teachers do not know why mathematically we can perform certain steps in a computation, they are less likely to be able to assist students in learning with understanding (Ma, 1999; Philipp et al., 2007).

An example of this type of problem is shown in Fig. 4.2. In this problem, PSTs explore a compensation strategy for addition in which both addends are adjusted in order to make an easier computation. PSTs often recognize that Fahmil adds a certain quantity to one addend and subtracts the same quantity from the other addend, but struggle to articulate why the expressions are equivalent. Justifying the validity of this strategy requires recognizing that adding and subtracting the same quantity is equivalent to adding or subtracting zero from the sum. Two new properties are often introduced at this time to provide PSTs with the precise mathematical language needed to help articulate their justification: additive inverses and additive identity.

4.5.1.3 Problem Type 3: Making Sense of Mathematical Structure

Making sense of mathematical structure and using it to solve problems is our third problem type. Mathematical structure is described in the CCSS-M Mathematical Practices and recognized by the MET II report (CBMS, 2012) as playing an important role in both the learning and teaching of mathematics. It entails decomposing mathematical objects into their component parts and looking for patterns. From a mathematical perspective, examining structure can support deeper understandings, pave the way for making connections between concepts and procedures, and shift learning away from memorization (Mason, Stephens, & Watson, 2009).

Figure 4.3 shows an example of this type of problem from the Addition Strategies lesson. In this example, PSTs examine the solution strategy of Karl, a hypothetical elementary school student who decomposed each addend by place value and then added corresponding place values together to get the sum. The problem also asks PSTs to consider the teacher's choice of model to represent Karl's strategy.

Karl's teacher recorded his steps for adding the numbers, 495 and 331. What did he do? Why did Karl's teacher model his strategy using this notation instead of on a number line?

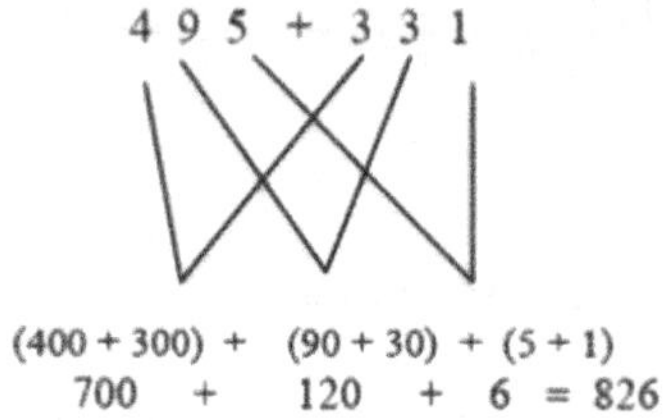

Fig. 4.3 *Making Sense of Mathematical Structure* problem from the Addition Strategies lesson

This problem attends to mathematical structure by focusing PSTs' attention on a computation strategy that leverages our place value system and its additive structure. Every number can be decomposed additively into its place value components. The representation chosen by the teacher for this problem is intended to help PSTs recognize this type of decomposition and that the order of addends does not affect the sum. Whereas not explicit, the commutative and associative properties of addition provide support for why Karl could reorder and regroup the decomposed quantities by place value. Using a number line model would not allow the teacher to easily represent Karl's thinking as it would not be clear that he is decomposing both addends by place value and recomposing by adding corresponding place value components.

4.5.1.4 Problem Type 4: Identifying the Mathematics in Elementary Students' Thinking

The fourth type of problem in the EMP curriculum focuses on identifying the mathematics in elementary students' thinking. Prior to making an instructional decision, a teacher must interpret students' thinking by first identifying the mathematical ideas, methods, and representations that students do and do not understand. However, Bartell, Webel, Bowen, and Dyson (2013) found that PSTs' content knowledge is not always sufficient for properly identifying evidence of students' understanding. Other studies have found that attending to children's mathematical thinking can help PSTs develop their own mathematical knowledge and productive beliefs (Philipp et al., 2007). Thus, we conceptualized these problems as providing additional opportunities for PSTs to develop their own understanding of concepts, procedures, and structure as they practice interpreting students' thinking (Fig. 4.4).

The example above as well as the examples given for problem types 2 and 3 could be classified as identifying the mathematics in elementary students' thinking.

One task of elementary teaching is for teachers to represent students' thinking in ways that help children make sense of a strategy. One representation that can be used is a number line. When a number line is used to represent thinking, you do not have to mark all numbers on the number line. Notice that jumps on the number line may not be accurate in terms of magnitude.

Examine the two number lines below for 495 + 331. Record numerically how each student decomposed, or split, one or both addends and what she did to find the sum. Circle the answer in the student work.

Clarissa

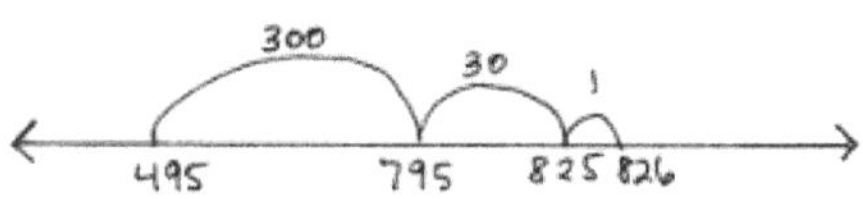

Molly

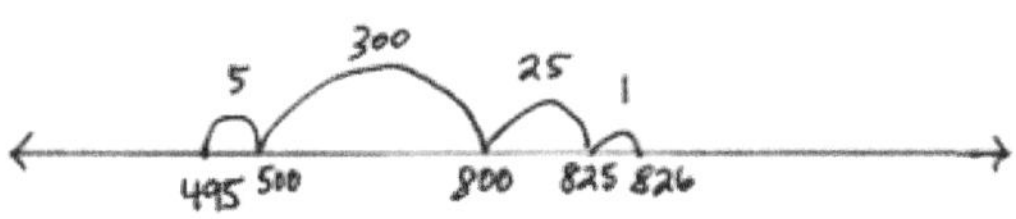

Fig. 4.4 *Identifying the Mathematics in Elementary Students' Thinking* problem from the Addition Strategies lesson

In all three cases, PSTs examine hypothetical students' strategies. In this example, there is the added feature of how a teacher represented each student's thinking on a number line. In this case, both students only decomposed the second addend, but in different ways. Clarissa decomposed 331 by place value, while Molly decomposed it in order to get friendlier numbers when adding. The purpose of this type of problem is to shift PSTs' attention to the mathematics in the response, a necessary first step in analysis. The emphasis is not to have PSTs consider pedagogical responses to students' thinking as this may distract from fully understanding what the students are doing mathematically. As they make sense of Clarissa's and Molly's thinking, PSTs also build their number sense flexibility and bolster their understanding of the decomposition strategy. In addition, we have found that this type of problem is highly motivating since PSTs' learning is situated in their future work as teachers.

4.5.2 Design Principle II: Engagement in Recurring Cycles of Collaborative Problem-Solving and Group Discussion

The EMP lessons follow a structure of recurring cycles of problems for small group and whole class engagement. Each lesson was designed to consist of one or two

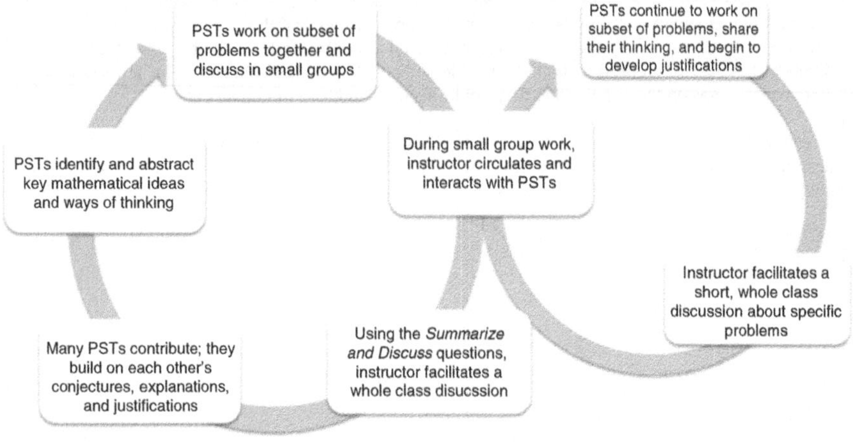

Fig. 4.5 Cycle of Enactment in an EMP Lesson

cycles—a cycle typically consists of 2–5 problems for PSTs to solve followed by a question designed for whole class discussion. The whole class discussion is intended to help PSTs identify and synthesize the big mathematical ideas inherent in the problems. PSTs share, refine, and extend their own thinking together as a community of learners.

A typical cycle, as shown in Fig. 4.5, starts with a launch by the instructor where the problem set is briefly introduced. PSTs then set to work on the mathematics, usually in groups of four. We expect PSTs to engage in collaborative work when solving problems in the cycle. This involves PST-to-PST talk that is initiated by them. They must ask each other questions and listen to each other's reasoning. During this time, the instructor monitors and supports PSTs' work by circulating from group to group, observing, questioning, and listening for different strategies and ways of thinking. The instructor interjects to help the PSTs refine their ideas by pushing them to be more clear and precise in their explanations. The instructor may also provide encouragement and some scaffolding or instruction to a group of PSTs, depending on the needs of the group members.

In every cycle, there also are whole class discussions. As shown in the cycle to the left in Fig. 4.5, this might be a culminating whole class discussion that occurs once all problems in the cycle have been completed. We have focused PSTs attention on the importance of together making sense of the underlying ideas explored in the problem set by labeling certain questions, "Summarize and Discuss." But just as often, there are mini-discussions prior to the culminating discussion, where the instructor interrupts small group work and brings the class together to discuss one or two of the problems. PSTs then return to group work and eventually discuss the

Summarize and Discuss problem. In our cycle image (Fig. 4.5) we have attempted to show how these mini-cycles (on the right) operate within the larger cycle. In these cases, the instructor decides that it would be beneficial for the whole class to come together to discuss a component of a problem, leading up to a whole class discussion.

During whole class discussions, PSTs use their co-constructed experiences solving the problems in small groups along with their prior knowledge to respond to the discussion questions. They justify steps in procedures or clarify concepts, but not always by presenting a formal mathematical explanation. We have found that PSTs focus on "how something works" usually before "why something works" and that generalization takes time. In our classes, over the course of a semester, PSTs become more focused on meaning and drive the conversations toward "why." The instructor, and the PSTs themselves, ask probing questions so that the mathematical meaning implicit in the cycle of problems and questions is explicated and thoroughly considered by the community.

4.6 Image of Instruction from Addition Strategies Lesson

We now share excerpts of transcript from enacting the EMP materials at our own university to illustrate both of our design principles. We selected video recordings from the Addition Strategies lesson, which is one of seven lessons in the *Addition and Subtraction* unit (see Appendix).

At Boston University, our elementary preservice teachers take three mathematics education courses, two sequenced mathematics content courses followed by one mathematics methods course during their sophomore and junior years (approximate ages are 19–21 years old), respectively. Below we examine a lesson from the first of the two content courses. The course met twice per week, for 1 h and 45 min each, for 14 weeks. Often during a class period, one lesson would be completed and another would be started, to be continued at the next meeting. There were 22 PSTs in this section of the course in fall 2017: 3 males and 19 females. Sixty-eight percent (68%) were White, 14% Hispanic, 14% Asian, and 4% African-American. Three PSTs speak a language other than English in their homes.

The EMP unit on Addition and Subtraction revisits these operations with the goal of understanding the underlying mathematics of several computation strategies and algorithms. It also introduces PSTs to various interpretations of these operations (e.g., comparison vs. takeaway), and uses common representations and manipulative materials used in the elementary grades to model actions and relationships. The unit builds on the lessons on place value from the *Number Concepts* unit as PSTs explore how features of the base ten number system enable us to manipulate expressions and perform calculations. The mathematics content of this episode focuses on how place value, decomposition, and arithmetic properties are used across a variety of addition strategies. In addition, PSTs must consider how these strategies are modeled using number lines, splitting notation, or mathematical symbols by teachers and children.

The transcript excerpts below illustrate a portion of a cycle; small group work followed by the instructor pulling the class together for short, whole class discussions about two problems. The cycle learning goal is for PSTs to deepen their understanding of decomposition and compensation strategies. To this end, PSTs work on problems that are categorized as Problem Types 2, 3, and 4. While this transcript does not explicitly highlight a Type 1 problem (connecting concepts and ideas), the other problems have subparts that ask the learner to explain the mathematical ideas that undergird the strategies. In addition, the final Summarize and Discuss question in the cycle is an example of a Problem Type 1 problem (see the Problem Type 1 example). In the previous lesson, PSTs examined addition strategies for basic facts and explored how addends could be decomposed and then recomposed to "make a 10" or "create doubles." That whole class discussion focused on how the commutative and associative properties of addition were used to perform these actions (decomposition/recomposition) and thus simplify the computations.

4.6.1 Addition Strategies, Small Group Work

We join a small group as they first discuss a problem about Clarissa and Molly's strategies (Problem Type 4 example) and then a problem about Karl's strategy (Problem Type 3 example). There are four female students in the group: Jordan, Alex, Noelle, and Dia. One student, Alex, gets up to use the restroom, so the discussion is mainly between the remaining three.

The EMP curriculum is designed to provide PSTs with many opportunities to become comfortable talking about new ideas and approaches. Ideas are revisited a number of times. Notice in this part of the transcript how the PSTs share their understanding and straddle the two learning communities of mathematical learner and future teacher. For example, by exploring a Type 4 problem in lines 1–6, PSTs engage in the common teaching practice of trying to make sense of others' mathematical thinking and representations (number lines). In doing so, PSTs simultaneously grapple with their own mathematical understandings of decomposition and place value. Type 3 problems, like the one in line 7, can also support PSTs' learning by attending to mathematical structure. Noelle's comment in line 12 helps the group better understand why Karl was able to decompose each addend and rearrange the parts to add by placevalue.

1.	Jordan	The first one, it was by place value. [referring to Clarissa's strategy]
2.	Alex	Yeah.
3.	Noelle	By place value.
4.	Jordan	Yeah, so the 331 was added to the 491 by place value.
5.	Noelle	And then in the second one [Molly's strategy], it was friendly numbers. But also by, no by friendly numbers. The 300 is kind of like a place value thing but also a friendly number. The primary way is by friendly numbers cause then she broke up the 26 to make it nicer.
6.	J and D	[both Jordan and Dia are agreeing with Noelle by nodding as she speaks.]
7.	Noelle	[She reads the problem about Karl aloud to the group.] *"Karl's teacher recorded his steps for adding the numbers, 495 and 331. What did he do? Why did Karl's teacher model his strategy using this notation instead of on a number line?"*
8.	Dia	Karl's extra [special], but I appreciate him.
9.	Jordan	Yeah, the lines that the teacher drew.
10.	Dia	I said that Karl did it by decomposing by place value but did an extreme version of this because he broke up each place value, the hundreds, the tens and the ones. Which, honestly, makes a lot of sense.
11.	Jordan	But it is confusing.
12.	Noelle	It's also kind of using the associative/commutative properties too, as you are not only just using the place values but you are also doing them in order.
13.	Dia	You are grouping. Because you are doing 400 plus 300, then the tens grouped together.

Small group work at its best has PSTs actively engaged with the content of the lesson; in lines 1–13, the PSTs are working together to articulate and generalize the ideas inherent in the problems. During this time, PSTs can resolve differences among themselves in how they have interpreted solution strategies. In addition, the small group format provides opportunities for an instructor, as she is circulating among the groups, to probe responses as well as provide instruction and information.

During the interchanges above, the instructor walked over to this small group and listened. She decided that the PSTs were making sense of decomposition and the commutative and associative properties of addition. So she asked about another problem in the set where Priyanka, a hypothetical student, uses compensation to simplify the computation, $57 + 34$. Priyanka solves the computation by adding 3 to one addend (57) to create a friendly number (60), adds 60 and 34, and then subtracts 3 from the sum (i.e., $57 + 34 \rightarrow (57 + 3) + 34 = 60 + 34 = 94$; $57 + 34 = 94 - 3$ or 91).

14.	Instructor	Explain to me how Priyanka solved the problem. What did she do there?
15.	Jordan	We said she used more friendly numbers rather than place value. So she added 3 to get to 60, from 57, and then jumped a whole 34 more which got her to 94. But then she realized that the 3 that she added, it's a little confusing, the 3 that she added from the 34 was already added at the beginning so she had to jump back to 91.
16.	Instructor	Okay. So where did that 3 that she added come from?
17.	Dia	From thin air. (laughter)
18.	Instructor	Yes, from thin air (laughter), she pulled it from the air, but where is it in the problem?
19.	Noelle	It's really when you subtract the 3 again, so rather than borrowing it from the 34, she just added it.
20.	Dia	She tried to get a friendly number by adding 3 to 57 to get 60, so 60 is a bit more malleable to work with. But we thought she should have broken down 34 and borrowed the 3 from 34 so it was 60 plus 31.
21.	Instructor	So let's use the word, "decompose." So she could have decomposed 34 and taken the 3 from the 34. You're right, that's another way she could have solved the problem, but she didn't (laughter). She computed to make a 60. So then we say, she "compensated" by subtracting 3 since at the end she knew she had added too much.

In lines 16 and 18, the instructor pushed the PSTs for clarity on Priyanka's steps and then in line 21 provided the term "decompose" to assist them in being more precise in their explanations. While she acknowledges that the PSTs might have solved the problem differently, she reminds them that as a community of future teachers they must focus on what *students actually say and do*. Problems like this in which PSTs are asked to make sense of students' thinking (Type 4) provide authentic opportunities for PSTs to take on the role of a teacher and interpret potentially unfamiliar ideas or approaches.

Next, the instructor facilitates a short whole class discussion about compensation using Fahmil's strategy (see Problem Type 2 example). The whole class discussions in the EMP curriculum provide the instructor with a venue for doubling down on the mathematics. The instructor can push PSTs to move beyond explanation to provide justification with evidence. However, sometimes this takes longer than anticipated. In the following transcript, notice that the PSTs are still clarifying how Fahmil's strategy works but not justifying it. Two important questions are explored, "Can the value you add and subtract to an expression be any number?" and "How do you decide what number to add/subtract to an expression?"

4.6.2 Addition Strategies, a Subsequent Whole Class Discussion

22.	Instructor	Let's take a look now at Fahmil's strategy which is a little different. I asked Ellery to come up and show one that her group made up.
23.	Ellery	[Ellery goes to document camera at front of room.] So I was trying to figure out if this one [strategy] always worked. [writes her calculations on document camera but the class can't see it] $47 + 16$ $(47 - 2) + (16 + 2)$ $45 + 18$ Which just became $45 + 18$ because I minused 2 from 47 and added 2 to 18.
24.	Instructor	Move your page over so we can see it. Let everybody take a look and see what you did. [pause of 6 s]
25.	Ellery	Which didn't really make the problem any easier so I tried it another way [again writes her calculations on the document camera] $47 + 16$ $(47 + 3) + (16 - 3)$ $50 + 13$ Which then you can easier see that it would be 63. Or you could also further break up the 13 to 10 plus 3.
26.	Instructor	So what are your thoughts on Ellery's first attempt at using Fahmil's strategy?
27.	Ada	I have a question about like Fahmil's strategy in general. Were we supposed to use a 3 or a 2 each time? Because my group didn't do that.
28.	Instructor	So what do you think? Do you have to use a 3 or a 2 each time? [pause as hands are raised] Call on someone, Ada.
29.	Ada	Maritza.
30.	Maritza	We did, we made up another one for instance, and it was 66 plus 34, and we used 4 for that one.
31.	Instructor	So you added 4 and subtracted 4?
32.	Martiza	Yes. And then we did one with 7, I did 23 plus 37 and I did that.
33.	Instructor	So you added and subtracted 7?
34.	Ada	[Directed to Ellery] So I'm just confused why you started with the 2?
35.	Ellery	While I was just trying, I was thinking what would be a number that would be like a friendly number for 47, so maybe 45. But I realized when I subtracted that 2 that it didn't really help to solve it easier. So I tried it the other way of making the 47 a friendly number to 50.
36.	Instructor	So could someone restate for me for why exactly Ellery first went with a 2, adding a 2 and subtracting a 2, but then changed her mind?
37.	Sophia	So from what I am understanding is you saw that 45 would be a friendly number, because sometimes it is when you are adding 10 or adding by 5s, so like if the number was 15. But you didn't realize that 16 plus 2 would not be a friendly number and wouldn't be easier to add. So then instead you realized that 47 plus 3 would make a friendly number of 50 and 16 minus 3 would just make a smaller number, and it is usually easier for kids to add groups of 10 with another number. I didn't think about breaking up the 13 to 10 plus 3 which you could also do and group the tens together by the associative property and add 3.

The EMP curriculum was not designed so that instructors show PSTs how to do mathematics or explain why steps or relationships make sense. The authors shifted these actions to the PSTs and then provided them with opportunities to practice both doing and explaining as they addressed the different problems. For example, in the transcript excerpt above, the instructor calls the whole class together to consider Fahmil's strategy using their own examples, giving the PSTs an opportunity to talk about why they selected particular quantities. In line 27, Ada asks if the strategy only works when you add and take away the quantities 2 or 3 (which were used in the examples provided to them). Over the next several turns of talk, several PSTs conclude that you can use any quantities, for the goal is to "help solve it [the problem] easier." At the prompting of the instructor, Sophia restates this idea (line 37). PSTs are collaboratively producing and negotiating new ideas as they make sense of what Ellery's group did and why, work they will do as a future teacher as they follow the reasoning of their students.

Since no justification for why the compensation strategy works was offered by the PSTs, the instructor next asked the class to work for 2 min in their small groups to figure out what was going on when the expression 47 + 16 was changed to 50 + 13, prior to adding. The instructor then reconvened the whole class discussion. She called one small group of PSTs to come to the board and present their thoughts. The group recorded the following expressions on the board and stated that this strategy works because "when you are adding 3 somewhere and subtracting 3 somewhere else, the numbers cancel each other out."

		47 + 16 47 + 3 + 16 − 3 50 + 13 63
38.	Instructor	So why does it make it the same? I agree, I see how you cancel them out, but what's going on there? Is there a way we can explain this a little more mathematically?
39.	Logan	If you use the commutative property and rearrange the numbers so it is 47 plus 16 plus 3 minus 3, you see that.
40.	Instructor	Wait a minute, let's write that [Juan, one of the PSTs at the board, records on the board, 47 + 3 + 16 − 3 = 47 + 16 + 3 − 3]. Okay, keep going.
41.	Logan	So if you just rewrite them in that order, then you see that you add 3 and then you immediately take away the 3 and essentially you didn't really do anything to it.
42.	Instructor	Who wants to add on to that? Lauren.
43.	Lauren	We looked at it more like compensation. So if you are adding 3 you have to compensate for that by subtracting the same amount as that is the way they cancel each other out.
44.	Instructor	What else could we say? [pause of 10 s] Juan, would you do me a favor and put parentheses around the 3 minus 3? [Juan adds parentheses so now the equation on the board is:] 47 + 3 + 16 − 3 = 47 + 16 + (3 − 3) Now stand back and I'd like everybody to just look at it. What's happening there?

45.	Class	Zero.
46.	Instructor	Right, zero. Can we add zero to anything?
47.	Class	Yeah, yes.
48.	Instructor	What's that property called?
49.	Juan	The associative property as we just grouped it differently.
50.	Instructor	The grouping of it would be the associative property but there is a special property about adding zero.
51.	Class	The additive property? The zero property?
52.	Instructor	Good guess. The identity property, the identity property of addition. The identity property says we can add zero to any addition sentence and not change the value.

In the transcript excerpt above, the instructor encourages the PSTs to justify Fahmil's compensation strategy. Logan, in lines 39 and 41, explains how by adding 3 and subtracting 3 "essentially you didn't really do anything to it [the expression]." Another PST in line 43 reiterates how the 3s cancel each other out. Next, the instructor provides scaffolded instruction in line 44, leading to naming the identity property of addition. Lines 22–52 illustrate the community of mathematical learners that has been created, where PSTs take each other seriously and move beyond the boundaries of their own thinking. They listen carefully to each other, reason, and think together about the mathematics.

4.7 Summary

In summary, the EMP curriculum was designed to ask PSTs to focus on the underlying mathematics of the elementary curriculum and at the same time make sense of and discuss hypothetical elementary students' ideas, strategies, and representations, activities that novice teachers will encounter in their future work as teachers. We provide one approach to assisting PSTs in deepening and connecting their own understanding of the content, both through our different problem types and our recurring cycles of group problem-solving and whole class discussions. We believe it is not enough for PSTs to solve problems in a way that makes sense to them; they must become flexible and competent in understanding and identifying how others also solve problems, as those activities are at the heart of their future work. By enculturating PSTs into dual but intersecting communities of mathematics learners and future teachers, we hope to also positively impact their confidence and mathematical identities.

From the authors' perspectives, our goal was to develop a curriculum that will help change what mathematics content elementary preservice teachers study as they train to become teachers. We designed the EMP curriculum so that it addressed our beliefs about what PSTs should learn and how they should learn mathematics content. In addition, we strived to better support instructors who are doing this work.

Appendix: EMP Units and Lesson Titles

Number Concepts	Addition and Subtraction
1. Base Ten Place Values	1. Addition and Subtraction Story Problems
2. Different Place Values	2. Addition Strategies
3. Large Numbers	3. Addition Algorithms
4. Decimals—Part 1	4. Problem-Solving: Consecutive Sums
5. Decimals—Part 2	5. Subtraction Strategies
6. Decimals—Part 3	6. Subtraction Algorithms
	7. Addition and Subtraction of Decimals

Multiplication and Division	Fractions
1. Multiplication and Division Situations	1. Fractions
2. Multiplication Strategies	2. Equivalent Fractions
3. Multiplication Algorithms	3. Fraction Comparisons
4. Multiplication Games	4. Addition and Subtraction of Fractions I
5. Making Sense of Division	5. Addition and Subtraction of Fractions II
6. Division Strategies	6. Multiplication of Fractions I
7. Division Algorithms	7. Multiplication of Fractions II
8. Multiplication and Division of Decimals	8. Division of Fractions I
	9. Division of Fractions II
	10. Fractions to Decimals

Number Theory
1. Factors
2. Prime Factorization
3. Divisibility
4. Divisibility Rules
5. Greatest Common Factor
6. Least Common Multiple

Geometry	Geometric Measurement
1. Attributes	1. Area Concepts
2. Angles	2. Area and Perimeter
3. Parallel and Intersecting Lines	3. Parallelograms and Triangles
4. Polygons	4. Trapezoids
5. Classifying Triangles	5. Circles
6. Constructing Triangles	6. Prisms
7. Classifying Quadrilaterals	7. Surface Area Part I
8. Interior Angles of Polygons	8. Surface Area Part II
9. Tessellations	9. Volume
10. Visualization	
11. Prisms	

References

Association of Mathematics Teacher Educators. (2017). *Standards for preparing teachers of mathematics*. Retrieved from amte.net/standards.

Ball, D., Thames, M., & Phelps, G. (2008). Content knowledge for teaching: What makes it so special? *Journal of Teacher Education, 59*, 389–407.

Bartell, T. G., Webel, C., Bowen, B., & Dyson, N. (2013). Prospective teacher learning: Recognizing evidence of conceptual understanding. *Journal of Mathematics Teacher Education, 16*(1), 57–79.

Bleiler, S. K. (2014). Increasing awareness of practice through interaction across communities: The lived experiences of a mathematician and a mathematics teacher educator. *Journal of Mathematics Teacher Education, 18*(3), 231–252.

Brown, J. S., Collins, A., & Duguid, P. (1989). Situated cognition and the culture of learning. *Educational Researcher, 18*(1), 32–42.

Browning, C., Edson, A., Kimani, P., & Aslan-Tutak, F. (2014). Mathematical content knowledge for teaching elementary mathematics: A focus on geometry and measurement. *The Mathematics Enthusiast, 11*(2), 333–383.

Callis, L. K. (2017). *Mathematics content courses for preparing elementary teachers: Curriculum and instruction* (Doctoral dissertation, Boston University). Retrieved from https://open.bu.edu/handle/2144/23303.

Carpenter, T. P., & Lehrer, R. (1999). Teaching and learning mathematics with understanding. In E. Fennema & T. A. Romberg (Eds.), *Mathematics classrooms that promote understanding* (pp. 19–32). Mahwah, NJ: Erlbaum.

Castro Superfine, A., Li, W., & Martinez, M. (2013). Developing preservice teachers' mathematical knowledge for teaching: Making explicit design considerations for a content course. *Mathematics Teacher Educator, 2*(1), 42–54.

Chapin, S., & O'Connor, C. (2007). Academically productive talk: Supporting students' learning in mathematics. In W. G. Martin, M. Strutchens, & P. Elliott (Eds.), *The learning of mathematics, 69th NCTM yearbook* (pp. 113–128). Reston, VA: National Council of Teachers of Mathematics.

Chapin, S. H., & O'Connor, C. (2012). Project Challenge: Using challenging curriculum and mathematical discourse to help all students learn. In C. Dudley-Marling & S. Michaels (Eds.), *High-expectation curricula: Helping all students succeed with powerful learning* (pp. 113–127). New York, NY: Teachers College Press.

Clements, D. (2007). Curriculum research: Toward a framework for "research-based curricula". *Journal for Research in Mathematics Education, 38*(1), 35–70.

Conference Board of the Mathematical Sciences. (2001). *The mathematical education of teachers*. Providence, RI and Washington, DC: American Mathematical Society and Mathematical Association of America.

Conference Board of the Mathematical Sciences. (2012). *The mathematical education of teachers II*. Providence, RI and Washington, DC: American Mathematical Society and Mathematical Association of America.

Eisenhardt, M., Borko, H., Underhill, R., Brown, C., Jones, D., & Agard, P. (1993). Conceptual knowledge falls through the cracks: Complexities of learning to teach mathematics for understanding. *Journal for Research in Mathematics Education, 24*, 8–40.

Elliott, P., & Garnett, C. (Eds.). (2008). *Getting into the mathematics conversation: Valuing communication in mathematics classrooms*. Reston, VA: National Council of Teachers of Mathematics.

Feiman-Nemser, S. (2001). From preparation to practice: Designing a continuum to strengthen and sustain teaching. *Teachers College Record, 103*(6), 1013–1055.

Feldman, Z. (2012). *Describing pre-service teachers' developing understanding of elementary number theory topics* (Doctoral dissertation, Boston University). Retrieved from http://gradworks.umi.com/35/29/3529017.html.

Forman, E. A. (1996). Learning mathematics as participation in classroom practice: Implications of sociocultural theory for educational reform. *Theories of Mathematical Learning, 115*, 130.

Forman, E. A. (2003). A sociocultural approach to mathematics reform: Speaking, inscribing, and doing mathematics within communities of practice. In J. Kilpatrick, W. G. Martin, & D. Schifter (Eds.), *A research companion to principles and standards for school mathematics* (pp. 333–352). Reston, VA: National Council of Teachers of Mathematics.

Gibbons, L. K., Feldman, Z., Chapin, S., Batista, L. N., Starks, R., & Vazquez-Aguilar, M. (2018). Facilitation practices in mathematics teacher education that promote productive identities in preservice elementary teachers. *Mathematics Teacher Education and Development, 20*(3), 20–40.

Greenberg, J., & Walsh, K. (2008). *No common denominator: The preparation of elementary teachers in mathematics by America's education schools*. Washington, DC: National Council on Teacher Quality.

Hart, L., Oesterle, S., & Swars, S. (2013). The juxtaposition of instructor and student perspectives on mathematics courses for elementary teachers. *Educational Studies in Mathematics, 83*(3), 429–451.

Hiebert, J., & Carpenter, T. P. (1992). Learning and teaching with understanding. In D. A. Grouws (Ed.), *Handbook of research on mathematics teaching and learning* (pp. 65–97). Reston, VA: National Council of Teachers of Mathematics.

Hill, H. C., Rowan, B., & Ball, D. L. (2005). Effects of teachers' mathematical knowledge for teaching on student achievement. *American Educational Research Journal, 42*(2), 371–406.

Hill, H. C., Sleep, L., Lewis, J. M., & Ball, D. L. (2007). Assessing teachers' mathematical knowledge: What knowledge matters and what evidence counts? In F. Lester (Ed.), *Second handbook of research on mathematics teaching and learning* (pp. 111–156). Charlotte, NC: Information Age Publishing.

Kastberg, S., & Morton, C. (2014). Mathematical content knowledge for teaching elementary mathematics: A focus on decimals. *The Mathematics Enthusiast, 11*(2), 311–332.

Lave, J., & Wenger, E. (1991). *Situated learning: Legitimate peripheral participation*. New York, NY: Cambridge University Press.

Lortie, D. C. (1975). *Schoolteacher: A sociological study*. Chicago, IL: University of Chicago Press.

Ma, L. (1999). *Knowing and teaching elementary mathematics: Teachers' understanding of fundamental mathematics in China and the United States*. Mahwah, NJ: Erlbaum.

Masingila, J., Olanoff, D., & Kwaka, D. (2012). Who teaches mathematics content courses for prospective elementary teachers in the United States? Results of a national survey. *Journal of Mathematics Teacher Education, 15*, 347–358.

Mason, J., Stephens, M., & Watson, A. (2009). Appreciating mathematical structure for all. *Mathematics Education Research Journal, 21*(2), 10–32.

McCrory, R., Zhang, C., Francis, A. P., & Young, S. (2009). Factors in the achievement of preservice elementary teachers in mathematics classes. In S. L. Swars, S. W. Stinson, & S. Lemons-Smith (Eds.), *Proceedings of the 31st Annual Meeting of the North American Chapter of the International Group for the Psychology of Mathematics Education* (Vol. 5, pp. 1179–1187). Atlanta, GA: Georgia State University.

National Governors Association Center for Best Practices, Council of Chief State School Officers. (2010). *Common core state standards (mathematics)*. Washington, DC: National Governors Association Center for Best Practices, Council of Chief State School Officers.

National Research Council. (2001). Adding it up: Helping children learn mathematics. In J. Kilpatrick, J. Swafford, & B. Findell (Eds.), *Mathematics Learning Study Committee, Center for Education, Division of Behavioral and Social Sciences and Education*. Washington, DC: National Academy Press.

Newton, K. J. (2008). An extensive analysis of preservice elementary teachers' knowledge of fractions. *American Educational Research Journal, 45*(4), 1080–1110.

O'Connor, C., Michaels, S., Chapin, S., & Harbaugh, A. G. (2017). The silent and the vocal: Participation and learning in whole-class discussion. *Learning and Instruction, 48*, 5–13.

Olanoff, D., Lo, J. J., & Tobias, J. (2014). Mathematical content knowledge for teaching elementary mathematics: A focus on fractions. *The Mathematics Enthusiast, 11*(2), 267–310.

Philipp, R., Ambrose, R., Lamb, L., Sowder, J. T., Schappelle, B., Sowder, L., et al. (2007). Effects of early field experiences on the mathematical content knowledge and beliefs of prospective elementary school teachers: An experimental study. *Journal for Research in Mathematics Education, 38*(5), 438–476.

Salinas, A., Feldman, Z., Callis, L. K., & Chapin, S. (2016, April). *A mathematics curriculum that supports preservice teachers' content knowledge for teaching mathematics*. Paper presented at the 2016 Annual Meeting of the American Educational Research Association (AERA), Washington DC.

Simon, M. (1993). Prospective elementary teachers' knowledge of division. *Journal for Research in Mathematics Education, 24*(3), 233–254.

Simon, M. (1994). Learning mathematics and learning to teach: Learning cycles in mathematics teacher education. *Educational Studies in Mathematics, 26*(1), 71–94.

Sowder, J. (2007). The mathematical education and development of teachers. In F. K. Lester (Ed.), *Second handbook of research on mathematics teaching and learning* (pp. 157–223). Reston, VA: National Council of Teachers of Mathematics.

TeachingWorks. (2015). *High-leverage content*. Retrieved from http://www.teachingworks.org/work-of-teaching/high-leverage-content.

Thanheiser, E., Whitacre, I., & Roy, G. (2014). Mathematical content knowledge for teaching elementary mathematics: A focus on whole number concepts and operations. *The Mathematics Enthusiast, 11*(2), 217–266.

U.S. Department of Education. (2008). *Foundations for success: The final report of the national mathematics advisory panel*. Washington, DC: U.S. Department of Education.

Walshaw, M., & Anthony, G. (2008). The teacher's role in classroom discourse: A review of recent research into mathematics. *Review of Educational Research, 78*, 516–551.

Wilhelm, A. G. (2014). Mathematics teachers' enactment of cognitively demanding tasks: Investigating links to teachers' knowledge and conceptions. *Journal for Research in Mathematics Education, 45*(5), 636–674.

Zazkis, R. (1998). Odds and ends of odds and evens: An inquiry into students' understanding of even and odd numbers. *Educational Studies in Mathematics, 36*, 73–89.

Zazkis, R., & Campbell, S. (1996a). Divisibility and multiplicative structure of natural numbers: Preservice teachers' understanding. *Journal for Research in Mathematics Education, 27*(5), 540–563.

Zazkis, R., & Campbell, S. R. (1996b). Prime decomposition: Understanding uniqueness. *Journal for Mathematical Behavior, 15*, 207–218.

Chapter 5
Mathematics and Mathematics Education at the University of Nebraska-Lincoln: Partnering to Implement Core Practices in Elementary Teacher Preparation

Michelle Homp and W. James Lewis

The mathematical education of future elementary teachers at the University of Nebraska-Lincoln (UNL) is an integrated learning experience that includes mathematical, pedagogical, and clinical learning experiences. The centerpiece of the program is the "Mathematics Semester" during which future teachers take a mathematics course, a math methods course, and a field experience (called a practicum at UNL). This article grew out of a presentation the authors gave at the "Mathematics Matters in Education" workshop held at Texas A&M University in 2017 and emphasizes the mathematical preparation of future elementary teachers at UNL.

Two aspects of the Mathematics Semester distinguish it from how future elementary teachers learn mathematics at other universities. One is a weekly assignment known at UNL as a "Habits of Mind" problem and the other is the partnership between UNL's Department of Mathematics (MATH) and Department of Teaching, Learning and Teacher Education (TLTE).

To address the importance of our partnership, we (1) provide a brief synopsis of the University of Nebraska-Lincoln's (UNL) journey to establishing what is now

W. James Lewis wishes to acknowledge the support provided by Math Matters (DUE-9981106), a National Science Foundation grant that made possible the development of the UNL program. In addition, this article and the presentation at Texas A&M is based upon work done while serving at the National Science Foundation. Any opinion, findings, and conclusions or recommendations expressed in this material are those of the author(s) and do not necessarily reflect the views of the National Science Foundation. W. James Lewis chaired the Steering Committee for MET I and was chair of the Writing Team for MET II. He was a member of the writing team for the *AMTE Standards*.

M. Homp (✉) · W. J. Lewis
Department of Mathematics, Center for Science, Mathematics and Computer Education, University of Nebraska-Lincoln, Lincoln, NE, USA
e-mail: mhomp@unl.edu; jlewis@unl.edu

Y. Li et al. (eds.), *Developing Mathematical Proficiency for Elementary Instruction*, Advances in STEM Education, https://doi.org/10.1007/978-3-030-68956-8_5

a firm partnership between MATH and TLTE, (2) describe how our elementary teacher education program (particularly the math content course taken as part of the Mathematics Semester) has evolved as a result of this partnership, and (3) offer comments on components needed to sustain such a partnership.

5.1 How Should Future Elementary Teachers Learn the Mathematics They Need to Know?

In the late 1990s, the second author (Lewis) was chair of the mathematics department at UNL. In that role he had a habit of reading the teaching evaluations of instructors in the department to be aware of problems that might arise in mathematics classrooms. From reading the evaluations obtained from classes for elementary education majors, two themes stood out. One was that the students did not believe that what they were studying in their mathematics classrooms had anything to do with their preparation to teach mathematics in elementary school.[1] The other prominent theme was that elementary education majors were often harshly critical of instructors who otherwise received outstanding evaluations.

At the same time, Lewis was chair of a Conference Board of the Mathematical Sciences (CBMS) committee that had been appointed to update the 1991 publication, *A call for change: Recommendations for the mathematical preparation of teachers of mathematics* (Leitzel, 1991). The committee's report, The Mathematical Education of Teachers (2001), stressed two themes: (1) the intellectual substance in school mathematics; and (2) the special nature of the mathematical knowledge needed for teaching (page xi). The report, now referred to as *MET I*, also included these recommendations:

- Prospective teachers need mathematics courses that develop a deep understanding of the mathematics they will teach.
- Along with building mathematical knowledge, mathematics courses for prospective teachers should develop the habits of mind of a mathematical thinker and demonstrate flexible, interactive styles of teaching.
- The mathematical education of teachers should be seen as a partnership between mathematics faculty and mathematics education faculty.

While the report was under development, discussions with Deborah Schifter and Judy Sowder, two of the report's authors, and members of the Steering Committee convinced Lewis that significant changes were needed in how UNL educated future elementary teachers of mathematics. His goal became to model the mathematical

[1]This is in agreement with past research findings which demonstrate these students believe they already possess the mathematical knowledge needed to successfully teach mathematics, making their math courses seem even less relevant to their future vocation (Thanheiser, Philipp, Fasteen, Strand, & Mills, 2013).

education of elementary teachers at UNL after the vision put forth by the report. In particular, he was convinced that in order to be successful teaching this audience, the mathematics instructor needed to be supported by a mathematics educator the students trusted.

In 1999, Lewis was also engaged in regular conversations with Dr. Ruth Heaton who, at the time, was a new mathematics educator at UNL. They concluded that prospective elementary teachers at UNL had weak understandings of mathematical content[2] that required an increase in the level of rigor of the courses required at the time for elementary education majors. Their discussions led to a successful Course, Curriculum and Laboratory Improvement NSF proposal (DUE #9981106) titled Math Matters with Heaton as the PI and Lewis as one of two co-PIs. As stated in the proposal, the goal was to "increase and deepen prospective teachers' mathematical content knowledge, and challenge and change their beliefs about the nature of mathematics, children's mathematical understandings, and what it means to teach mathematics".[3] Influences on the development of Math Matters included ideas for program design, course content, and materials obtained from NSF-funded projects at Kansas State University (Wright, 1997) and the Oregon Collaborative for Excellence in the Preparation of Teachers at Portland State University. With support from the grant, they developed an innovative partnership for elementary mathematics teacher preparation build around the vision:

- Create a mathematician-mathematics educator partnership with the goal of improving the mathematical education of future elementary school teachers.
- Link field experiences, pedagogy, and mathematics instruction.
- Create math classes that are both accessible and useful for future elementary school teachers.

A decade after the 2001 CBMS report, Lewis was chair of the writing team that produced the report known as *MET II* (2012), which also included recommendations for the mathematics that teachers should know and how they should come to know that mathematics. While both reports are intended to be resources for those who teach mathematics and statistics, the authors express the opinion that the reports will be a resource for mathematics educators. MET II also asserts (2012, p. xi) that:

> Mathematics teacher education is a complex, interdisciplinary enterprise requiring knowledge of teaching and learning as well as knowledge of mathematics. This argues strongly for a partnership between mathematics educators and those that teach mathematics.

Recommendation 1 in MET II states that teachers need "mathematics courses that develop a solid understanding of the mathematics they will teach," and argues

---

[2]Studies that influenced changes to the UNL program include Simon (1993) and Simon and Blume (1994).

[3]Wideen, Mayer-Smith, and Moon's (1998) analysis of research on teacher development and Shulman's (1986) work on pedagogical content knowledge support a focus on a program that attends to the intersection of mathematical content, how children learn the content, and teaching.

that this knowledge is "substantial yet quite different from that required for other mathematical professions" (2012, p. 17).

More recently, the Association of Mathematics Teacher Educators released their report, *AMTE Standards for Preparing Teachers of Mathematics*[4] (2017) which acknowledges the contributions of MET II and develops standards for mathematics teacher preparation programs. It too stresses partnerships. Standard P.1. Partnerships states the following:

> An effective mathematics teacher preparation program has significant input and participation from all appropriate stakeholders.

An article discussing the first decade of their partnership was published in the *Notices of the American Mathematical Society* (Heaton & Lewis, 2011). This article extends the information in the initial report (Heaton & Lewis, 2008) by addressing key components of sustaining a partnership beyond the development phase and implementing a program with long-term success. The recommendations of the MET II and the AMTE on the importance of partnerships continue to serve as a framework for the teacher preparation program at UNL.

5.2 Establishing the Partnership

Originally, Heaton and Lewis envisioned developing a math specialist program that they hoped would attract about one-sixth of the students in UNL's elementary teacher preparation program. During AY 2000–2001, 16 future elementary teachers volunteered to be part of Math Matters, taking a math course, a pedagogy course, and a field experience (practicum) each semester. After about 2 years, a summative evaluation of Math Matters was quite positive and UNL's teacher education program decided to make a one-semester version of the program a required part of elementary teacher preparation for all students. Thus, starting in Fall 2003 and continuing to the present time, all students majoring in elementary education participate in the Mathematics Semester which includes the following:

- A number and number sense mathematics course
- A math methods course
- A two-credit field experience that meets 2 days per week in a local school.

Together this represents an 8-hr block of courses that students complete in cohorts, taking all three courses with the same group of peers. In addition to the Mathematics Semester, all elementary education majors take three additional mathematics courses, two of which are specifically designed for students in elementary education.

[4] Lewis also had the opportunity to be a part of the writing team for the AMTE report.

The mathematics course that is part of the Mathematics Semester focuses on the development of a profound understanding of the fundamental mathematics (Ma, 1999) which comprise elementary mathematics curriculum and the attributes of mathematicians seriously engaged in doing mathematics, i.e., the "habits of mind of a mathematical thinker" (CBMS, 2012, p. 8, 19; Cuoco, Goldenberg, & Mark, 1996). Simultaneously, the methods course emphasizes how students can use their deepened understanding of mathematics to find the mathematical content in the many tasks of teaching mathematics (Loewenberg Ball, Thames, & Phelps, 2008). This helps prospective teachers develop productive habits of pedagogy (Rolle, 2008) and understand mathematics from the child's point of view. Additionally, because the students are simultaneously spending time in K–5 classrooms, the future teachers also see the relevance of their course work in managing the realities of mathematics teaching and learning. Indeed, the authors want to stress the contributions made by the pedagogy course and the field experience to convincing undergraduate students of the complexity of teaching (National Mathematics Advisory Panel, 2008) and the importance of the mathematics class to their preparation to be an elementary teacher.

In the beginning, there were unexpected challenges to working across department lines in a large university. For example, at first it was difficult to get the mathematics course and the methods course scheduled back-to-back in the same classroom. After a few semesters, however, the two departments learned to work together cooperatively. Another challenge concerned the sequence in which elementary education majors take this and the other two math courses specifically designed for elementary education majors. TLTE uses a portfolio admission process where students apply for admission to the elementary education program near the end of their second year of college. Once admitted to the program, students' first semester is the Mathematics Semester. MATH believes that student performance in the other math courses for elementary education majors is much better if taken after the Mathematics Semester while advisers in TLTE believe that there is not sufficient "room" in students' schedules to take the courses after the Mathematics Semester. Sustaining a long-term partnership requires compromise among both parties, so the current solution is to have students take one of the courses as a general elective in their first 2 years and the other course after the Mathematics Semester.

It should be noted that originally the Mathematics Semester included a fourth course titled "The Learner Centered Classroom" which Heaton designed to focus on mathematics content and using engaging tasks to achieve classroom management. This afforded additional time to focus on mathematics and teaching children mathematics. Few things remain the same, of course, so, in part because concepts of management and student-focused learning are cross disciplinary, about 5 years ago TLTE moved The Learner Centered Classroom course to a different semester. Now, 15 years (since program-wide implementation) and several department chairs later, and with new faculty in both departments, the long-term partnership and the Math Semester continue to be successful. So much so that current TLTE faculty are considering ways to match the math–math education courses with science content and science methods courses to extend the model and transition to a STEM Semester.

5.3 Habits of Mind Problems

In many ways, the mathematics course taken as part of the Mathematics Semester is similar to other "number and number sense" or arithmetic courses for elementary education majors. Our primary goal is to "study the mathematics they (will) teach in depth, and from the perspective of a teacher" (CBMS, 2012, p. 23). Currently, we use the fifth edition of Sybilla Beckmann's *Mathematics for Elementary Teachers* (2017) and cover the material in the first six chapters.[5] A visitor to one of our classes would observe an active learning environment with heavy use of the activities that accompany Beckmann's book and student presentations of homework problems.

One aspect of our course that is not typical is the weekly assignment of a challenging problem that students work on outside of class in groups. We feel no obligation to have these problems connect to the material being studied in class at the time. Instead, we view these problems as contributing to the development of the Mathematical Practices found in the *Common Core State Standards for Mathematics* (NGA, 2010) or the "habits of mind of a mathematical thinker" discussed in *MET II*. Originally, we viewed these problems as just good opportunities to learn to solve problems, but when we began to use the problems in one of our NSF-funded professional development programs, the teachers in the program dubbed them to be "Habits of Mind Problems" and began adapting them for use in their own classrooms.

A well-designed problem will be sufficiently challenging for our students so that solutions are not immediately obvious, but are accessible to those who demonstrate the perseverance we hope to develop. Ideally, the types of problems selected have multiple pathways toward a solution, allowing students to experience a sense of pride or even a smile once they have solved the problem. It is not unusual for good solutions to require 2–4 pages of explanation, and students typically find it challenging to explain their reasoning. However, they do quickly learn the value of working with a group.

One example of a Habits of Mind Problem is the following that Lewis first learned from Deborah Ball.

43 Cents
What is the fewest number of coins that it will take to make exactly 43 cents if you have available pennies, nickels, dimes, and quarters? After you have solved the problem, provide an explanation which proves that your answer is correct. How

[5]Our algebraic thinking course covers Chaps. 7–9 and our geometry and measurement course covers Chaps. 10–14.

does the answer (and the justification) change if you only have pennies, dimes, and quarters available? Again, explain why your answer is correct.

We have frequently assigned this problem the first day of class. Students adopt a "greedy algorithm" approach to the first part, using a quarter, dime, nickel, and three pennies. The first time we assigned the problem, we were quite surprised that nearly 50% of the class used the same approach for the second part of the problem and argued that for the second part of the problem, it takes ten coins, a quarter, a dime, and eight pennies. When we got similar results semester after semester, we gained a new respect for the challenge of developing the capacity of our students to reason mathematically. After returning papers to our students, the results afford an opportunity to ask students whether the "greedy algorithm" approach to the first part of the problem constitutes proof of the correct answer if the same approach leads to an incorrect answer to the second part of the problem.

Another problem assigned early in the semester uses Crossing the River (Glencoe/McGraw Hill, 1991), a problem Lewis first observed at a conference organized by Ira Papick (who at the time was in Missouri). An edited version of the problem is as follows:

Crossing the River
A group of adults go on a camping trip with a group of fourth grade students. They come to a river that is too deep to wade across. They find a boat, but it isn't very big. The adults are rather big and only one adult can fit in the boat at one time, but the boat can hold any two fourth grade students. The students have experience boating and each can safely row across the river by themselves.

If there are four adults and two students on the trip, is it possible to get all of them across the river? If yes, how many one-way trips across the river will it take? What if there were five adults and only one student? What if there were 5 adults and 2 students or 4 adults and 6 students? How can the problem be generalized?

Because the problem is assigned early in the semester, some students struggle with explaining their reasoning or with stating and solving the problem for a adults and s students. Over time, our students gain confidence in their ability to solve challenging problems and in their ability to construct mathematical arguments. By the end of the semester, our students are correctly solving problems that are notably more challenging, they are developing the ability to produce high-quality written solutions, and they are gaining an appreciation for the value of writing carefully crafted explanations.

5.4 Math Content and Pedagogy: Building Course Connections

The first major achievement of the Mathematics Semester was having the courses offered back-to-back in the same room and to the same group of students. When their schedules permitted, Heaton and Lewis would sit in each other's classroom. Heaton was particularly good at affirming the importance of the work done in the content course by finding opportunities to say "As Jim was discussing last week . . . ," during the methods course.

Semester after semester, Heaton and Lewis looked for ways to strengthen the connections between mathematics content and mathematics pedagogy, both between their own courses, and in the minds of the students. Over time, the goals and tasks from the mathematics and pedagogy courses and the field experience were integrated into a single syllabus. While both courses include assignments which are distinct, this syllabus also includes several assignments which require students to apply what they learn from one course setting to another. These assignments are evaluated by both the mathematics and the methods instructors, and receive a single grade which counts in both courses. A few such integrated assignments are described below.

5.4.1 Teaching and Learning Project

One major integrated assignment contributing to both courses is the Teaching and Learning Project. Early in the semester, students are assigned a homework problem in the mathematics class that many find challenging (*Crossing the River* is one example of such a problem). Later that semester, the future teachers are asked to work in pairs to plan a lesson for a child based on that same mathematics problem. Part of the planning process includes adapting the problem so that it is accessible to elementary students in the grade level of future teachers' field experience classroom. Since the goal is to help the child have a successful learning experience, they need to consider vocabulary, instructional representational tools, a sequence of tasks, and possible questions to move the child along in his or her thinking about the problem. The elementary education majors then select a K–5 student from their field experience classroom, have that student work on the problem, and then submit a written report about the experience.

When instructed to adapt the problem for an elementary student, the future teachers may think it is an unreasonable problem for a child in second grade—or even in fifth grade—to solve. Often, in their reflections, the future teachers will write about their surprise that young children can be creative and successful with such challenging mathematical tasks. Thus, the Teaching and Learning Project proves valuable in many ways. The students come to appreciate that (1) they cannot teach mathematics successfully unless they understand it themselves; (2) they must rely on their pedagogical knowledge to plan how to present information and prepare appropriate manipulatives; (3) they need to anticipate the difficulties children will have and prepare questions accordingly; and finally (4) they should not underestimate children's abilities to solve problems. This fully integrated learning experience helps students better comprehend the complexity of the task of teaching (National Mathematics Advisory Panel, 2008). Furthermore, a direct look at children's thinking strategies in relationship to learning mathematics has been shown to have a positive impact on students' beliefs about mathematics and about the teaching and learning of mathematics (Philipp, Clement, Thanheiser, Schappelle, & Sowder, 2003). This integrated approach to the Teaching and Learning Project is just one of the many benefits of our mathematician/educator/teacher partnership.

5.4.2 Curriculum Project

Another major integrated assignment in which students are asked to synthesize knowledge gained from their field experiences, methods, and mathematics courses involves an analysis of standards (both content and process) and the elementary curriculum. In this project students investigate a content domain in elementary mathematics and consider what teachers and children need to know in order to teach and understand the topic in deep and meaningful ways. Guided by documents such as the *Common Core State Standards* (and the *Nebraska Mathematics Standards*), students work in small groups of 3–5 to complete the analysis, submit a report, and create a presentation for their peers in which they summarize their findings. (Originally, the assignment was built around the NCTM's *Principles and Standards for School Mathematics* (NCTM, 2000) and NSF-funded reform curricula.) Students typically take great pride in their report. One group that produced a 69-page report even printed T-shirts saying "I survived the Curriculum Report" that they wore the day they presented their report.

An abbreviated version of the assignment description is below.

Curriculum Project

*This curriculum project asks you to investigate an area of the elementary mathematics curriculum and consider what teachers need to know to teach it well and what children need to learn to understand the topic in the deep and meaningful ways suggested by the Common Core State Standards in Mathematics (*http://www.corestandards.org*).*

You will work in groups on this project to investigate a grade band (K–2 or 3–5), a CCSS Content Domain, and a set of curriculum materials from your practicum classroom and school. Possible content domains include Counting and Cardinality, Operations and Algebraic Thinking, Number and Operations in Base Ten, Geometry, Number and Operations-Fractions.

1. *Read, analyze, and synthesize the Common Core State Standards for Mathematics (CCSS) content domain you have chosen, first focusing on one single grade, then the entire grade band. What does a child need to understand the topic in the ways suggested by the standards?*
2. *Read, analyze, and synthesize CCSS Mathematical Practices relevant to your content domain. What should be taught and what does a child need to begin to understand to engage in these mathematical practices?*
3. *Study the curriculum used by the local school district. How well do the materials address your content domain according to the CCSS? Do the curriculum materi-*

als afford an opportunity to teach the Common Core Standards for Mathematical Practice?

4. *Select and work three sample math problems* ***for future teachers*** *that address your domain. The problems should engage teachers in learning mathematics they need to know to teach as envisioned by the CCSS and be appropriate for college students preparing to be teachers.*
5. *Select and work out three sample math problems* ***for children*** *that address your domain and help children learn mathematics as envisioned by the CCSS. Explain how working on the problem provides children opportunities to learn the mathematics in deep and meaningful ways.*

5.4.3 Professional Writings

A final example of an integrated assignment involves what are referred to as "professional writings." These assignments are designed to make connections within and across mathematical, pedagogical, and field experiences through written reflections. These writing tasks engage students in considering issues related to mathematics education fueled by their own experiences and/or things they have read. Here are two examples of professional writing prompts.

Professional Writing #1

Read: "What do Math Teachers Need to Be?" by Herb Clemens, Teaching academic subjects to diverse learners (pp. 84–96), Clemens, 1991.

Questions:

- *Where does your own practice of teaching mathematics stand in relationship to what Clemens says mathematics teachers need to be: unafraid, reverent, humble, opportunistic, versatile, and in control of their math.*
- *If he came to your practicum classroom and watched you teach a math lesson tomorrow, how would he answer:* ***Can this teacher teach [math] with conviction, and with some feeling for its essence?*** *Explain.*

Professional Writing #2

*Based on the scenario by Liping Ma (*1999*).*

For this professional writing, assume that you are the fourth grade math team leader at your elementary school. The instructor of one fourth-grade class, Miss X, asks you to help her respond to a situation that came up in her class.

Miss X explains that her students have been working in groups to study properties of rectangles by designing a rectangular fenced-in yard for a new pet and have been encouraged to share their findings and observations. When Miss X asked the groups to share their observations, one group declared the following:

"If you use more fencing, you enclose more area."

Your task is to compose a letter to Miss X to help her address this situation. In your response you should:

Discuss the validity of the student's assertion and explain your reasoning.
Suggest a strategy for Miss X to help her class investigate the assertion (using pictures or diagrams as needed) and arrive at a correct conclusion.

Some things to consider as you compose your response: How do you make sense of the mathematics in this problem? How might the group have come to this conclusion? What pedagogical issues are important as you decide what to write? What mathematical ideas should you discuss in your letter? How can you explain these mathematical ideas in ways that will support Miss X and be a positive learning experience both for her and for her students?

5.5 A Sustained Partnership: Strengthening Course Connections

Now 19 years after piloting the first "Math Matters" semester, The Mathematics Semester and the partnership between UNL's Department of Mathematics and Department of Teaching, Learning and Teacher Education has continued, even as new faculty have transitioned into the positions of leadership. Following Lewis and Heaton, Dr. Michelle Homp (MATH) and Dr. Amanda Thomas (TLTE) have assumed leadership roles in sustaining The Mathematics Semester. As a mathematician with considerable experience working with K–12 teachers, Homp serves as the (long-term) coordinator for the mathematics component, while Thomas coordinates the math methods courses and field experiences. Together they continued the practice of administering (and refining) assignments common to both the content and methods courses, and further integrating the instruction between the content and methods courses. In addition, they are responsible for mentoring graduate students from each of the departments who teach in the program. Several graduate students who have taught in the program report that the experience was helpful as they applied for academic positions upon completing their Ph.D. degrees.

5.5.1 Support for Instructors

One of the early challenges in implementing the Math Semester resulted from the sometimes conflicting cultures of the mathematics faculty and the faculty in mathematics education. As previously mentioned, student evaluations of mathematics faculty teaching content courses for future elementary teachers tend to be critical, even for faculty who generally receive outstanding student evaluations. Consequently, it is not uncommon for tenure-track mathematics faculty to resist teaching these courses, leaving them to be taught by graduate students or part-time lecturers.

The reasons for the criticism vary. One reason, is that coming into the content course students equate understanding mathematics with knowing how to follow the algorithms used to perform operations (Graeber, 1999). Thus, they perceive themselves as already possessing the mathematical knowledge needed to be successful at teaching math (Thanheiser et al., 2013), and therefore without reason (other than fulfilling a requirement) to have to take the course. By the time students are enrolled in the Math Semester, many of them have had one or two courses directly related to their preparation to be elementary teachers. This can make it difficult for some to see the relevance in the mathematics being studied in content courses. Additionally, there are cultural differences between the math and education departments in how instruction is delivered and students are assessed. As a student moves through the Elementary Teacher Education Program, the evaluation for their work as students becomes more dependent on projects than on homework and tests. Yet homework and tests are still useful means of assessment for mathematics faculty. This often results in a stark contrast between the kinds of expectations students encounter in their math courses and in their education courses, making the mathematics expectations seem particularly unyielding and overwhelming.

Grading is also an issue. On UNL's campus, College of Education and Human Sciences faculty gives among the highest grades on campus, while the mathematics department in the College of Arts and Sciences gives among the lowest.

Though these challenges remain, as the program has matured one strategy developed for addressing them is to ensure that support for the Mathematics Semester extends beyond the involvement of two individuals. Full support from the Department of Mathematics has led to the practice of selecting instructors for the math content courses (generally 2–5 of them) from among only veteran graduate teaching assistants (GTAs) with solid teaching credentials and experience with a variety of pedagogical strategies. With additional support from TLTE, prior to the start of the academic year, Thomas and Homp lead a mini-workshop for the Mathematics Semester instructors to provide an overview of the program, to discuss the collaborative nature between the math content and the methods courses, and to prepare them for the unique challenges and rewards of working with future elementary teachers. The Mathematics Semester conveners share all course materials (including sample lesson plans) with the other instructors via file-sharing, and all exams are written collaboratively to ensure consistent expectations. Taken together, these steps help to ensure the Mathematics Semester experience, especially in the math classroom, is more likely to be a positive one, both for the students and the GTAs serving as instructors.

5.5.2 Further Integration of Math Content and Methods Instruction

As the program has continued, more coordination between the mathematics content and methods courses has been implemented. Thomas and Homp continue to work together to update and revise common assignments. Following the guidelines of ATME Standard P.3. Opportunities to Learn to Teach Mathematics (2017), which states "mathematics, practices for teaching mathematics, knowledge of students as learners, and the social contexts of mathematics teaching and learning are integrated," they have made efforts to strengthen the interplay between experiencing mathematics as a learner and as a teacher. One such revision is the reordering of the topics in the two courses so that they are in alignment whenever possible. (Because of the sequential nature of mathematics, this largely meant adjusting the order of topics in the methods course to align with those in the mathematics course.) For example, addressing methods instruction related to multiplication (representations, learning facts, alternative algorithms, etc.) now take place on the same day that the meaning of multiplication and the reasoning behind the standard algorithm is taught in the math content course. Similarly, attributes of addition and subtraction are addressed in the content and methods courses on the same day. After making the observation that the methods course did very little to address instruction in geometry (which is taught in a required geometry course outside of the Mathematic Semester), the conveners coordinated with the instructors of the geometry course, Geometry for Elementary Teachers, to plan a class period in which concepts of pedagogy as they relate to geometry instruction are brought as a workshop into the mathematics class.

Additional revisions framed by the AMTE standards have included even further collaboration between the two departments to develop two, fully integrated lessons in which math and pedagogical content is intertwined throughout the consecutive class periods. One of these lessons addresses the various structures of addition and subtraction problems (e.g., take away vs. comparison, etc.) and is featured on the AMTE website featuring supplementary materials for enacting the Standards for Preparing Teachers of Mathematics (Thomas & Homp, 2019). Another fully integrated lesson addresses the meaning of division (partitive and measurement models) and understanding the division algorithm. Though currently anecdotal, student response to the integrated lessons has been overwhelmingly positive, even candidly cited as among the most meaningful experiences within the Mathematics Semester.

5.6 Gains in Mathematical Knowledge for Teaching

As a way of determining whether the Math Semester was achieving its goal of preparing well-started beginning elementary teachers, we began using questions from the Learning Mathematics for Teaching (LMT) assessment (Hill, Schilling,

& Ball, 2004) to measure students' growth from the beginning to the end of the semester. We use the questions on the LMT pre- and posttests which focus on number and number sense. We administer the pretest on the first day of the semester and the posttest during the second to last week of the semester (week 14) to allow time for the results to be analyzed and shared with the class. Data for 37 sections since 2012 reveal the average pretest score is −0.46, the average posttest score is 0.46, yielding an average gain of 0.92. These increases are noteworthy since an increase of 0.5 is considered statistically significant.

5.7 Sustaining the Partnership

While a sustained partnership between MATH and TLTE has been instrumental in strengthening UNL's elementary teacher preparation program, the need for such a partnership is not limited to the context of teacher training. Many current K–12 teachers have needs which can be best addressed through the combined expertise that a mathematician and a mathematics educator can bring to the table. We have evidence that this kind of partnership works; at Nebraska, our partnership resulted in two large NSF grants for designing professional development through Math Science Partnerships (MSP), Math in the Middle Institute Partnership, and NebraskaMATH (see http://scimath.unl.edu/).

Yet, a sustained partnership between two departments does not happen of its own volition. It takes time to develop and the resulting changes to the teacher preparation program will occur gradually. Commitment to the partnership must extend beyond individuals. Leadership matters so that turnover in personnel does not disrupt full departmental support for the continuation of the Mathematics Semester. TLTE, in particular, must continue its advocacy so that systemic changes in teacher training requirements do not result in the "crowding" or dismantling of the Mathematics Semester. The discussion around adopting a STEM Semester is viewed as very positive evidence that UNL's teacher education program wishes to be known for its emphasis on mathematics and science in the elementary education program. The Mathematics Department needs to be selective when first asking a mathematician to serve in a (long-term) capacity as liaison/coordinator for the department's contribution to the Mathematics Semester, and then support and reward his or her work in the field of K–12 education.

In particular, continued collaboration requires both partners to

- Value the integration of content and pedagogy while respecting the contributions of each.
- Commit to a long-term partnership, regardless of leadership/faculty turnover.
- Build on existing relationships and capitalize on strengths.
- Appreciate mathematics for teaching as rigorous.
- Mediate expectations for learning mathematics for teaching.
- Support one another's goals.

- Understand and respect differences in how instruction is delivered.
- Accept cultural differences in how students are assessed.

5.8 Conclusion

Looking back on the partnership that began in Fall 2000, we are convinced that our collaboration has been the key to our success. The methods courses at UNL are more mathematical than they were two decades ago. The mathematics courses have a much stronger connection to the work of teaching elementary mathematics than previously. By supporting each other, we are able to hold our students to high standards, to help them make important progress learning both mathematics and how to teach mathematics, and to have them end the semester with a positive attitude toward mathematics.

References

Association of Mathematics Teacher Educators. (2017). *Standards for preparing teachers of mathematics*. Retrieved from amte.net/standards.

Beckmann, S. (2017). *Mathematics for elementary teachers* (5th ed.). London, UK: Pearson.

Clemens, H. (1991). What do math teachers need to be? In M. Kennedy (Ed.), *Teaching academic subjects to diverse learners* (pp. 84–96). New York, NY: Teachers College Press.

Conference Board of the Mathematical Sciences. (2001). *The mathematical education of teachers* (Vol. 11). Providence, RI and Washington, DC: American Mathematical Society and Mathematical Association of America.

Conference Board of the Mathematical Sciences (2012). The mathematical education of teachers II. Providence, RI and Washington, DC: American Mathematical Society and Mathematical Association of America.

Cuoco, A., Goldenberg, E. P., & Mark, J. (1996). Habits of mind: An organizing principle for mathematics curricula. *The Journal of Mathematical Behavior, 15*(4), 375–402.

Glencoe/McGraw-Hill. (1991). *MathScape: Seeing and thinking mathematically: Course 1 Patterns in numbers and shapes*. New York, NY: Glencoe/McGraw-Hill.

Graeber, A. O. (1999). Forms of knowing mathematics: What preservice teachers should learn. *Educational Studies in Mathematics, 38*(1–3), 189–208.

Heaton, R. M., & Lewis, W. J. (2008). A mathematics educator and mathematician work in partnership to prepare elementary mathematics teachers: A local response to a national imperative. In *Conference Proceedings for the Penn GSE US-China Math Education Exchange*. Philadelphia, PA: University of Pennsylvania.

Heaton, R. M., & Lewis, W. J. (2011). A mathematician-mathematics educator partnership to teach teachers. *Notices of the AMS, 58*(3), 394–400.

Hill, H. C., Schilling, S. G., & Ball, D. L. (2004). Developing measures of teachers' mathematics knowledge for teaching. *The Elementary School Journal, 105*(1), 11–30.

Leitzel, J. R. (1991). *A call for change: Recommendations for the mathematical preparation of teachers of mathematics. An MAA Report*. Washington, DC: Mathematical Association of America.

Loewenberg Ball, D., Thames, M. H., & Phelps, G. (2008). Content knowledge for teaching: What makes it special? *Journal of Teacher Education, 59*(5), 389–407.

Ma, L. (1999). *Knowing and teaching elementary mathematics: Teachers' understanding of fundamental mathematics in China and the United States*. New York, NY: Routledge.
National Council of Teachers of Mathematics. (2000). *Principles and standards for school mathematics*. Reston, VA: National Council of Teachers of Mathematics.
National Governors Association Center for Best Practices, Council of Chief State School Officers. (2010). *Common core state standards mathematics*. Washington, DC: National Governors Association Center for Best Practices, Council of Chief State School Officers.
National Mathematics Advisory Panel. (2008). *Foundations for success: The final report of the National Mathematics Advisory Panel*. Washington, DC: US Department of Education.
Philipp, R. A., Clement, L., Thanheiser, E., Schappelle, B., & Sowder, J. T. (2003, April). Integrating mathematics and pedagogy: An investigation of the effects on elementary preservice teachers' beliefs and learning of mathematics. In *Research presession of the Annual Meeting of the National Council of Teachers of Mathematics*. San Antonio TX: National Council of Teachers of Mathematics.
Rolle, Y. A. (2008). *Habits of practice: A qualitative case study of a middle-school mathematics teacher*. Lincoln: The University of Nebraska-Lincoln.
Shulman, L. S. (1986). *Those who understand: Knowledge growth in teaching*. Educational researcher, *15*(2), 4–14.
Simon, M. A. (1993). Prospective elementary teachers' knowledge of division. *Journal for Research in Mathematics Education, 23*(3), 233–254.
Simon, M. A., & Blume, G. W. (1994). Building and understanding multiplicative relationships: A study of prospective elementary teachers. *Journal for Research in Mathematics Education, 25*(5), 472–494.
Thanheiser, E., Philipp, R. A., Fasteen, J., Strand, K., & Mills, B. (2013). Preservice-teacher interviews: A tool for motivating mathematics learning. *Mathematics Teacher Educator, 1*(2), 137–147.
Thomas, A., & Homp, M. (2019, June). *Structures of addition and subtraction: Supplementary materials for AMTE standards for preparing teachers of mathematics*. Retrieved from https://amte.net/sptm/supp.
Wideen, M., Mayer-Smith, J., & Moon, B. (1998). A critical analysis of the research on learning to teach: Making the case for an ecological perspective on inquiry. *Review of Educational Research, 68*(2), 130–178.
Wright, E. (1997). *NSF Elementary Science, Mathematics, and Technology Preparation Project Final Report, 1990–1996*. NSF Award Number 9050015. Manhattan, KS: Kansas State University.

Chapter 6
Focusing on Mathematical Structure *and* Sensemaking in Courses for Future Teachers

Sybilla Beckmann and Andrew Izsák

Mathematics courses for future teachers have been recognized as worthy of significant attention and careful design (e.g., Conference Board of the Mathematical Sciences, 2012), but the field needs examples of these courses together with discussions of their features and what the future teachers in these courses are able to achieve. We are interested in these questions: How can courses for future teachers (1) focus on important mathematics the teachers are responsible for teaching, (2) help future teachers see this mathematics as coherent, logically structured, and organized by a small set of core ideas, and (3) draw on future teachers' own sensemaking to reason about the mathematics? We approach these questions, and the goals they imply, as a kind of engineering design problem: to identify or formulate definitions, statements, proofs, explanations, and problem-solving methods that we want future teachers to use or develop, and which they can use and develop in practice.

In this chapter, we report on our mathematics courses for future middle-grades and secondary-grades teachers designed from this perspective. We discuss some central, distinctive features of our courses and explain why we think those features are important to emphasize. We give examples and discuss the kind of reasoning and explanations future teachers are able to produce with our approach and where we would like to do better. Course design requires balance and humility: the goals we describe above compete with each other, and we discuss how we address and balance all of them. In practice, this means we have to aim for something that is good enough, not perfect. We do not claim that our courses are better than others;

S. Beckmann (✉)
Department of Mathematics, University of Georgia, Athens, GA, USA
e-mail: sybilla@uga.edu

A. Izsák
Department of Education, Tufts University, Medford, MA, USA
e-mail: Andrew.Izsak@tufts.edu

Y. Li et al. (eds.), *Developing Mathematical Proficiency for Elementary Instruction*, Advances in STEM Education, https://doi.org/10.1007/978-3-030-68956-8_6

we are just trying to design courses that work. We also cannot claim that any of the features we think are important are actually what make our courses work—it is the course as a whole that achieves its effect on the students who are enrolled in it. But we will show that future teachers who take our courses are able to solve problems, give explanations, and make connections that the research literature has repeatedly documented are difficult. As such, this chapter provides an existence proof of what can be accomplished in mathematics courses for future teachers and a discussion of what we have emphasized in our courses and why.

6.1 A Focus on the Multiplicative Conceptual Field

The interrelated topics of multiplication, division, fractions, proportional relationships, and linear relationships provide a critical foundation for mathematics and science and are therefore a reasonable focus of courses for future teachers. Gérard Vergnaud (1988) described the *multiplicative conceptual field* as "all situations that could be analyzed as simple or multiple proportion problems and for which one usually needs to multiply or divide" (p. 141). Vergnaud advocated studying such comprehensive systems because he viewed concepts as developing in relationship with each other rather than in isolation. Vergnaud was interested in analyzing and classifying the situations of the multiplicative conceptual field and describing the ideas and procedures students use to deal with them.

We are interested in an additional aspect of the multiplicative conceptual field: its coherence. We see this coherence as deriving from quantitative definitions for fractions and for multiplication. As we have discussed elsewhere (Beckmann & Izsák, 2015; Izsák & Beckmann, 2019), such definitions can serve as central organizing ideas for multiplication, division, ratio, and proportional relationships with whole numbers as well as nonnegative fractions and decimals. In particular, our definition of multiplication presents multiplication as a single arithmetic operation with one meaning that applies across all nonnegative numbers and relevant word problem types. But designing a mathematics course for coherence—by returning repeatedly to a few definitions and central ideas—will not necessarily make the course understandable or useful to future teachers. We therefore formulate definitions that strike a balance between mathematical precision and being usable by future teachers in practice. We present versions of our definitions in subsequent sections, but first we turn to the issue of balancing the expert mathematical perspective with future teachers' productive reasoning.

6.2 Balancing Logical Structure and Sensemaking

In discussing how coherence was a guiding principle in writing the Common Core State Standards for Mathematics (National Governors Association Center for Best

Practices & Council of Chief State School Officers, 2010), McCallum (2018) contrasted a *sensemaking* stance toward mathematics with a *making-sense* stance. The sensemaking stance views mathematics as pattern-seeking and problem-solving, and it foregrounds processes such as communication. We view the sensemaking stance as focusing on the learner and their thinking. According to McCallum, the danger with this sensemaking stance is that critical mathematics content could become a backdrop to the action—for instance, learners might not recognize or look for mathematical structure. In contrast, the making-sense stance views mathematics as something to be actively structured in such a way that helps people to makes sense of it. As McCallum pointed out, although this structuring is constrained by the logic of mathematics, it does not tell you how to make mathematics make sense. We view the making-sense stance as focusing on experts' views of mathematical ideas and how to organize the ideas coherently to foreground mathematical structure. A danger of not attending both the making-sense stance and to the sensemaking stance is that

> ... attending to logical structure alone can lead to overly formal and elaborate structuring of mathematical ideas. Just as it is a risk of the sense-making stance that the mathematics gets ignored, it is a risk of the making-sense stance that the sense-maker gets ignored. (McCallum, 2018)

Thus, McCallum (2018) argued that we must attend to both stances when supporting learners.

In this chapter, we try to convey how our courses for future teachers blend a making-sense stance with a sensemaking stance. We motivate the need for the definitions we use, but we expect future teachers in our courses to accept these definitions as given pieces of mathematics and to use them in constructing arguments. We use the definition of multiplication as a central organizer for our discussions of multiplication, division, and proportional relationships. But we also foster future teachers' own sensemaking about quantities in ordinary situations in the world. We ask future teachers to derive various standard calculation procedures (such as how we multiply and divide fractions) from such reasoning, and we ask them to explain how the definition of multiplication applies to a range of types of problems about everyday situations.

In the next section, we describe and discuss some distinctive features of our approach to fractions and multiplication in our mathematics courses for future middle-grades and secondary teachers.

6.3 Distinctive Features of Our Making-Sense Stance

In our courses for future teachers, we take a making-sense stance by traversing a significant portion of the multiplicative conceptual field in a logically connected way using a few central ideas repeatedly. In this section, we discuss four distinctive features of our making-sense stance and our approach to coherence. The first main aspect of our making-sense stance is to give future teachers explicit definitions of fractions and of multiplication to use in their reasoning and explanations throughout the course. We have developed these definitions from our expert point of view, and we see them as like Vygotsky's (1986) notion of scientific concepts, the understanding of which evolves through interactions with the teacher and classmates. Over many semesters, we have refined and revised the wording of our definitions. On the one hand, we have sought to bring forth and emphasize core mathematical ideas. The future teachers in our courses will be returning to these definitions repeatedly, so the ideas within them should be worth emphasizing. Recent versions of our definition of fractions and multiplication (see Izsák & Beckmann, 2019) have grounded fractions and multiplication explicitly in concepts of measurement—a foundation that we suspect needs much more emphasis in courses for teachers and students.

Second, because we take a definition of multiplication as a central organizer for the multiplicative conceptual field, and this definition applies to whole numbers as well as to fractions and decimals, we organize our courses to highlight the unity of this definition. This leads to traversing the material in a different order than some traditional treatments. Before we discuss multiplication or division in our courses, we first discuss how to understand whole numbers and nonnegative fractions and decimals as a coherent set of numbers, namely as results of measurement. We then discuss multiplication as a coherent operation that has the same meaning across whole numbers and nonnegative fractions or decimals. We go on to discuss division as multiplication with an unknown factor and also as a coherent operation across whole numbers and nonnegative fractions and decimals.

Our way of organizing the material may seem counterintuitive: Should not a discussion of division precede a discussion of decimals and fractions? Our answer is no, not necessarily, and for several reasons. First, although students in elementary school will begin to learn about multiplication and division with whole numbers before they work with fractions, it does not follow that courses for future teachers must traverse the material in this same order. Instead, we have opted to highlight the unity of multiplication as a single coherent concept across whole numbers and fractions by drawing on measurement (see Izsák & Beckmann, 2019). We thus aim to help future teachers develop a new, more advanced, and more coherently unified perspective on material they learned in elementary and middle school.

Second, we discuss multiplicative and additive structure in terms of modeling with equations of the form $A \bullet B = C$ or $A + B = C$ rather than in terms of physical or mental actions, or taxonomies of problem types. As explained by Izsák and Beckmann (2019), we distinguish between the physical or mental operation of *partitioning*—cutting an object or stuff up into equal portions—from the arithmetical operation of *division*—something we do with numbers. Likewise, we distinguish the physical or mental operation of *iterating*—replicating an object or stuff multiple times—from the arithmetical operation of *multiplication*. We also do not use the various different means for classifying word problems into types that have been discussed in the literature to characterize multiplicative (or additive) structure (see Izsák & Beckmann, 2019).

Third, we combine our making-sense stance, in which we give future teachers carefully formulated definitions and deliberately traverse the material in a certain order, with a sensemaking stance, in which we support students' own sensemaking to reason with the definitions we give them about quantities in the world. We see our approach of combining our expert-generated definitions with future teachers' own reasoning as reminiscent of Vygotsky's (1986) idea that scientific concepts and spontaneous or everyday concepts are not antagonistic or mutually exclusive forms of thinking but rather are interdependent and evolve together. Just as McCallum (2018) noted a dichotomy in views of mathematics that are associated with taking either a sensemaking stance or a making-sense stance to the exclusion of the other, Kozulin (1986) noted various researchers' and educators' tendencies to exclude either scientific concepts or everyday concepts from the focus of learning.

In our courses, we aim to deepen future teachers' understanding of numbers and multiplicative structure. To do so, we think that neither expert-generated definitions nor future teachers' (or students') own everyday ideas about numbers, multiplication, and division will suffice alone. We therefore try to create opportunities for productive intertwining of scientific and everyday ideas.

In the next sections, we illustrate the distinctive features we have discussed in this section with specific examples from the first semester of a two-semester sequence for future grades 4–8 teachers that was taught by the first author in the fall semester of 2018. The course webpage, which can be found at https://www.sybillabeckmann.com/arithmetic-for-middle-grades-teache, provides the topic of each day's lesson and homework assignment. The data we use come from a multi-year project in which we have been studying the multiplicative reasoning of future secondary (grades 6–12) teachers and future middle-grades (grades 4–8) teachers. These future teachers were enrolled in one or two mathematics courses taught by the authors. The data we collected included all of the future teachers' written work that was part of the regular business of the course.

6.4 Reasoning with the Common Core Definition of Fractions

In this section, we discuss how we typically introduce the Common Core definition of fractions in our courses for future teachers. We then discuss how the future middle-grades teachers in the fall of 2018 were able to use this definition productively: to solve problems they were not able to solve at the beginning of the course, to solve problems by reasoning about math drawings that they made, and to explain fraction addition. We also discuss two aspects of fraction reasoning that remained difficult for the future teachers even after instruction, both of which concern interpreting the role of numbers that arise in real-world situations. In one case, the difficulty concerns how the numbers in the situation relate to solving a problem by partitioning a math drawing. In the other case, the difficulty concerns determining how the numbers should be involved in an addition or subtraction equation that models the problem. This difficulty brings up an important distinction between viewing situations structurally, in the very precise way that they are modeled by equations, versus in terms of actions, such as combining, iterating, and

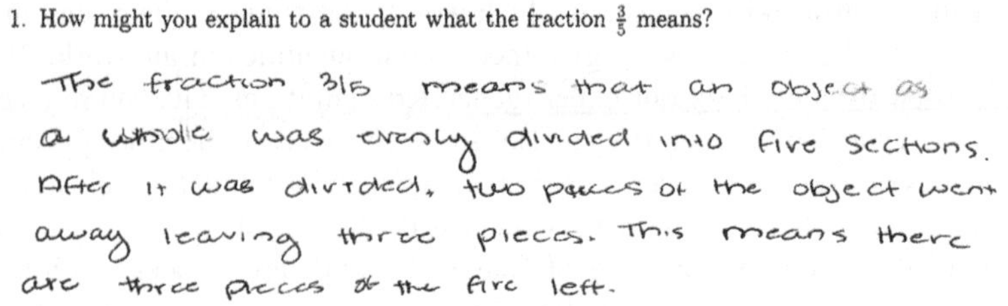
1. How might you explain to a student what the fraction $\frac{3}{5}$ means?

The fraction 3/5 means that an object as a whole was evenly divided into five sections. After it was divided, two pieces of the object went away leaving three pieces. This means there are three pieces of the five left.

Fig. 6.1 A future teacher's initial explanation of what 3/5 means

partitioning. At the end of this section, we briefly discuss how we explicitly connect the Common Core definition of fractions to other ways of thinking about fractions.

6.4.1 Introducing the Definition of Fractions

In our courses, we often introduce fractions by asking future teachers how they would explain the meaning of a fraction such as 3/5. Many of the future teachers describe 3/5 as "3 out of 5." For example, Fig. 6.1 shows a typical response to the question "How might you explain to a student what the fraction 3/5 means?" on a fraction survey that was given as an initial homework assignment before instruction in fractions (see Appendix A for the survey questions). In a subsequent class discussion, we typically ask the future teachers if they see limitations to describing fractions as "*A* out of *B*," and we introduce questions such as "what about improper fractions such as 7/5?" and "what about plotting fractions on number lines?" if the future teachers do not bring these limitations up on their own. The class comes to the conclusion that viewing 3/5 as "3 out of 5" is not helpful for plotting 3/5 on a number line and that describing 7/5 as "7 out of 5" is not reasonable. Therefore, we need another definition of fractions besides "*A* out of *B*."

The definition of fractions that we use during the rest of the course is very similar to the one found in the Common Core State Standards for Mathematics and proceeds in two steps. Given some amount or quantity, which we call a unit amount, a whole, a referent unit, or just a referent, we define 1/*B* of the unit amount as the new amount formed by 1 part when the unit amount is partitioned into *B* equal parts. We then define *A*/*B* of the unit amount as the amount formed by *A* parts, each of size 1/*B* of the unit amount. We intentionally define "*A*/*B* of the unit amount" rather than just "*A*/*B*" because unit amounts are essential when we use fractions to describe quantities in the world. For example, some rice could be simultaneously one-third of a cup and half of a serving.

Future teachers in our courses often find this fraction definition wordy, but in practice, in the midst of an explanation, we usually shorten it to the form "7 parts, each of size 1/5" or "7 parts, each 1/5." In fact, we could view these ways of describing 7/5 as elaborations on what "seven fifths" literally means: There are 7

The strip below is $\frac{2}{3}$ as long as another strip.

(a) Is the other strip longer or shorter than the strip below?

(b) Draw the other strip. Explain your reasoning in a sentence or two.

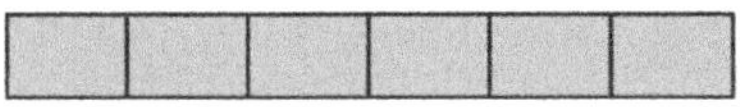

Fig. 6.2 A fraction "reversibility" task on a pre-instruction fraction survey

things, and each of them is a fifth. Thus, we ask the future teachers to understand fractions in a way that is closer to the literal meaning of fraction words instead of re-interpreting fractions in an "out of" sense.

6.4.2 *Difficult Fraction Reasoning That the Definition Makes Accessible*

Does the Common Core definition of fractions help future teachers reason about fractional quantities? In this section, we present evidence that instruction that emphasizes the Common Core definition of fraction can help future teachers reason about quantities shown in math drawings and thereby solve difficult fraction problems that they were not able to solve at the beginning of the course.

To see how much facility future teachers had with reasoning about quantities shown in math drawings before instruction, we gave a fraction survey (see Appendix A) that included the task show in Fig. 6.2. Of the 19 future teachers in the class, 6 (32%) produced a correct drawing and gave a valid explanation for their solution. Based on percentages correct, this was the hardest task on the survey. Several aspects of this task may make it difficult. First, if a future teacher thinks of the fraction 2/3 as "2 out of 3," then they may be puzzled if they do not see that a 6-piece strip can show 2 things out of 3 things. Also, the task asks for reversing one's thinking about the fraction 2/3: Instead of starting with a whole, partitioning it into 3 equal parts, and then considering 2 of those parts, one must consider the given 6-piece strip to be the outcome of such a process. Furthermore, to solve the task, one reverses the usual roles of the numerator and denominator: One partitions the given 6-piece strip into 2 parts and then makes 3 such parts to create the original whole.

Figure 6.3 shows another task on the survey that was difficult for many future teachers. It is similar to tasks used by Lovin, Stevens, Siegfried, Wilkins, and Norton (2018, Fig. 6.2c p. 215, Fig. 6.8 p. 228), which these authors found were difficult for future teachers. Of the 19 future teachers in the class 8 (42%) produced a correct

4. The strip below is $\frac{9}{7}$ as long as another strip.

   (a) Is the other strip longer or shorter than the strip below?

   (b) Draw the other strip. Explain your reasoning in a sentence or two.

Fig. 6.3 An improper fraction "reversibility" task on a pre-instruction fraction survey

drawing by partitioning the given strip into 9 parts and creating the desired strip from 7 such parts. This task might be difficult both because it requires considering an improper fraction (9/7), which is greater than the whole, and because it requires "reversing" one's thinking about fractions as discussed above for the task in Fig. 6.2. Only 4 (21%) of the 19 future teachers in the class produced correct answers on both tasks in Figs. 6.2 and 6.3.

The Common Core definition of fraction that we used in the course proved to be a tool that the future teachers could use in practice. On the first test, which was a take-home test about 5 weeks into the semester, and after about 4 weeks of instruction in reasoning about fractions, 100% of the 19 future teachers produced a correct drawing and gave a valid explanation for their drawing on the task shown in Fig. 6.4, which is similar to the pre-instruction survey tasks in Figs. 6.2 and 6.3. Mabel,[1] whose (correct) work is shown in Fig. 6.5, gave incorrect answers to the tasks in Figs. 6.2 and 6.3 on the pre-instruction survey. Although the instruction included discussions of tasks like the improper fraction "reversibility" task in Fig. 6.3, it did not include discussions of tasks like the one in Fig. 6.2. This result shows that, with 4 weeks of instruction, this group of 19 future middle-grades teachers was able to use the Common Core definition of fraction to reason about quantities in real-world situations that extended tasks discussed in class.

In summary, for the future teachers we have taught, the definition makes explaining improper fractions (such as 4/3) and solving many problems that involve "reversibility" (producing a whole from a fraction, as in the tasks in Figs. 6.2, 6.3, 6.4 and 6.5) readily accessible. In fact, in our experience, even though future teachers often stumble over the wording of the Common Core definition of fractions at first, most are able to use the definition to explain fractions and solve problems within a lesson or two.

[1]All names are pseudonyms.

1. The strip below shows a runway that is $\frac{4}{3}$ of a mile long. Using the runway below as a guide, draw another runway that is 1 mile long. Explain your reasoning. As part of your discussion, explain what $\frac{4}{3}$ means according to our (Common Core) definition of fraction.

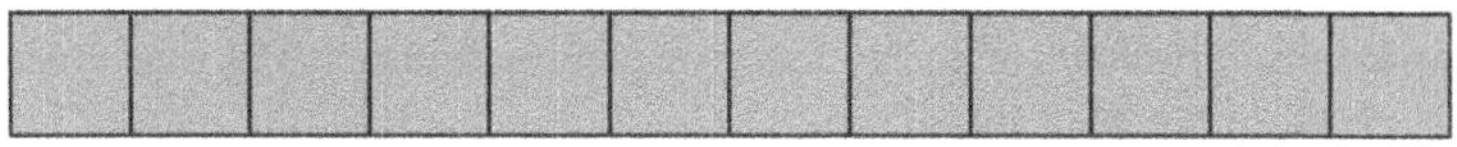

Fig. 6.4 An improper fraction "reversibility" problem on the first test

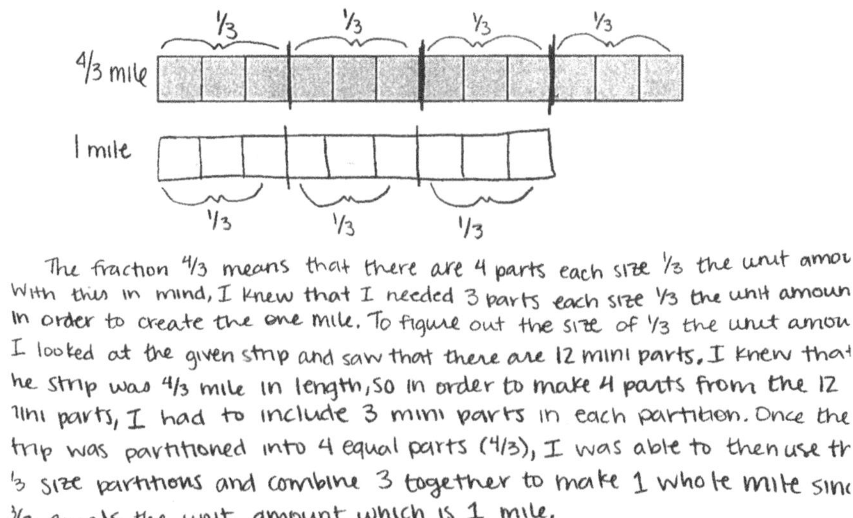

Fig. 6.5 Mabel's solution to the test problem in Fig. 6.4

6.4.3 Solving Fraction Word Problems by Reasoning About Partitioning and Iterating

We expect future teachers in our courses to engage in sensemaking as they reason with the Common Core definition of fractions. To encourage and develop this sensemaking, we give word problems that involve ordinary situations about quantities in the world, and we ask the future teachers to make math drawings that represent the quantities and to explain how their solutions derive from reasoning about these quantities shown in their math drawings. When they come to our classes, future teachers do not seem to be familiar with solving problems by reasoning with math drawings. Although most of the future teachers we work with develop such reasoning in our courses, as we illustrate in this section, there is a common difficulty: explaining the detailed logic of how and why to partition quantities into smaller parts.

We purposely pose problems in which future teachers practice reasoning about fractional quantities that they show in math drawings before we introduce modeling problems with equations. In our experience, we have found that understanding and explaining the connection between equations and word problems is a significant challenge, even for future teachers who can solve word problems by reasoning about math drawings. For example, consider the following Workout Problem:

> Your workout is to swim 5/6 of a mile. Today you only want to do 2/3 of your workout. How far should you swim (in miles)?
>
> Explain how to solve the problem by reasoning about quantities that you show in carefully structured math drawings. Be sure to explain the reasoning or logic of the partitioning that you do in the process of solving the problem.

Notice that the Workout Problem could be modeled by the multiplication equation $2/3 \cdot 5/6 = ?$. However, the Workout Problem was on the first test (a take-home test 5 weeks into the semester), before the future teachers were asked to explain how to model such word problems with equations. At this point in the semester, the goal was for the future teachers to engage in sensemaking about quantities. In her sound solution, Hillary made no mention of multiplying fractions (Fig. 6.6).

During class, the future teachers had solved and discussed problems like the Workout Problem as well as word problems that could be considered fraction division problems. To solve these problems by reasoning about quantities shown in math drawings requires that future teachers not only rely on what fractions mean but also (1) recognize and carefully distinguish between different referent units (wholes) in the situation and (2) determine how and why to create common partitions (see Izsák, Beckmann, & Kulow, 2018 for a discussion of partitioning in the case of fraction division word problems). Class discussions emphasized both of these points. On the test, 17 (89%) of the 19 future teachers in the course made a suitable math drawing that appropriately represented the different referent units (wholes) in the situation and used the drawing to find a solution to the Workout Problem, although one of those 17 did not provide a written explanation to accompany her drawing.

In solving the Workout Problem, Hillary showed that 2/3 of 5/6 is 10/18. In fact, her solution could be part of an explanation for why we multiply fractions the way we do. We see the reasoning that Hillary (and other future teachers) used to solve the Workout Problem as a foundation for discussions later in the semester, after introducing our definition of multiplication, during which future teachers modeled fraction multiplication word problems with equations and explained why we can multiply fractions by multiplying numerators and multiplying denominators.

Hillary's solution of the Workout Problem also shows that the logic of partitioning to solve the problem is more subtle than one might anticipate. Because the Workout Problem involves the fractions 5/6 and 2/3, which have different denominators, one might predict that the rationale for partitioning into mini-parts involves "common denominators." This is not the case, but many future teachers in our courses make such a judgment initially. In her solution, Hillary relied on the idea that 5 parts, each of which is 1/6 of a mile, are aligned with 3 parts, each

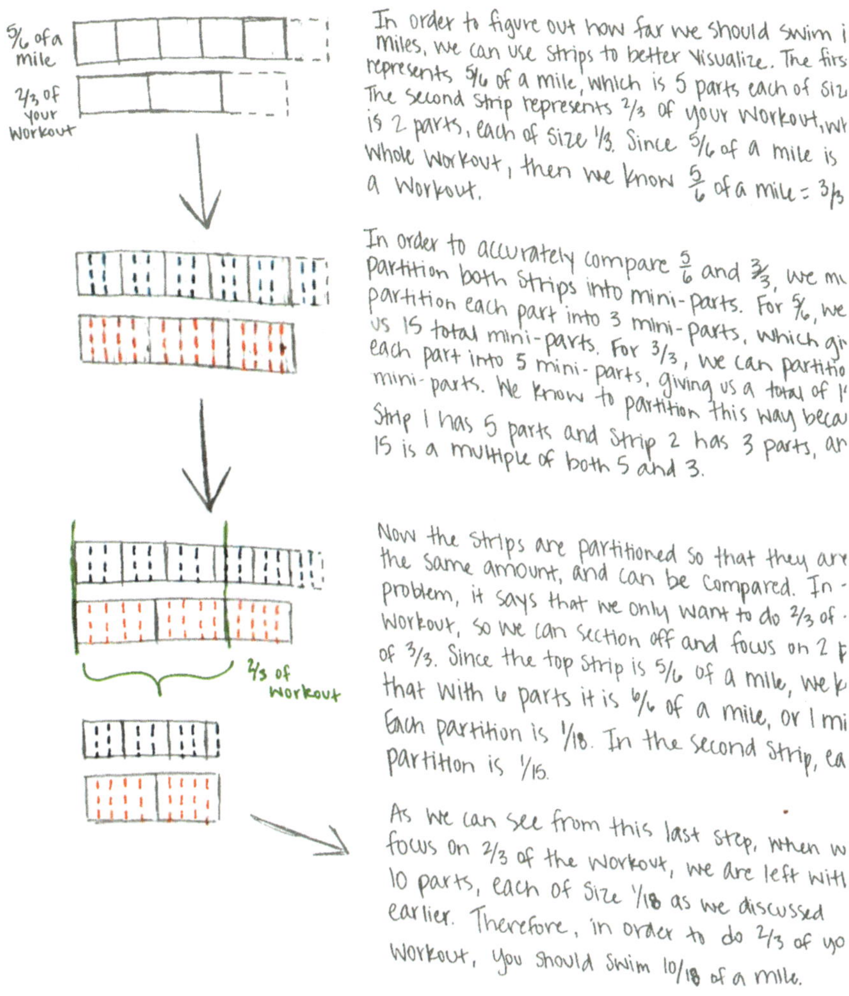

Fig. 6.6 Hillary's solution to the Workout Problem

of which is 1/3 of a workout. In order to coordinate the 5 parts with the 3 parts, she created 15 mini-parts by partitioning each of the 5 parts into 3 mini-parts and each of the 3 parts into 5 mini-parts. When she extended this partitioning to the entire mile, the entire mile was made of 18 mini-parts (6 groups of 3). Two-thirds of the workout consists of 10 mini-parts (2 groups of 5). Thus, 2/3 of the workout is 10/18 of a mile. Although the denominator of the answer 10/18 *is* the product of the denominators of 5/6 and 2/3, Hillary's careful reasoning in Fig. 6.6 shows that the rationale for partitioning into mini-parts—finding a smaller part that simultaneously partitions two given amounts—actually involves the numerator 5 from 5/6 and the denominator 3 from 2/3.

In our experience, one of the most challenging aspects of solving problems by reasoning about quantities shown in math drawings is explaining the exact logic for *how and why* to partition into mini-parts. This logic remains difficult for many future teachers, even after instruction and class discussions. On the Workout Problem, 6 (32%) of the 19 future teachers faltered in explaining how and why to partition into mini-parts, for example, by saying that they needed to find a common multiple between 6 (from 5/6 of a mile) and 3 (from 2/3 of a workout) to create 18 mini-parts in a whole mile. Although almost all the future teachers partitioned their drawings appropriately so that they could derive a correct answer, a significant number did not have a valid mathematical explanation for why they partitioned that way and not some other way. In particular, they did not recognize that it was the numerator 5 from 5/6 and the denominator 3 from 2/3 that governed the partitioning into mini-parts. Thus, attending to the precise roles that numbers in a real-world situation play in producing a solution to a problem and capturing those precise roles in a nuanced mathematical explanation seemed to be difficult for these future teachers.

In summary, based on the Workout Problem and other problems, including ones that could be considered fraction division problems, the class of future teachers was adept at sensemaking with fractional quantities after about 4 weeks of reasoning with the class definition of fractions. Research has shown that future teachers often struggle with fraction problems (for a review see Olanoff, Lo, & Tobias, 2014), but the future teachers in this class were able to make math drawings to show quantities in real-world situations and were able to derive solutions to problems about these situations by reasoning about these math drawings.

6.4.4 Applying the Definition of Fractions to Fraction Addition and Subtraction

In this section we discuss two aspects of fraction addition and subtraction that we think are worth discussing in courses for future teachers. First, the Common Core definition of fractions is helpful for explaining why we add and subtract fractions the way we do in a way that other ways of thinking about fractions are not. Second, by attempting to model situations with fraction addition and subtraction equations one sees that these equations apply in a precise, constrained way, which seems to be invisible when working with whole number situations. To model a situation with addition or subtraction, by placing numbers from the situation into an addition or subtraction equation, requires identifying a very precise structure that is more specific than that the situation involves combining, taking away, or comparing. Furthermore, examining this precise structure brings up an important point of contrast between how addition and subtraction equations model situations (or word problems) and how multiplication and division equations model situations (or word problems).

4. (a) Discuss how viewing the fractions $\frac{3}{5}$ and $\frac{4}{5}$ as "3 out of 5" and "4 out of 5" might lead students to add $\frac{3}{5} + \frac{4}{5}$ incorrectly.
   (b) Write a simple word problem for $\frac{3}{5} + \frac{4}{5} = ?$.
   (c) Explain how to reason with our (Common Core) definition of fractions and math drawings to solve your word problem in part (b).

Fig. 6.7 A final exam problem about fraction addition

One important application of the Common Core definition of fractions is to explain why we add and subtract fractions the way we do. Figure 6.7 shows a final exam problem that asked the future teachers to discuss why it is problematic to view fractions as "*A* out of *B*" in the context of addition. The final exam was a 3-h in-class exam, with no books or notes allowed, and covered material from the entire semester. Of the 19 future teachers in the class, 18 (95%) discussed either the error of adding the numerators and adding the denominators or that it would be problematic to interpret the answer to 3/5 + 4/5 as "7 out of 5" (or both). These same 18 future teachers also used the Common Core definition of fractions to explain why it makes sense to add 3/5 + 4/5 the way we do. Figure 6.8 shows Steve's work on this final exam problem, including his explanation, "if we add 3 parts, each size 1/5, to 4 parts, each size 1/5, we end up with a total of 7 parts, each size 1/5."

Writing suitable word problems for fraction addition and subtraction is surprisingly challenging. It brings up the importance of attending to referent units for fractions and reveals that expressions and equations model situations (or word problems) in ways that are more precise and constrained than simply whether the situation involves combining, taking away, or comparing. As Fig. 6.8 shows, Steve included a suitable word problem for 3/5 + 4/5 for part b of the final exam problem. However, 6 (32%) of the 19 future teachers wrote a word problem for 3/5 + 4/5 that could be considered flawed. For example, several future teachers formulated a situation in which there is 3/5 of one thing, 4/5 of another identical thing, and asked how much of the two things one has, which could legitimately be answered with the fraction 7/10 instead of 7/5. Similarly, the Wire Problem in Fig. 6.9 looks like a 1/3 − 1/4 problem, but it is not because the referents for 1/3 and 1/4 are different. On the final exam, 100% of the future teachers identified that the Wire Problem is not a 1/3 − 1/4 problem and 17 (89%) explained why not, although 5 (26%) wrote another word problem that was not appropriate for 1/3 − 1/4.

That teachers can have difficulty writing appropriate word problems to illustrate an operation has been widely discussed (for a review see Olanoff et al., 2014), but some of the issues involved in writing word problems have not. Although addition and subtraction concern situations of combining, taking away, or comparing, and can be defined in terms of those ideas, precise mathematical definitions of addition and subtraction lead to exacting requirements on when an addition or subtraction equation $A + B = C$ or $A - B = C$ models a situation. In particular, those equations

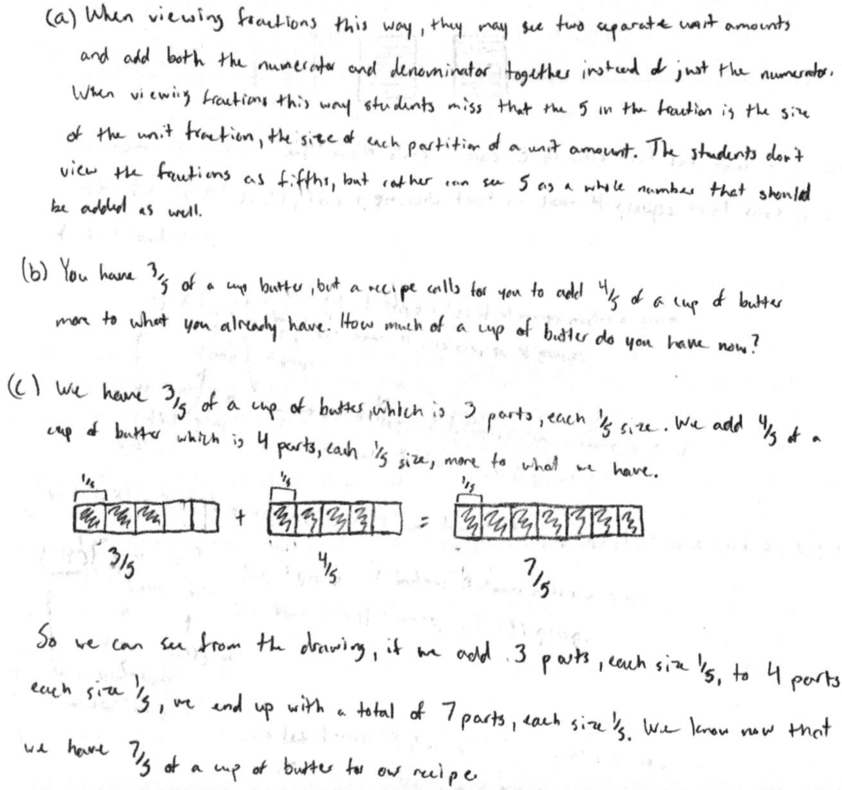
(a) When viewing fractions this way, they may see two separate unit amounts and add both the numerator and denominator together instead of just the numerator. When viewing fractions this way students miss that the 5 in the fraction is the size of the unit fraction, the size of each partition of a unit amount. The students don't view the fractions as fifths, but rather can see 5 as a whole number that should be added as well.

(b) You have 3/5 of a cup butter, but a recipe calls for you to add 4/5 of a cup of butter more to what you already have. How much of a cup of butter do you have now?

(c) We have 3/5 of a cup of butter, which is 3 parts, each 1/5 size. We add 4/5 of a cup of butter which is 4 parts, each 1/5 size, more to what we have.

So we can see from the drawing, if we add 3 parts, each size 1/5, to 4 parts, each size 1/5, we end up with a total of 7 parts, each size 1/5. We know now that we have 7/5 of a cup of butter for our recipe.

Fig. 6.8 Steve's solution to the fraction addition problem in Fig. 6.7

) **Wire Problem A**: You had a 1 meter long piece of wire that weighed $\frac{1}{3}$ of a gram. You used $\frac{1}{4}$ of that wire for a project. Now how much wire do you have left?

Is Wire Problem A a word problem for $\frac{1}{3} - \frac{1}{4}$? If so, explain why it is. If not, explain why not and write a new wire problem for $\frac{1}{3} - \frac{1}{4}$. (You do not need to solve the problem.)

Fig. 6.9 A final exam problem that seems like fraction subtraction

can only model a word problem when *A*, *B*, and *C* all reference an identical unit (e.g., each references 1 Liter). But this requirement for identical units may be invisible when working with whole number addition and subtraction. For example, we can legitimately say that 3 apples plus 2 pears make 5 pieces of fruit, and model this situation with the equation $3 + 2 = 5$. But in the equation, the 3 and the 2 reference fruit, not apples or pears, a technicality that may seem inconsequential when working with whole numbers, but becomes more clearly important when working with fractions. Thus, viewing addition and subtraction as structural models expressed with equations is different from viewing these operations as models of actions such as combining, taking away, or comparing. Because addition and subtraction are introduced so early in children's education, it may not be clear to future teachers (and others) that there even are mathematical definitions for these operations that need to be maintained as one passes from whole numbers to fractions and decimals. Another subtlety is that although A, B, and C all reference an identical unit in the case of addition and subtraction, that is not the case in multiplication and division, as will be discussed later.

6.4.5 Connecting the Definition of Fractions to Other Ways of Thinking About Fractions

For any concept in mathematics, there is generally more than one good way to think about it. So too with fractions. Research has identified various distinct interpretations or "subconstructs" of fractions, including viewing a fraction as a ratio, an indicated division (a quotient), an operator, or a measure (e.g., Behr, Lesh, Post, & Silver, 1983), the latter of which isncludes the Common Core definition of fraction. Connections among different ways of interpreting fractions have long been recognized as important (e.g., Kieren, 1988). Because our making-sense stance foregrounds coherence, our courses develop arguments that explain how the Common Core definition of fraction connects to other fraction subconstructs. We explicitly develop arguments to relate the Common Core definition of fraction to division—namely that $A \div B$ is equal to A/B, what we call the Fundamental Theorem of Fractions (see Wu, 1998 for a proof)—and to different ways of thinking about the rates A/B and B/A that are associated with quantities in a fixed ratio of A to B.

Figure 6.10 shows Carla's explanation on the final exam for why $3 \div 5 = 3/5$. Her explanation relies on the Common Core definition of fractions, the definition of multiplication that was used in the course, and division defined as multiplication with an unknown factor. In the next section, we discuss our approach to multiplication and its central role in our making-sense stance that foregrounds coherence.

6.5 Reasoning with a Definition of Multiplication

In this section, we discuss how we typically motivate and introduce a definition for multiplication in our courses for future teachers. We then discuss specific features of our definition, which we have formulated from an expert perspective, as part of our making-sense stance. As with the Common Core definition of fractions, our definition is based in measurement. In our definition, multipliers and multiplicands play different roles so that modeling situations with multiplication (and division)

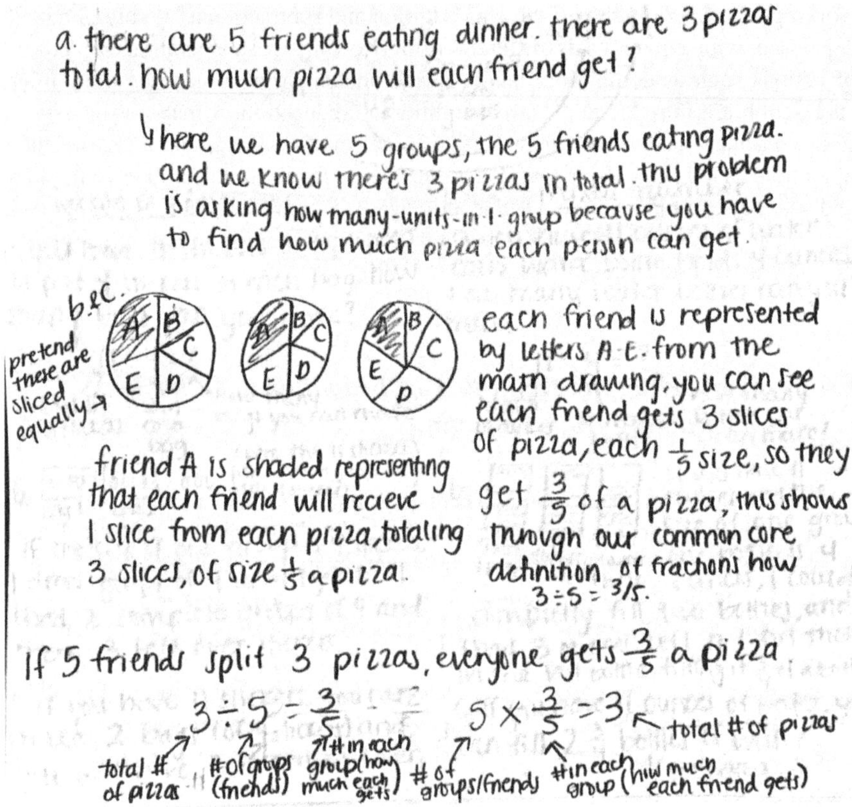

Fig. 6.10 Carla's explanation for why 3 ÷ 5 = 3/5

requires attending to this structural distinction and not just to actions such as replicating or stretching or to types or taxonomies of word problems. In our view, this structural distinction between multipliers and multiplicands plays an important role in viewing multiplication coherently, as a single concept that retains its meaning across whole numbers and fractions. We discuss how, on a pre-instruction survey given in the fall 2018, most future middle-grades teachers did not draw analogies between whole number and fractional multipliers, but, after instruction, most were able to discuss multiplication as a single concept that retains its meaning across whole numbers and fractions. To conclude the section, we briefly discuss applying our definition of multiplication to a range of types of multiplication word problems. Even after instruction, it remained challenging for future teachers to make a fully precise connection between features of a real-world situation and the exact structure of the definition of multiplication.

6.5.1 The Class Definition of Multiplication: Its Features and How We Introduce It

In this section, we discuss how we typically motivate the need for a definition of multiplication and describe some key features of the definition we use. Before introducing a definition of multiplication, we ask future teachers how they might explain what multiplication means—for example what does 3•5 mean?—and to write a simple word problem to illustrate the meaning. The future teachers usually bring up repeated addition, such as $5 + 5 + 5$ or $3 + 3 + 3 + 3 + 3$. We ask the class to consider whether there might be limitations to defining multiplication as repeated addition. If the case of fraction multiplication does not come up during the discussion, we ask about it. The class as a whole comes to the conclusion that repeated addition is not adequate for defining fraction multiplication. We then introduce a class definition of multiplication that we use throughout the remainder of the instruction.

We have used various versions of a definition of multiplication across many iterations of our courses (the first author has taught the courses for middle-grades teachers for over 15 years). We have tinkered with the definition to strengthen it from an expert perspective, for example, in recent versions we have clarified how the concept of measurement undergirds multiplication. We have also tinkered with the definition to make it more understandable and useable for future teachers. Key features of every version are (1) it is based on quantities in situations, (2) it can be used to identify any situation that involves multiplication or division of nonnegative real numbers as such and to model such situations with an equation (where division is interpreted as multiplication with an unknown factor and a known product), and (3) it applies to cases involving not only whole numbers but also nonnegative fractions and decimals. In the last few years, we have defined multiplication explicitly in terms of measurement.

Our definition applies to situations involving quantities in the world, and it involves identifying some quantity in the situation as constituting 1 group, consisting of N base units, where a "base unit" is some unit of measurement (standard or non-standard), and N is some nonnegative number, which could be a whole number, fraction, or decimal. We can consider this 1 group, which is N base units, as another unit with which we could measure quantities. Then if M is a nonnegative number (whole number, fraction, or decimal), we can consider a quantity formed from M of this unit, namely we can consider M of an N-base-unit group. We define $M \cdot N$ as how many base units it takes to make these M N-base-unit groups exactly. We write P for this number, called the product. We call M the multiplier (which we sometimes define as how many groups make the product amount, P base units) and we call N the multiplicand. Therefore, we can summarize the definition of multiplication with the annotated equation in Fig. 6.11. Appendix B provides the two-page handout that was given to the fall 2018 class of future middle-grades teachers to summarize this definition of multiplication and to indicate how to use it. We note that in future iterations of our courses, we plan to refer to the multiplicand

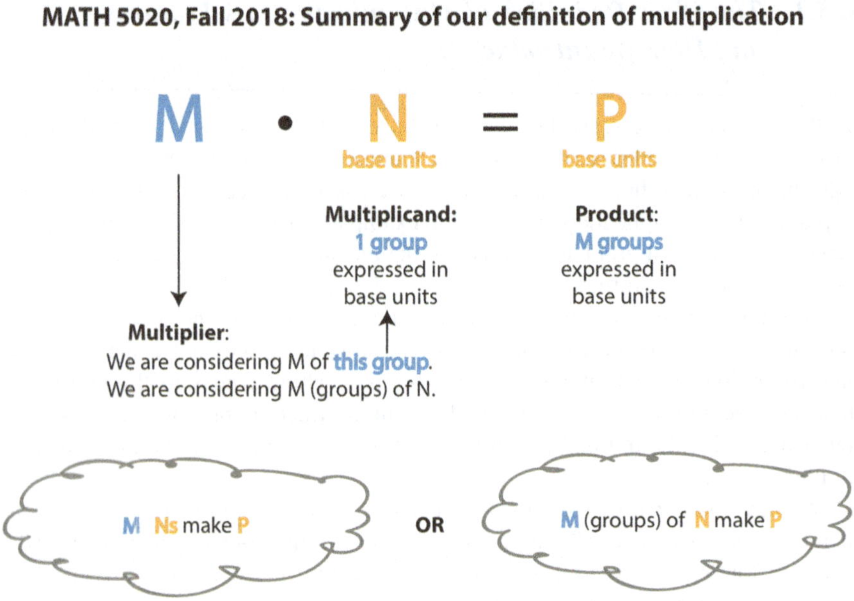

Fig. 6.11 An annotated equation summarizing one version of our definition of multiplication

and the product as "*measured* in base units" rather than "*expressed* in base units," to draw better attention to the measurement perspective.

Elsewhere (Izsák & Beckmann, 2019), we have discussed how our definition of multiplication is related to other ways that multiplication has been conceptualized in the research literature. In our courses for future teachers, we place a high priority on a logically coherent development of arithmetic that is rooted in reasoning about quantities in the world. We see this as a blend of sensemaking and making sense. We enact our making-sense stance by constructing our courses around an explicit meaning for multiplication. We see a definition of multiplication as a central organizer for the multiplicative conceptual field because it can support reasoning about division and proportional relationships with nonnegative whole numbers, fractions, and decimals (Beckmann & Izsák, 2015; Izsák & Beckmann, 2019). As with the definition of fractions, we expect future teachers to use the definition of multiplication in sensemaking about quantities in the world. We ask future teachers to explain why situations are multiplication situations by identifying appropriate base units and groups in the situation and arguing that the situation has the multiplicative structure given in the definition. It is not enough for future teachers to simply assert that a problem can be solved by multiplication because it is of a certain type, such as an area problem or Cartesian product problem. Instead, they must argue that the definition of multiplication applies.

In our definition of multiplication, the multiplier M and the multiplicand N play different roles. For example, suppose we want to make 4 servings of rice, where each serving is 6 ounces, and we want to know how many ounces that is in all. If we let 1 group be a 6-ounce serving, then we want to know how many ounces make four 6-ounce groups, which we model with the equation $4 \cdot 6 = ?$. Here the multiplier, M, is 4 and the multiplicand, N, is 6. In contrast, suppose we want to make 6 small servings of rice, where each small serving is 4 ounces, and we want to know how many ounces that is in all. If we let 1 group be a small 4-ounce serving, then we want to know how many ounces make six 4-ounce groups, which we model with the equation $6 \cdot 4 = ?$. Here the multiplier, M, is 6 and the multiplicand, N, is 4. In both cases there are 24 ounces of rice in all, but the two situations—four 6-ounce servings and six 4-ounce servings—are different. Therefore, with our definition of multiplication, $6 \cdot 4$ and $4 \cdot 6$ model two distinct situations and multiplication is not obviously commutative—or we might say that multiplication is not "situationally commutative"—which some future teachers and some of our colleagues have found disturbing. Part of using our definition of multiplication involves *deducing* that multiplication is numerically commutative, for example, by seeing that arrays can be structured according to their rows or their columns. But many multiplicative situations that elementary- and middle-school students work with are not inherently or situationally commutative, and we think it is important for future teachers to understand that numerical commutativity is therefore not obvious and should not simply be taken for granted.

One final comment about our definition of multiplication is that it provides a criterion for when a real-world situation can be modeled with a multiplication equation, which involves identifying a specific structure in the situation. The criterion is flexible, which explains why there are equivalent ways of calculating. For example, to identify a situation as multiplicative, one must be able to identify a suitable "base unit" and "group" in the situation, such as 1 ounce and one 6-ounce serving above. However, the choice of base unit and group is not unique. If we took a base unit to be 1 ounce and a group to be four 6-ounce servings combined, then we could interpret one 6-ounce serving as ¼ of this group. In this case, we can model the situation in which we are asking how many ounces are in four 6-ounce servings with the equation $\frac{1}{4} \cdot ? = 6$. This way of reasoning can explain why dividing 6 by ¼ is equivalent to multiplying 6 by 4, and more generally, why we can divide fractions by multiplying by the reciprocal of the divisor. Notice, however, that modeling the situation with the equation $\frac{1}{4} \cdot ? = 6$ involved using the number ¼, which was not stated in the description of the situation. In real-world situations that concern multiplication or division with two nonnegative real numbers, if one wants to model the situation with an equation that uses the numbers given in the situation, then there will be a unique choice of base unit and group.

6.5.2 *Understanding Multiplication as a Single Concept Across Whole Numbers and Fractions*

In this section we discuss how future middle-grades teachers in the fall 2018 course understood the roles of whole numbers and fractions (which were multipliers and multiplicands from our point of view) in real-world multiplicative situations before instruction in multiplication. Before instruction, most of the future teachers did not consistently interpret fractional multipliers as analogous to whole number multipliers. After instruction, almost all of the future teachers were able to discuss multiplication as a single concept that retains its meaning across whole numbers and fractions.

To see if future teachers were sensitive to the distinct roles of multipliers and multiplicands in situations, we gave them the survey in Appendix C before instruction in the definition of multiplication. The survey asked the future teachers to compare the quantities in various situations to the quantities in a Soccer Ball Problem:

> There are 5 soccer balls in each bag. The coach has 2 bags. How many soccer balls does the coach have?

For example, the first task on the survey asked about a Cheese Problem:

> You bought 8 packages of cheese. The cheese in each package weighs 1/3 of a pound. How many pounds of cheese did you buy?

In the Cheese Problem, the multiplicand is a fraction (1/3), but the multiplier is a whole number (8). Of the 18 future teachers who completed the survey before instruction in the definition of multiplication, 15 (83%) identified the 8 packages in the Cheese Problem as analogous to the 2 bags in the Soccer Ball Problem. In contrast, three other tasks on the survey had a fractional multiplier, including the Cracker Problem in which the multiplier is ¼:

> One serving of crackers contains 24 grams of carbohydrates. If you eat ¼ of a serving, how many grams of carbohydrates will you eat?

On this problem, only 7 (39%) of the 18 future teachers identified ¼ of a serving as analogous to 2 bags in the Soccer Ball Problem. Overall, only 4 (22%) of the 18 future teachers consistently identified the multiplier as analogous to the multiplier in the Soccer Ball Problem across the entire survey.

From an expert perspective, multipliers and multiplicands can have consistent meanings across whole numbers and fractions (e.g., the meanings expressed in our definition of multiplication). Based on the responses to the survey, before instruction in the definition of multiplication, most future teachers in this class seemed not to take an expert view, in which multipliers and multiplicands have consistent meanings across whole numbers and fractions. Perhaps this is because fractional multipliers are harder to understand than whole number multipliers (e.g., Simon, Kara, Norton, & Placa, 2018), which we have found across multiple iterations of our courses for future teachers. To foreshadow one result, later in this section we show

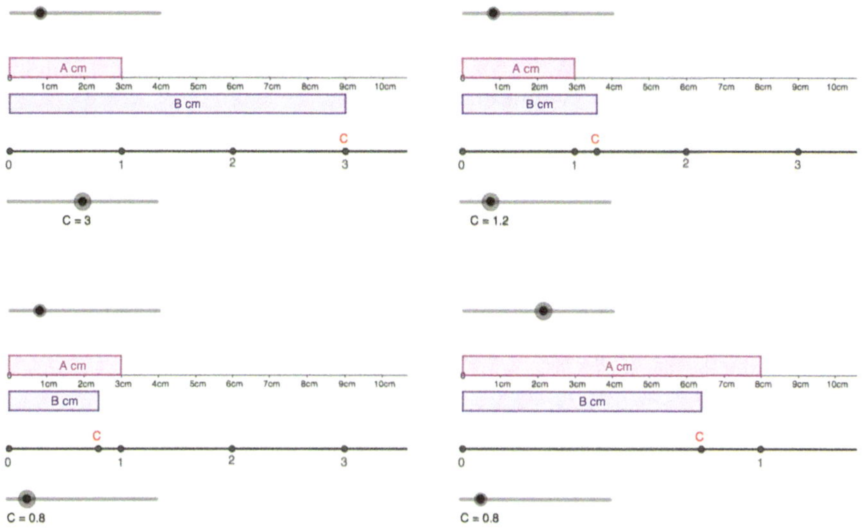

Fig. 6.12 Screen shots of a Geogebra sketch at https://ggbm.at/bmz9sgpx

that after instruction almost all the future teachers were able to express a coherent view of multiplication across whole numbers and fractions and, in particular, were able to discuss a consistent meaning of the multiplier in their own words.

After introducing the class definition of multiplication, but before discussing multiplication with fractions, we wanted to prepare the future teachers to think about how the distinct roles of the multiplier and multiplicand can be understood consistently across whole numbers and fractions or decimals. To do so, we asked the future teachers to work with the dynamic Geogebra sketch at https://ggbm.at/bmz9sgpx, which shows an *A* cm strip, a *B* cm strip, and a value *C*, which varies on a number line. The screen shots in Fig. 6.12 show that one button changes the value of *C* and the other button changes the coordinated lengths of the *A* cm and *B* cm strips while leaving *C* fixed.

We asked the future teachers to move *C* in the Geogebra sketch to various different values, including whole numbers and decimals greater than 1 and less than 1, and to answer the question:

> How many of the *A* cm strip does it take to make the *B* cm strip exactly?

In each case, the answer is *C*. We also asked the following question for various different values of *C*, such as $C = 1.2$:

> What does the 1.2 tell you about how *A* and *B* are related? The 1.2 is 1.2 of what?

We posed these questions to help future teachers develop a measurement sense of *C* in the Geogebra sketch situation. Specifically, we wanted them to see the *A* cm strip as a unit with which to measure the *B* cm strip, and to see that *C* of the *A* cm strip makes the *B* cm strip exactly, whether or not *C* is a whole number. Because of

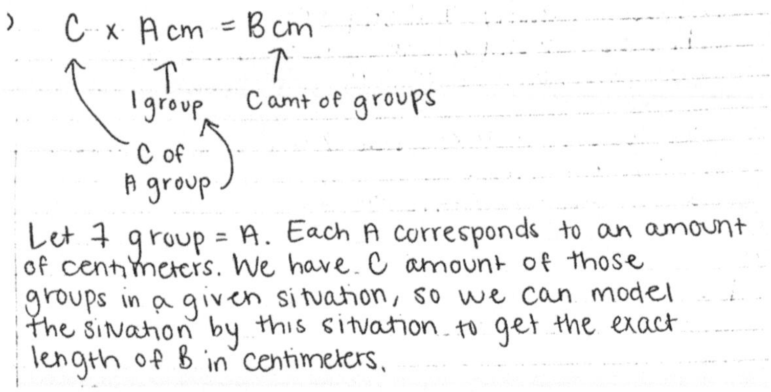

Fig. 6.13 Becky's explanation of the relationship among *A*, *B*, and *C* in the Geogebra sketch

this relationship, if we let 1 group be *A* cm, then it takes exactly *B* cm to make *C* *A*-cm groups. So, according to our definition of multiplication, $C \cdot A = B$. To see if the future teachers could formulate such a relationship before we discussed fraction multiplication explicitly, we asked them to respond to this question:

> Identify a relationship among *A*, *B*, and *C* that always holds, no matter how you move the buttons. Explain this relationship in terms of our definition of multiplication.

Of the 19 future teachers in the class, 13 (68%) formulated a multiplication equation and explained it in a way that fit with the class definition of multiplication (see Fig. 6.13 for an example), although 2 wrote their equation as $A \bullet C = B$. Two more future teachers described the situation in terms of the ideas of the class definition but did not actually formulate an equation.

The Geogebra sketch activity was intended to prime and preview how the measurement ideas involved in our definition of multiplication can be used with fractions and decimals. We conjecture that understanding multiplication in such a measurement way is productive for sensemaking with multiplicative structures, especially in cases of fractional multipliers. Although the entire class of future teachers did not immediately interpret the strips in the Geogebra sketch in terms of the class definition of multiplication, more than half of the class did.

After about 6 weeks of instruction in multiplication, approximately 3 of which focused on fraction multiplication, the future teachers took a take-home test that included the problem in Fig. 6.14. This problem asked the future teachers to discuss how to understand multiplication as having a consistent meaning for both whole numbers and fractions. Figure 6.15 shows part of April's response to that problem (her drawing shows 15ths, which were not necessary).

One goal for instruction in multiplication was for future teachers to use their own sensemaking as they used the definition throughout the course. We saw evidence that the future teachers used their own sensemaking in discussing the coherence of multiplication across whole numbers and fractions. On the test problem in Figs. 6.14 and 6.18 (95%) of the 19 future teachers provided math drawings and corresponding

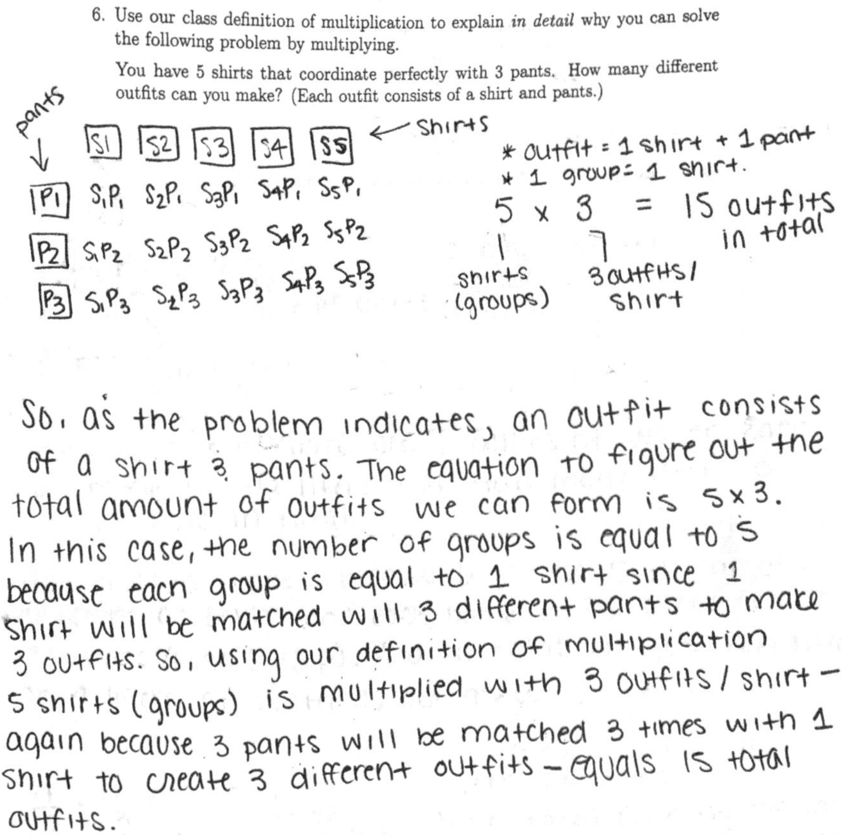
6. Use our class definition of multiplication to explain *in detail* why you can solve the following problem by multiplying.

You have 5 shirts that coordinate perfectly with 3 pants. How many different outfits can you make? (Each outfit consists of a shirt and pants.)

So, as the problem indicates, an outfit consists of a shirt & pants. The equation to figure out the total amount of outfits we can form is 5 x 3. In this case, the number of groups is equal to 5 because each group is equal to 1 shirt since 1 shirt will be matched will 3 different pants to make 3 outfits. So, using our definition of multiplication 5 shirts (groups) is multiplied with 3 outfits / shirt – again because 3 pants will be matched 3 times with 1 shirt to create 3 different outfits – equals 15 total outfits.

Fig. 6.19 A Cartesian product problem and Jenny's explanation for why the definition of multiplication applies

connected way and by drawing repeatedly on a few core ideas: measurement meanings for numbers and multiplication. We thus foreground seeing multiplicative situations in the world in terms of a specific mathematical structure that applies across types of situations and types of numbers. This program reflects our making-sense stance, in which experts organize a domain to help the field as a whole understand it better and to help teachers develop a deeper understanding of the material they will teach. We explained how we integrate our making-sense stance with our sensemaking stance, in which teachers and students use their own powers to reason, solve problems, and communicate about quantities in the world. We gave examples to illustrate the kind of reasoning and explanations the future teachers in our courses are able to engage in, including cases of fractions and multiplication that the research literature has documented as being difficult. We also discussed cases that remain difficult for future teachers even after instruction.

Our program to bring coherence to the multiplicative conceptual field extends beyond what we were able to present here. Elsewhere, we discuss how we implement other aspects of our coherence program in courses for future teachers, including how we treat division as multiplication with an unknown factor and how we extend multiplication and division to proportional relationships and to generating and explaining equations for lines (Beckmann, Izsák, & Ölmez, 2015; Beckmann & Kulow, 2018; Izsák et al., 2018; Izsák, Kulow, Stevenson, Ölmez, & Beckmann, 2019).

In our experience teaching future middle-grades and secondary teachers at a large public research university, our courses are successful at deepening future teachers' understanding of the multiplicative conceptual field. The future teachers in our courses tell us they appreciate what they learn. We have been fortunate to be able to offer courses that allow us to spend most of one or two semesters on the ideas of the multiplicative conceptual field. Even though virtually all of the future teachers in our courses have taken at least one semester of calculus, and many have taken more advanced mathematics courses, the time we spend on the multiplicative conceptual field feels productive and needed.

Acknowledgments This research was supported by the National Science Foundation under Grant No. DRL-1420307. The opinions expressed are those of the authors and do not necessarily reflect the views of the National Science Foundation.

Appendix A: Fraction Survey

1. How might you explain to a student what the fraction 3/5 means?
2. Draw a strip that is 7/5 as long as the strip below.

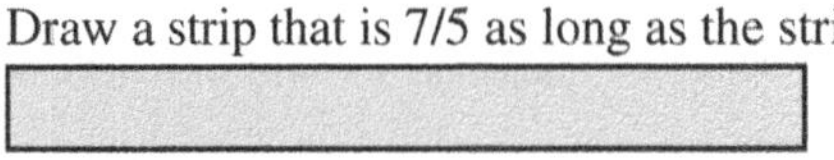

3. In the context of problem above, how might you explain what the fraction 7/5 means?
4. The strip below is 9/7 as long as another strip.
   (a) Is the other strip longer or shorter than the strip below?
   (b) Draw the other strip. Explain your reasoning in a sentence or two.

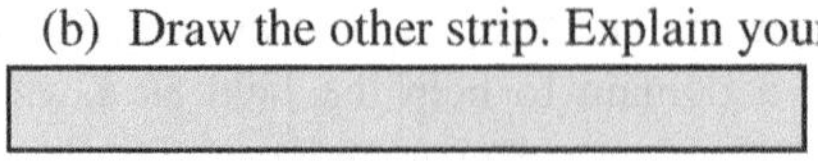

5. The strip below is 2/3 as long as another strip.
   (a) Is the other strip longer or shorter than the strip below?
   (b) Draw the other strip. Explain your reasoning in a sentence or two.

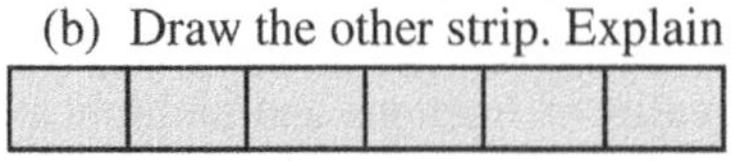

6. How many of Strip A does it take to make Strip B exactly? Explain how you know.

Strip A:

Strip B:

7. How many of Strip B does it take to make Strip A exactly? Explain how you know.

Strip A:

Strip B:

8. The drawing below shows a runway that is 1/3 of a mile long and is 1/5 as long as another runway. Draw the other runway and determine how long it is in miles. Explain your reasoning in a sentence or two.

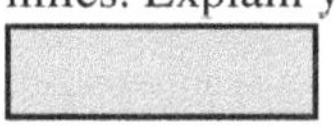

9. The drawing below shows a runway that is 3/4 of a mile long and is 3/7 as long as another runway. Draw the other runway and determine how long it is in miles. Explain your reasoning in a sentence or two.

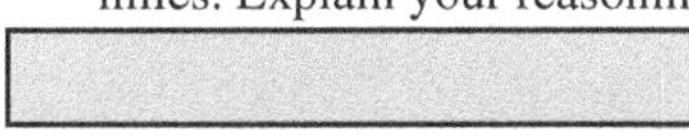

10. The drawing below shows a track that is 5/8 of a mile long. Draw another track that is 3/8 of a mile long. Explain your reasoning in a sentence or two.

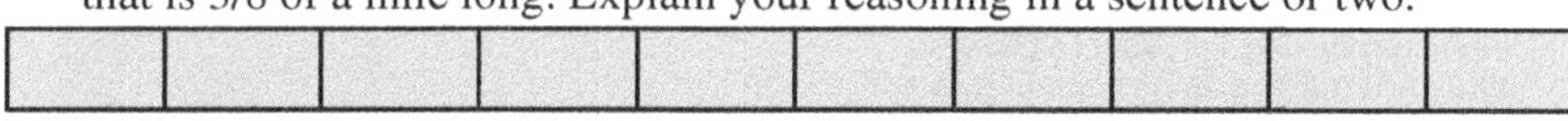

11. The drawing below shows a sidewalk that is 4/25 of a mile long. Draw another sidewalk that is 9/25 of a mile long. Explain your reasoning in a sentence or two.

12. The drawing below shows a track that is 8/15 of a mile long. The track will be paved in sections that are 2/15 of a mile long. How many sections will be needed? Explain your reasoning in a sentence or two.

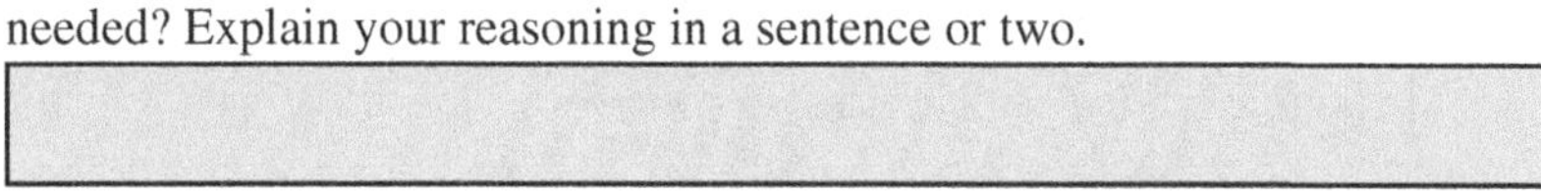

Appendix B

A 2-page handout on the class definition of multiplication.

MATH 5020, Fall 2018: Summary of our definition of multiplication

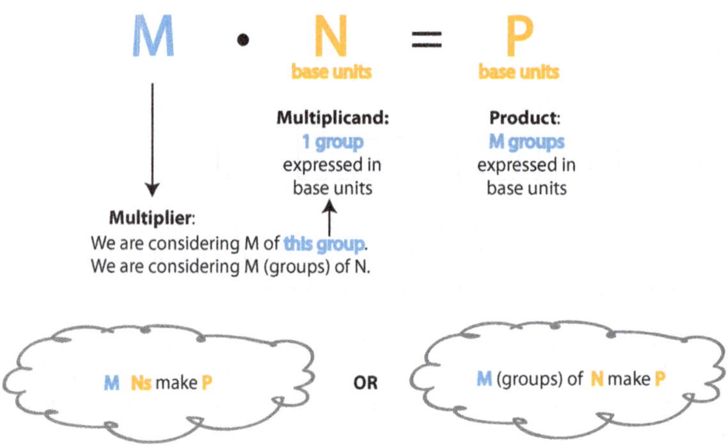

Example: We have 3 packages of cookies. Each package has 5 cookies. How many cookies do we have?

Let 1 group = 5 cookies. A package makes 1 group.
We are considering 3 groups of 5 cookies.
What are those 3 groups in terms of cookies?

Note: The 3 groups are the same in terms of number. The different packages could be different flavors as long as each package has 5 cookies.

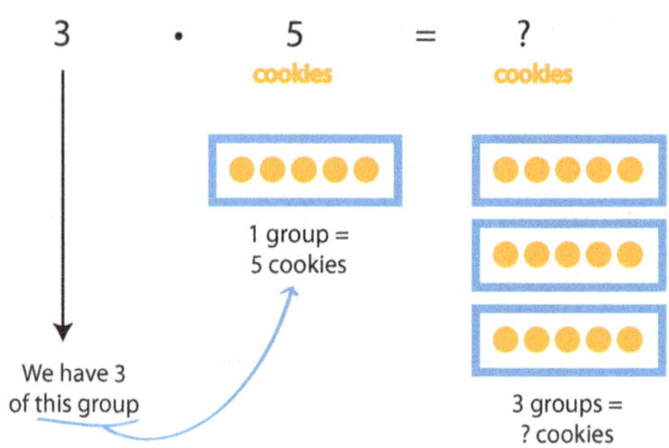

MATH 5020, Fall 2018: Summary of our definition of multiplication

Example If 1 serving of a food is 6 ounces, then how many ounces are in 4 servings of the food?

Let 1 group = 6 ounces (a serving).
We are interested in 4 groups.

4 servings = ? ounces

↓

We are interested in 4 of 1 group.

4 (1 serving) = ? ounces

↓

We are interested in 4 of 6 ounces.

4 (6-ounces) = ? ounces

↓

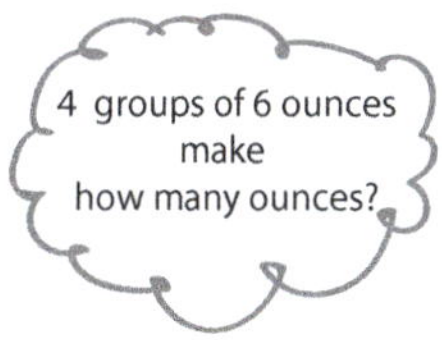

$4 \cdot 6 = ?$

6: ounces, 1 serving; ?: ounces, 4 servings

Multiplication involves describing quantities in two ways:
(1) in terms of "groups"
(2) in terms of "base units"

This example:
1 group = 6 ounces
a group is a serving

Groups		Base units
1 serving	⟶	6 ounces
4 servings	⟶	? ounces

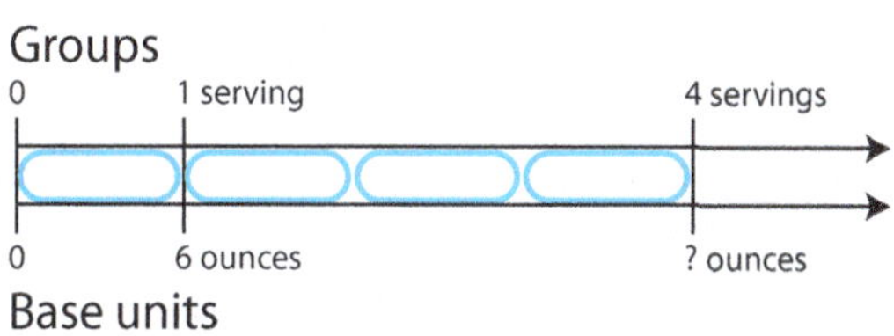

1 serving		2 servings		3 servings		4 servings
6	+	6	+	6	+	6
ounces		ounces		ounces		ounces

(This works for whole numbers of servings)

Appendix C: Equations for Word Problems Survey

Soccer Ball Problem: There are 5 soccer balls in each bag. The coach has 2 bags. How many soccer balls does the coach have?
We could write either of the following two equations for the soccer ball problem:
$5 \times 2 = ?$ or $2 \times 5 = ?$

1. **Cheese Problem**: You bought 8 packages of cheese. The cheese in each package weighs 1/3 of a pound. How many pounds of cheese did you buy?
   (a) Write an equation for the Cheese Problem, using a question mark for the unknown amount.
   (b) If you think about mathematical structure, would you say that the 8 packages, the 1/3 of a pound, and the unknown number of pounds in the Cheese Problem are analogous to any of the quantities in the Soccer Ball Problem (the 5 balls, 2 bags, or ? balls)? Explain.
2. **Tricycle Problem**: Jacinda is making toy tricycles. She has 21 wheels. If each tricycle gets 3 wheels, how many tricycles can Jacinda make?
   (a) Write an equation for the Tricycle Problem, using a question mark for the unknown amount.
   (b) If you think about mathematical structure, would you say that the 21 wheels, the 3 wheels, and the unknown number of tricycles in the Tricycle Problem are analogous to any of the quantities in the Soccer Ball Problem (the 5 balls, 2 bags, or ? balls)? Explain.
3. **Cracker Problem**: One serving of crackers contains 24 g of carbohydrates. If you eat 1/4 of a serving, how many grams of carbohydrates will you eat?
   (a) Write an equation for the Cracker Problem, using a question mark for the unknown amount.
   (b) If you think about mathematical structure, would you say that the 24 g, the 1/4 of a serving, and the unknown number of grams in the Cracker Problem are analogous to any of the quantities in the Soccer Ball Problem (the 5 balls, 2 bags, or ? balls)? Explain.
4. **Orange Juice Problem**: One serving of orange juice is 1/5 of a liter. Chelsea drank 3/4 of a serving. How many liters of orange juice did Chelsea drink?
   (a) Write an equation for the Orange Juice Problem, using a question mark for the unknown amount.
   (b) If you think about mathematical structure, would you say that the 1/5 of a liter, the 3/4 of a serving, and the unknown number of liters in the Orange Juice Problem are analogous to any of the quantities in the Soccer Ball Problem (the 5 balls, 2 bags, or ? balls)? Explain.

5. **Chili Problem**: Nick used 2/3 of a can of tomato paste to make chili. The can contained 4 ounces of tomato paste. How many ounces of tomato paste did Nick use to make chili?
   (a) Write an equation for the chili problem, using a question mark for the unknown amount.
   (b) If you think about mathematical structure, would you say that the 2/3 of a can, the 4 ounces, and the unknown number of ounces in the Chili Problem are analogous to any of the quantities in the Soccer Ball Problem (the 5 balls, 2 bags, or ? balls)? Explain.
6. In your math education so far, have you had experiences thinking about what whole number multiplication and fraction multiplication mean, and whether their meanings are connected?

References

Beckmann, S., & Izsák, A. (2015). Two perspectives on proportional relationships: Extending complementary origins of multiplication in terms of quantities. *Journal for Research in Mathematics Education, 46*(1), 17–38.

Beckmann, S., Izsák, A., & Ölmez, İ. B. (2015). From multiplication to proportional relationships. In X. Sun, B. Kaur, & J. Novotná (Eds.), *Proceedings of the Twenty-Third ICMI Study: Primary Mathematics Study on Whole Numbers* (pp. 518–525). Macau, China: University of Macao.

Beckmann, S., & Kulow, T. K. (2018). How future teachers reasoned with variable parts and strip diagrams to develop equations for proportional relationships and lines. In Y. Li, J. Lewis, & J. Madden (Eds.), *Mathematics matters in education: Essays in honor of Roger E. Howe* (pp. 117–148). Cham, Switzerland: Springer. https://doi.org/10.1007/978-3-319-61434-2

Behr, M. J., Lesh, R., Post, T. R., & Silver, E. A. (1983). Rational number concepts. In R. Lesh & M. Landau (Eds.), *Acquisition of mathematics concepts and processes* (pp. 91–125). New York, NY: Academic Press.

Conference Board of the Mathematical Sciences. (2012). *The mathematical education of teachers II*. Providence RI and Washington DC: American Mathematical Society and Mathematical Association of America.

Izsák, A., & Beckmann, S. (2019). Developing a coherent approach to multiplication and measurement. *Educational Studies in Mathematics, 101*, 83–103.

Izsák, A., Beckmann, S., & Kulow, T. (2018). Uncommon partitions: Partitioning as a problem-solving activity. *Mathematics Teaching in the Middle School, 24*(3), 164–171.

Izsák, A., Kulow, T., Stevenson, D., Ölmez, B., & Beckmann, S. (2019). Using coordinated measurement with future teachers to connect multiplication, division, and proportional relationships. *Mathematics Teacher Educator, 8*, 49.

Kieren, T. E. (1988). Personal knowledge of rational numbers: Its intuitive and formal development. In J. Hiebert & M. Behr (Eds.), *Number concepts and operations in middle grades* (pp. 162–181). Reston, VA; Hillsdale, NJ: National Council of Teachers of Mathematics; Erlbaum.

Kozulin, A. (1986). Vygotsky in context (1986). In L. Vygotsky (Ed.), *Thought and language* (pp. xxv–lxxii). Cambridge, MA: MIT Press.

Lovin, L. H., Stevens, A. L., Siegfried, J., Wilkins, J. L. M., & Norton, A. (2018). Pre-K-8 prospective teachers' understanding of fractions: An extension of fraction schemes and operations research. *Journal of Mathematics Teacher Education, 21*, 207–235.

McCallum, W. (2018). Making sense of mathematics and making mathematics make sense. In Y. Shimizu & R. Vithal (Eds.), *ICMI Study 24 Pre-Conference Proceedings: School mathematics curriculum reforms: Challenges, changes and opportunities* (pp. 1–8). Tsukuba, Japan: University of Tsukuba.

National Governors Association Center for Best Practices & Council of Chief State School Officers. (2010). *Common core state standards for mathematics*. Washington, DC: Author.

Olanoff, D., Lo, J., & Tobias, J. (2014). Mathematical content knowledge for teaching elementary mathematics: A focus on fractions. *The Mathematics Enthusiast, 11*(2), 267–310.

Simon, M., Kara, M., Norton, A., & Placa, N. (2018). Fostering construction of a meaning for multiplication that subsumes whole-number and fraction multiplication: A study of the Learning Through Activity research program. *Journal of Mathematical Behavior, 52*, 151–173.

Vergnaud, G. (1988). Multiplicative structures. In J. Hiebert & M. Behr (Eds.), *Number concepts and operations in middle grades* (pp. 141–161). Reston, VA; Hillsdale, NJ: National Council of Teachers of Mathematics; Erlbaum.

Vygotsky, L. (1986). *Thought and language*. Cambridge, MA: MIT Press.

Wu, H. (1998). *Teaching fractions in elementary school: A manual for teachers*. Retrieved from https://math.berkeley.edu/~wu/fractions1998.pdf.

Chapter 7
The Work of Coaching in Rehearsals to Enlist Mathematical Knowledge for Teaching

Hala N. Ghousseini

Research in mathematics education has established that high-quality instruction requires mathematical knowledge for teaching (MKT)—a form of unpacked mathematical knowledge that enables teachers to make features of content visible to and learnable by students. There is growing evidence that mathematical knowledge for teaching is integral to the ways in which teachers teach mathematics (Ball, Thames, & Phelps, 2008; Hill et al., 2008; Wilson, Floden, & Ferrini-Mundy, 2001) and positively associated with students' mathematics achievement (Hill, Rowan, & Ball, 2005). Scholars have argued, in particular, that to be effective, teachers not only need to have MKT but also learn it for use in "the dynamic interplay of content with pedagogy in teachers' real-time problem solving" (Ball & Bass, 2000, p. 88). Teachers draw on MKT often in non-routine situations, requiring a capacity to improvise in the midst of contingent interactions, relying on professional judgment to marshal knowledge and skill in the service of professional goals. The mathematical preparation of mathematics teachers, accordingly, needs to create opportunities for them to learn MKT *in, from, and for practice*, where it can be conceived as a form of sensemaking or "knowing" in practice; not only as knowledge to be applied to practice (Cook & Brown, 1999). The present study depicts the teacher education pedagogy of rehearsals as a context for learning MKT as a form of sensemaking in practice, and describes the work that the teacher educator does to support this work.

Typical approaches to the mathematical preparation of teachers rely on the use of rich mathematics problems in methods and content courses to support teachers in developing the mathematical knowledge they need for teaching (Charalambous, Hill, & Ball, 2011; Steele & Hillen, 2012; Thanheiser, 2015). These rich problems engage teachers in unpacking and building connections among mathematical ideas,

---

H. N. Ghousseini (✉)
Department of Curriculum and Instruction, University of Wisconsin-Madison, Madison, WI, USA
e-mail: Ghousseini@wisc.edu

Y. Li et al. (eds.), *Developing Mathematical Proficiency for Elementary Instruction*, Advances in STEM Education, https://doi.org/10.1007/978-3-030-68956-8_7

and in considering alternative representations and solution methods. In other approaches, teacher educators are also integrating work on these rich mathematics problems with work on pedagogy through the use of video or narrative cases and samples of student work (Silver, Clark, Ghousseini, Charalambous, & Sealy, 2007; Steele & Hillen, 2012). Underlying these approaches is the assumption that teachers' nascent mathematical knowledge for teaching can find grounding in practice through the examination of other teachers' records of practice or through teachers' own attempts to take these understandings to their own classrooms. While this assumption appears to have some bearing on teacher learning (Silver et al., 2007), it suggests that MKT is mainly learned through investigations of mathematics or teaching and then applied to one's own practice. It also leaves unanswered the question of how teachers learn to embed MKT and can be supported to do it in the interactive work of teaching—*in, from, and for practice*.

In this study, I argue that the professional preparation of novice teachers (NTs) can address this problem by moving beyond the use of rich tasks and cases of teaching to also focus on contexts of learning that engage novices directly in the interactive work of teaching. A current teacher education pedagogy that serves this purpose is rehearsals of teaching, which provide novice teachers (NTs) with guided opportunities to both simulate and analyze manageable chunks of contingent, interactive practice before enacting them with students in classrooms (Lampert et al., 2013). During rehearsal, one or more NTs participate as teachers, while other NTs participate as students, exhibiting their understanding of how children think about mathematical ideas, and simulating the multiple relationships with students and content that might be in play. Through their participation, NTs often contribute insights and questions related to the routine and spontaneous instructional interactions that must be managed in teaching. Guiding participation during the simulation, the teacher educator acts as both coach and simulated student, enabling them to put into play a range of issues related to students, subject matter, and instructional decision-making. Brief discussions can also take place among the participants in which decisions are weighed and alternatives considered.

The next section lays out a brief conceptual framing of MKT and a set of theoretical considerations grounded in a situative perspective on learning (Greeno, 1998). Then following the description of data collection and methods, I share findings in the context of one analytic vignette to demonstrate in rich detail the kind of support that the teacher educator provides in rehearsal to enlist MKT in the interactive work of teaching.

7.1 Mathematical Knowledge for Teaching

Based on their analysis of the tasks of teaching that require mathematical insights and reasoning, Ball et al. (2008) propose that MKT could be conceptualized in various domains that fit under two main distinguishable categories: subject matter knowledge and pedagogical content knowledge. Subject matter knowledge, they suggest, comprises knowledge of how curricular content is situated in and connected to the broader disciplinary territory. It also includes both basic knowledge of concepts and procedures (enabling one to solve problems or judge the correctness of a solution) and a specialized knowledge "uniquely needed by teachers in the conduct of their work" (p. 34). This specialized knowledge, they argue, allows teachers, for example, to reason about which representations to choose and to interpret students' mathematical errors—mathematical work not necessarily needed in professions other than teaching. Under pedagogical content knowledge, Ball et al. include knowledge that combines knowing about students and their thinking with particular understandings about mathematics, as well as knowledge that combines knowing about pedagogical issues that affect learning and specific mathematical understanding.

7.2 A Situative Perspective for Learning MKT

The present study frames MKT as situative and dynamic in nature (Fennema & Franke, 1992; Hodgen, 2011), and as located social practice (Adler, 1998; Cook & Brown, 1999; Hodgen, 2011; Shulman, 1986).[1] Shulman (1986) proposes that one aspect of teacher knowledge gets manifested in a form of *strategic knowledge*—or "strategic *knowing*" (p. 14). He explains, "Strategic knowledge comes into play as the teacher confronts particular situations or problems whether theoretical, practical, or moral, where principles collide and no simple solution is possible" (p. 13). MKT can be instantiated in the strategic knowing that takes place in rehearsal when the learner is provided with opportunities to both experience *and* reflect on corresponding changes between enactments of the skill and resulting consequences (Hatano & Inagaki, 1986). Repeated cycles of experience and reflection in rehearsal, coupled with guidance and feedback from a knowledgeable other, become a form of deliberate practice through which performance can be improved (Ericsson, Krampe, & Tesch-Römer, 1993). The design elements of rehearsal—the back and forth between simulation and analysis—provide opportunities to attend to MKT as form of strategic knowing inside the relational work of teaching. The present study will highlight how the teacher educator can infuse considerations for MKT into the continuous back and forth that takes place in rehearsals.

---

[1]A point that justifies looking at MKT inside the work of teaching rather than measuring it using surveys.

7.3 Study Context and Methods

The context of this study is mathematics "methods" courses in Master's level teacher certification programs for elementary teachers at three public universities in the United States. These courses typically start with 3–4 days of focused work on tasks that allow novice teachers to develop an understanding of important mathematical ideas and of children's mathematical thinking (e.g., exploring the meaning of the traditional algorithm for addition and subtraction). The design of the courses is part of a collaborative effort to use recurring Cycles of Enactment and Investigation centered on common Instructional Activities to enable novice teachers not only to develop knowledge about teaching and mathematics but also use the knowledge in actual interactions with children. These features are described in the following.

Instructional Activities: At the center of each cycle is an instructional activity (IA) that is common to the elementary mathematics curriculum, chosen or designed to be a context where novices work on principled instructional practices and mathematical knowledge in integrated ways. The instructional practices include eliciting and responding to student contributions, representing student thinking verbally and on the board, and orienting students to one another. The principles that guide the implementation of these practices include treating students as sensemakers and designing instruction for all children to have equitable access to rigorous academic work in school. The mathematics in these activities often draws on the understandings that the teachers developed during the 3–4 days of focused work on mathematical tasks at the beginning of the course. Strategy Sharing is an example of an IA that will be featured in this analysis. In Strategy Sharing, the teacher poses a problem and then solicits several ideas after the students had solved it. The teacher elicits students' explanations for how they solved the problem and encourages students to listen to each other's ideas, asking them to repeat and track their classmates' strategies. The purpose of Strategy Sharing is to allow students to contribute in different ways to solve the same problem while moving across a broad terrain of concepts, procedures, representations, and explanations (Kazemi & Hintz, 2014).

Cycles of Enactment and Investigation: Cycles of Enactment and Investigation begin with the novice teachers observing and analyzing the teaching of an IA, such as Strategy Sharing, in a classroom context, either live or on video. This phase aims to help the novices identify teaching practices and discuss their principled use in practice. It also aims to engage the novices in the mathematical content being addressed in the IA. Following the observation and analysis of the IA, the novice teachers next prepare to teach it to elementary students, rehearsing it first publicly in front of their peers and the teacher educator. Guiding participation in this setting, the teacher educator acts as both coach and simulated student, enabling both the rehearsing novice and others in the group to study the range of actions a teacher might take in response to particular student performances. All novice teachers then enact the IA with students in elementary school classrooms, video-recording their work. Continuing the cycle, the teacher educator guides a collective analysis using records of the enactments to examine how the mathematical content and instructional practices designed into the activity played out in particular classroom contexts. These cycles typically occur weekly over the duration of the methods course.

7.3.1 Data Sources

The data consist of video-recorded rehearsals, taken from a larger pool of 90 rehearsal sessions collected across the three teacher education programs. These rehearsals varied along several dimensions: They were led by different teacher educators (TEs), involved different cohorts of novice teachers, focused on different IAs, varied in length (averaging 15 min in length), and number of exchanges between the teacher educator and novice teachers (average of 14 exchanges per rehearsal). Using this initial data set, 10 representative rehearsals were sampled from each teacher education program for a total data set of 30 rehearsals. The sampling took into consideration the variations described above by keeping a balanced representation of the different dimensions (IAs, TEs, length of rehearsals, etc.) in the smaller data set.

7.3.2 Analytic Procedures

7.3.2.1 Segmenting Rehearsals

The video-analysis software Studiocode©[2] was used to segment rehearsals into two kinds of instances: (a) instances where a novice teacher (NT) was leading an IA as a teacher with other novice teachers and the teacher educator participating as students (*NT Rehearsing*), and (b) instances where there were pauses in the simulation for exchanges between the TE and novice teachers (*TE/NT Exchange*). An NT Rehearsing instance was defined from the moment an NT starts leading a segment of the IA to the time when the simulated lesson is paused for feedback or coaching. Across all 30 rehearsals, there were 296 total instances of NT Rehearsing. In comparison, there was a total of 430 instances of TE/NT Exchanges. These exchanges were defined from the moment the TE or NT pause the simulation to offer/request guidance about what to do next until the rehearsal is resumed.

---

[2]Studiocode is a software package that connects codes directly to segments of video. The "timeline" feature of Studiocode enables a chronological representation of all codes using multi-layered, graphical representations, descriptors, and narrative comments attached to a particular video.

7.3.2.2 Analysis of the Teacher Educator's Work

In order to understand how the teacher educator supported novices' work on tasks that enlisted MKT, I performed thematic analyses at two levels: Identifying the problems of practice that prompted a pause in the rehearsal, and then analyzing the teacher educator's interventions during these moments.

Problems of Practice

To characterize the kinds of problems of practice that arose for novice teachers in the context of rehearsing the teaching of an IA, I looked inside the TE/NT exchanges for events that precipitated a pause in the rehearsal. I recognized such events based on moves or cues in the exchanges that signaled an aspect of classroom interaction experienced by the NT(s) or perceived by the TE as being worthy of comment, troublesome, unclear, or challenging (Horn & Little, 2010). For instance, an NT may pause the simulation to ask for guidance about a particular issue. The TE, perceiving an aspect of the NT's performance as either promising or problematic, may also pause the simulation to give feedback. I looked for patterns across these events and then grouped them into three main categories, featured in Table 7.1.

The Work of the Teacher Educator

Following this first level of analysis, I conducted a thematic analysis of the teacher educator's interventions that followed the events identified above, focusing on the way these interventions supported the work that enlisted MKT. I then grouped

Table 7.1 Categories of events that initiated work on problems of practice and co-opted TEs' interventions

Precipitating events	Description	Example
NT struggles or solicits support	TE provides support to NT after observing that she/he was facing difficulty with some aspects of teaching. Or NT rehearsing asks for clarification or guidance about particular aspects of practice.	NT asks TE what to do if she asks a student how he solved, "9 + 9," and the student mainly says, "I just knew it."
TE signals a problematic teaching move	TE intervenes to note a problematic NT move.	TE points to a problematic aspect of the NT's representation of a student strategy on the whiteboard.
TE inserts a possible situation	TE uses an opening in the simulation to invite NTs to consider how they would respond to a possible situation.	TE plays the role of the student and provides a non-standard solution to a problem.

the different themes that emerged from the analysis of the teacher educator's interventions under broader analytic categories (Saldaña, 2013), and then used them in a more in-depth analysis of three rehearsals, each representing a different teacher education site and a different instructional activity. For the purposes of this chapter, I will focus in my findings on an analytic vignette taken from one representative rehearsal.

7.4 Results

The findings reveal that the TE's interventions to support work that enlists MKT during rehearsals of teaching happened in the context of three main types of events (see Table 7.1): The NT struggles with a task of teaching or solicits support from the TE, the TE signals a problematic teaching move, and the TE inserts a possible situation to initiate work on a problem of practice. These events initiated a pause in the simulation, allowing the TE to support novices' work on tasks that enlist MKT by (a) providing insight into student thinking or the mathematics, (b) facilitating novices' unpacking of the mathematics, (c) noting affordances and constraints of instructional moves in relation to MKT, and (d) guiding instructional decisions to attend to the mathematics. Table 7.2 provides a description of these thematic categories with illustrative examples. It is important to note that several of these interventions happened concurrently in practice. For example, at times the teacher educator would guide decision-making about instructional moves by noting their affordances and constraints in relation to the mathematics.

The in-depth examination of how the TE used these interventions in rehearsal revealed that the first three types of interventions (a, b, and c) appeared to guide the NTs in *interpreting problems of practice* in light of considerations for the mathematics and student thinking—a finding also revealed in Lampert et al. (2013). The fourth type of intervention (d) was about guiding decision-making and appeared to guide NTs in *connecting the interpretation of problems to possible courses of action*. Together these patterns of the TE's interventions suggest how novice teachers can be supported in embedding MKT in the interactive work of teaching. This happened through a continuous back and forth between NTs enacting various tasks of teaching that enlist MKT and the TE helping them consider aspects of their performance and its consequences in relation to a body of knowledge that includes particular mathematical ideas and goals as well as student mathematical thinking. I illustrate how this process comes together in the following analytic vignette.

Table 7.2 Teacher educator interventions during tasks of teaching that enlist MKT

Category of TE intervention	Thematic description	Example
(a) Noting the affordances (or constraints) of instructional moves	TE notes the way an instructional move attends to (or detracts from) the mathematics and/or students' thinking about the mathematics.	During a Strategy Sharing activity, a rehearsing novice teacher (Reh-NT) asks what language she should use to represent a student's strategy when the student indicates that she imagined transferring a dot from one ten frame to another. She asks whether she should call it "borrowing." The TE indicates that this would be misleading because borrowing indicates that one will give something back, which is not what happens in this case. She suggests saying "move" instead.
(b) Providing insight into the mathematics or students' thinking	TE provides insight into student thinking by offering responses (including errors) that students are likely to give or discussing student thinking in general. TE provides insight into the mathematics by commenting on the mathematical goal, elaborating a mathematical idea, or making connections to other mathematical ideas.	A Reh-NT is leading a skip count by 8, and wonders whether she should stop after the first eight counted numbers to ask students what the next number is, and how they know. The TE says that stopping the count at that point may not give the students enough numbers to look for patterns. However, she suggests that if students struggle to add 8 each time, it might be a good idea to stop and ask how students are figuring out the next number that early in the count. The TE goes on to say, "Students might say well it's different. When it lands on an 8, I do 2 and 6. When I land on a 6, I do 4 and 4 . . . "
(c) Facilitating the unpacking of the mathematics	TE helps the novices explain their thinking about the mathematics and engage in each other's mathematical ideas.	During a Strategy Sharing activity, the TE presses the class to consider the similarities and differences between two strategies.
(d) Guiding instructional decisions to attend to the mathematics	TE guides decisions about what question to ask, what representation to use, what problem to pose, and when to elicit student thinking. She may suggest a course of action or help novices reason about one.	During a skip count that is recorded on the board, the TE suggests that if there's a particular pattern that the novice teachers want to surface from the activity, they might want to think about what types of questions would support students in noticing that pattern. She then supports them in considering particular questions.

7.4.1 Launching a Subtraction Problem and Attending to a Student Strategy

Molly,[3] the NT in the role of the teacher, starts launching the activity by stating, "We are going to work on a problem today: 76 – 27." She then writes the problem on the board vertically, placing 27 under 76. As she prepares to continue with her instructions, the TE intervenes: "Okay, actually, one thing I want you to do differently is, don't write the problem vertically like that. Write [the numbers] horizontally." The TE explains that stacking the numbers vertically encourages students to "imagine the standard algorithm for subtraction in their head," rather than producing a range of informal strategies, which is the goal of the instructional activity. Molly follows the TE's feedback and rewrites the problem horizontally. Next, Molly offers a story context to help students make sense of the problem, telling the "students" (played by the TE and other NTs in the audience): "To help you visualize this problem, let's say that Maria has 76 dollars that she has saved up, and she wants to buy a 27-dollar video game. How much money will she have once she buys that video game?" Molly gives the students a little time to solve the problem and then asks them for their responses. She solicits three different answers: 49, 43, and 51, and writes all three of them on the board without making any judgment on their correctness. She then starts to elicit students' strategies, starting with Nora's who gave the response of 49.

Nora shares her approach for arriving at 49, explaining, "I broke 27 up into 20 and a 7, and then I thought that 76 – 20 is 56, and then 56 – 7 is 49". Then Molly continues, addressing the group:

3	Molly: Okay. So, can anyone repeat what Nora just said in his own or her own words?
4	TE [in role of student]: I think I know where she started. She said she went 76 minus [she hesitates to continue, exhibiting uncertainty about what Nora did].
5	Students: 20.
6	TE: 20.
7	Molly: She broke down [and then stepping out of her teacher role, she addresses the TE] Like, so, should I write down, like, what she said as she was saying it, or... should I, like, ask somebody else to repeat what she said, before I wrote it down?
8	TE: So, that's a lot of what [Nora] just said, right? So, let's say, "Okay. Let's see if we can recreate how Nora solved this. Did anyone catch where she started?" And if nobody [can recreate that], you can say, "Nora, where did you start again? Tell us."

Continuing with the rehearsal, Molly takes up the TE's suggestion to unpack Nora's strategy from the point where she started. She asks, "Can anyone remember where Nora started this problem? Larry?" Larry explains that Nora started from 27, breaking it down into 20 and 7. Molly represents the strategy on the board while

[3] All names are pseudonyms.

prompting other students to reconstruct its steps. In the end, Molly turns back to Nora and asks,

11	Molly: And what number did you get?
12	Nora: 49
13	Molly: And how did you get 49 here? [Writes 49 at the end of the number sentence]
14	Nora: Because I knew that 56 minus 7 is 49.
15	Molly: [Looking skeptical] You just knew it? Like, how did you count it?
16	Nora: I knew that, if I took a 10 from the 5, and I had a 4, I'd have a 16 minus 7 is 9, and then I brought down my 4.
	Molly pauses at this point, looking perplexed by Nora's explanation. The TE intervenes and asks Nora, "Really, is that how you did it?"
17	Nora: I think so. I just know that 16 minus 7 is 9.
18	Another NT: She's breaking down the 56 into 16 and 40 and then subtracting.

Next, Molly represents Nora's strategy on the board. She decomposes 56 into 40 and 16, and then writes $16 - 7 = 9$ (see Fig. 7.1). Her action prompts another NT in the audience to exclaim, "That is a complicated way to do it!" In response to that, the TE intervenes, suggesting that $56 - 7$ is, in and of itself, a good problem to give to students, because it taps into their invented strategies for subtracting across a decade, "You just go back 6 and then you go back 1 more." She also explains that most adults tend to use the traditional algorithm for such a problem, while kids often find more efficient ways to solve it. "Kids would count backwards, 55, 54, 53, …" She suggests to the NTs that it would be good for them to try to practice alternative strategies to the traditional algorithm in their head. A novice teacher then inquires,

19	NT: So, do you think kids will not do [the traditional algorithm] anymore, like, do they not regroup anymore, like…?
20	TE: It just depends how much experience they've had with standard algorithms. You're more likely to see more variety with the kids than what we were able to produce in the classroom here.

7.4.1.1 Guiding NTs' Participation with Considerations for MKT

Several problems of practice emerge in this excerpt, opening opportunities for the TE to intervene and infuse the problem space with considerations for MKT. To elaborate on this, I focus on three problems related to the way Molly represented the numerical problem when launching it, and the way she responded to Nora's strategy.

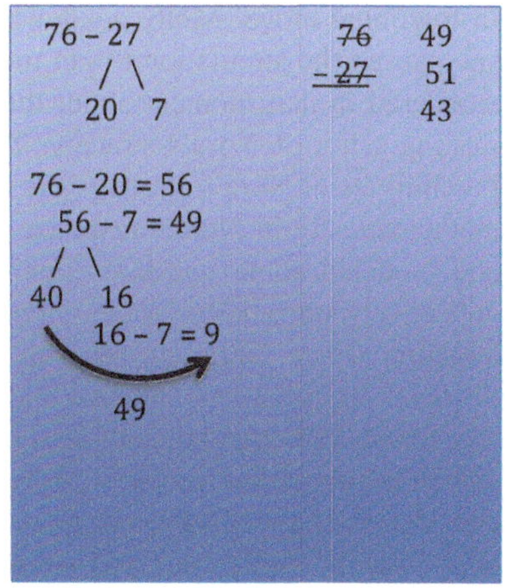

Fig. 7.1 Molly's use of the board space to represent students' work

Representing the Numerical Problem

In her first intervention, following Molly's representation of the math task on the board, the TE noted a constraint in the NT's instructional move, by signaling a problem of practice that could result from writing the subtraction problem vertically. Given that the purpose of the Strategy Sharing activity is to allow students to contribute in different ways to solve the same problem, this vertical representation could limit the variety of solutions that the students would present, consequently narrowing the richness of the mathematics that the class can work on. The TE drew the NTs' attention to this problem by providing insight into how students might respond to such a representation, highlighting a general tendency in students to rely on the traditional algorithm when the numbers are stacked in this manner. In her intervention, the TE not only attended to knowledge of student thinking in relation to particular content but also connected the choice of a particular representation to the mathematical goals of the activity. Her intervention set the stage for the novice in the teacher role to continue with the intended goals of the activity, making it more likely that she will get a variety of strategies from the students.

Responding to Nora's Strategy

After Nora shared her strategy, both the TE and NTs, respectively, contributed in ways that enlisted MKT in the interactive work of teaching. In her first response to Nora's strategy, Molly invited other students to revoice it (Turn 3). The TE intervened at this point to highlight the demands that students might face in repeating an unconventional strategy that relies on decomposing the subtrahend into tens and ones, then taking away the tens from 76. She used an opening in the rehearsal to simulate a student's confusion in recalling the details of this strategy (Turn 4), and accordingly communicated to the NTs a potential constraint in Molly's instructional move. Molly acknowledged her recognition of this constraint when she paused the simulation to get more guidance on how to proceed (Turn 7). She wondered in that moment about the kind of move that would support students in making sense of Nora's strategy. The TE used this opportunity to point to some possibilities in Nora's strategy and emphasized that there is much to be unpacked mathematically in Nora's response, "That's a lot of what she just said" (Turn 8), and modeled some questions that the novice could use to engage the class in collectively unpacking Nora's strategy. In her modeling, she communicated that targeting where Nora *started* with her strategy would be an important first step (Turns 8–9) in helping students make sense of it because focusing on where Nora started can key other students into how she used decomposition in this example to solve the subtraction problem. In both aspects of her intervention, the TE weaved considerations for MKT in her guidance of NTs' participation: She provided insight into students' difficulties and suggested a course of action that has potential to support students in understanding the central mathematical relationships in Nora's strategy. This feedback supported Molly in continuing with her work on eliciting the reasoning behind Nora's strategy, inviting participation from different students and scaffolding it in accordance with the way MKT was enlisted in the TE's feedback.

Continuing with the unpacking of Nora's strategy provided additional opportunities for the group of NTs to tap into aspects of MKT, specifically considering the mental strategies that students might use in subtracting a single-digit number from a two-digit number. When Nora said that she "just knew" that $56 - 7$ is 49, Molly pressed her to explain her reasoning, and with her body language exhibited her perception of something problematic with Nora's answer (Turn 15). Her move—exhibiting how MKT plays out in practice—suggested her perception that making the leap from 56 to 49 may not be as straightforward to students as Nora suggested. When Molly pressed Nora to explain her thinking, it opened up a space for the group to consider the likelihood that third-grade students (to whom this instructional activity is targeted) would reason about this subtraction problem by using the traditional algorithm. The TE used this problem space to attend to the NTs' knowledge of content and teaching, noting that mathematical tasks that require students to subtract across a multiple of 10 on the number line could help support students' number sense. She also provided insight into how third graders may approach such a subtraction task by noting two strategies that students are likely to use: (a) counting back and (b) bridging across a multiple of 10 to subtract (first subtracting 6 from 56 to get to 50, and then taking away 1). Both of the TE's interventions supported the NTs in attending to both knowledge of content and students and knowledge of content and teaching *in* practice—in the moment to moment problem-solving that teachers face in their work.

7.5 Discussion

The analysis illustrates the opportunities to enlist MKT in the interactive work of teaching that the teacher educator can support inside rehearsals. In the rehearsal of instructional activities, novice teachers had opportunities to consider and unpack mathematical ideas while carrying out particular tasks of teaching like posing questions to launch a problem and using representations to attend to student thinking. The analysis of the vignette illustrates how the teacher educator's interventions—providing insight into the mathematics or student thinking, noting affordances and constraints of instructional moves in relation to students' mathematical learning, engaging the novices in the unpacking of mathematical ideas, and guiding pedagogical decision-making—enlisted MKT in the work of teaching.

A key aspect of the way the teacher educator enlisted MKT in the work of teaching during rehearsals is through the repeated process of helping novices interpret problems of practice, guiding their pedagogical reasoning, and then their enactment. The interpretation of problems of practice supported novices' opportunities to learn MKT in the way it framed the problem space and what matters in it around MKT and its different domains. Over time this framing would help novices in identifying problems of practice and deciding on situationally relevant, appropriate action (Horn & Little, 2010). Horn and Little propose that when teachers work collaboratively on practice, they develop "shared frames of reference—shared concepts, principles, and terminology—to structure their talk, justify their decisions, and guide their interpretations of classroom-based problems of practice" (p. 209).

One may argue, however, that helping novices interpret what matters mathematically in a particular teaching context may not be enough for mobilizing this knowledge for use in practice. Research in teacher education, in fact, attests to the limitations of mainly teaching novices to be analytic about practice, and suggests that they need to be supported in using what they know in the complex interactions of classrooms (Ebby, 2000; Lloyd, 2005). However, the results of this study reveal how the teacher educator does not stop at the framing of problems of teaching, but also prompts novices to use mathematical insights to address these problems in practice. The teacher educator guides pedagogical decision-making by either suggesting particular moves or prompting novices to reason about a possible course of action (Lampert et al., 2013). The rehearsing novice teacher then experiments with these actions in practice. Through repeated cycles of deciding on pedagogical moves, testing them out, and then inquiring into their resulting consequences, novices are supported in developing professional judgment (Hatano & Inagaki, 1986) regarding the problems they need to address and possible ways to work on them. Hatano and Inagaki argue that the development of such professional judgment is central to adaptive expertise.

This study approached the learning of MKT from a situative perspective, where rather than treating this knowledge only as a form of declarative knowledge that can be acquired and then retrieved in particular contexts (including surveys of teacher knowledge), it attended to it as dynamic in nature and connected to the contexts of its use (Adler, 1998). Research on complex human performance suggests that key aspects of the usability of knowledge are its organization and connectedness. Developing expertise in a field requires "reorganizing pieces of prior knowledge" (Hatano & Inagaki, 1986, p. 265) and knowing the "conditions of applicability" of one's knowledge (Bransford, Brown, & Cocking, 2000, p. 237). Novices' engagement in tasks that enlist MKT inside rehearsal provided them with opportunities to reorganize their MKT and learn about the conditions of its applicability. These opportunities emerged through a process of repeated practice with opportunities for both reflection and action. Hatano and Inagaki (1986) describe the role of repeated practice of procedural skills in daily life in equipping novices with efficiency not only in enacting these skills but also "in understanding the meaning and nature of their object" (p. 263), which can include content knowledge for teaching.

MKT is also treated in rehearsal as a form of knowing in the way it gets embedded in deliberate practice—starting with the interpretation of a problem of practice that emerges in the context of tasks that enlist MKT, then figuring out the next move(s) a teacher might make to address it, and ending with acting on this move(s) and reflecting on its consequences. This process is consonant with the way MKT—as a form of knowing—is used in the online problem-solving that practitioners do in the actual work of teaching (Shulman, 1986; Stein, Engle, Smith, & Hughes, 2008), where teachers interpret situations and act on them using mathematical insights and sensibilities. This in-the-moment problem-solving that practitioners do is simulated in the deliberate practice that is part of rehearsal by slowing down a segment of teaching, looking at some of its details that can be hard to detect in the normal course of action, and then using mathematical considerations to figure out how to enhance aspects of it through decision-making and experimentation. In this process, MKT is not treated in isolation from principles of teaching and learning, but looked upon as part of the "horizon of observation" (Hutchins, 1996, p. 52) that informs teachers' practice.

This study mainly focuses on the role of the teacher educator in supporting novices' opportunities to develop MKT in rehearsals; however, it is important to note that the instructional activities (IAs) used in rehearsal and throughout the cycle of enactment and investigation (described in the context of this study) play an equally important role. The IAs have structures that give school children multiple entry points and promote participation that is not solely dependent on talk (Lampert, Beasley, Ghousseini, Kazemi, & Franke, 2010). Accordingly, these structures enable novice teachers to engage their MKT in the context of managing student contributions. For instance, the vignettes featured in this study evidence how the activity of Strategy Sharing engaged the novices in doing mathematics, anticipating student thinking, eliciting multiple student conjectures, and responding to students' ideas using verbal and written representations. All of these tasks of teaching, built into the structure of the IA, provide opportunities to enlist MKT in the work of teaching. Furthermore, the embeddedness of the IAs throughout the cycle of enactment and investigation provides novices additional opportunities to develop and use MKT as they prepare for rehearsal and enactment. The reader can recall that the first stages of the cycle of enactment and investigation begin with novice teachers observing and analyzing an enactment of an IA in a classroom context, either live or on video. Following this phase, novices also prepare for rehearsing the IA by anticipating student thinking and crafting ways to connect particular content to different learners. Hence, novices get repeated opportunities to engage in tasks that enlist MKT across the cycle. Circling back to the work of the teacher educator, it is important to note the centrality of her role in supporting these opportunities. It would be naïve to think that we can simply give novice teachers instructional activities to practice together without teacher educator guidance and expect them to learn MKT for use.

Pedagogies like rehearsals represent a shift in teacher education toward a practice-centered curriculum organized around core practices that can help develop professional knowledge and skills (Grossman, Hammerness, & McDonald, 2009). This shift has been met with some caution from scholars who see a focus on core practices as reminiscent of previous efforts in teacher education to focus on "the visible practices of teaching" with potentially not enough attention to purpose (Kennedy, 2015). Kennedy, for example, states, "the notion of core practices still has weaknesses when we move to teaching preservice teachers. Even though the labels do a better job of conveying the meaning and significance of individual practices, efforts to teach novices about specific core practices can still become so procedural that their ultimate purposes are overlooked." (p. 4). The present study attends to these concerns by illustrating how a pedagogy of enactment like rehearsals could prepare novices to enact core practices with attention to knowledge of content and students.

7.6 Conclusion

Preservice teachers need to understand mathematical content in ways that would enable them to effectively help others learn mathematics with understanding. This means that they must not only develop MKT as a form of propositional knowledge but also learn it for use in the dynamic interplay of content with pedagogy in teachers' real-time problem-solving (Ball & Bass, 2000). The present study shows that pedagogies of teacher education like rehearsals of teaching provide novices opportunities to engage in work that enlists MKT inside the interactive work of teaching, with guidance from the teacher educator. One of the clear implications of this study is that methods courses can be conceived as sites for developing novices' content knowledge. This is important given that the responsibility for developing preservice teachers' mathematics content knowledge is typically relegated to content courses housed in mathematics departments. With content knowledge needs for novice teachers being so high, pedagogies such as rehearsals of teaching can provide additional opportunities to develop MKT.

References

Adler, J. (1998). Lights and limits: Recontextualising Lave and Wenger to theorise knowledge of teaching and of learning school mathematics. In A. Watson (Ed.), *Situated cognition and the learning of mathematics* (pp. 161–177). Oxford: Centre for Mathematics Education Research, University of Oxford.

Ball, D. L., & Bass, H. (2000). Interweaving content and pedagogy in teaching and learning to teach: Knowing and using mathematics. In J. Boaler (Ed.), *Multiple perspectives on the teaching and learning of mathematics* (pp. 83–104). Westport, CT: Ablex.

Ball, D. L., Thames, M. H., & Phelps, G. (2008). Content knowledge for teaching: What makes it special? *Journal of Teacher Education, 59*(5), 389–407.

Bransford, J. D., Brown, A. L., & Cocking, R. R. (2000). *How people learn: Brain, mind, experience, and school*. Washington, DC: National Academy Press.

Charalambous, C. Y., Hill, H. C., & Ball, D. L. (2011). Prospective teachers' learning to provide instructional explanations: How does it look and what might it take? *Journal of Mathematics Teacher Education, 14*, 441–463.

Cook, S. D., & Brown, J. S. (1999). Bridging epistemologies: The generative dance between organizational knowledge and organizational knowing. *Organization Science, 10*(4), 381–400.

Ebby, C. B. (2000). Learning to teach mathematics differently: The interaction between coursework and fieldwork for pre-service teachers. *Journal of Mathematics Teacher Education, 3*, 69–97.

Ericsson, K. A., Krampe, R. T., & Tesch-Römer, C. (1993). The role of deliberate practice in the acquisition of expert performance. *Psychological Review, 100*(3), 363–406.

Fennema, E., & Franke, M. L. (1992). Teachers' knowledge and its impact. In D. A. Grouws (Ed.), *Handbook of research on mathematics teaching and learning* (pp. 147–164). New York, NY: Macmillan.

Greeno, J. (1998). The situativity of knowing, learning, and research. *American Psychologist, 53*(1), 5–26.

Grossman, P., Hammerness, K., & McDonald, M. (2009). Redefining teaching, re-imagining teacher education. *Teachers and Teaching: Theory and Practice, 15*(2), 273–289.

Hatano, G., & Inagaki, K. (1986). Two courses of expertise. In H. Stevenson, H. Azuma, & K. Hakuta (Eds.), *Child development and education in Japan* (pp. 263–272). New York, NY: Freeman & Co..

Hill, H. C., Blunk, M., Charalambous, C., Lewis, J., Phelps, G. C., Sleep, L., & Ball, D. L. (2008). Mathematical knowledge for teaching and the mathematical quality of instruction: An exploratory study. *Cognition and Instruction, 26*, 430–511.

Hill, H. C., Rowan, B., & Ball, D. L. (2005). Effects of teachers' mathematical knowledge for teaching on student achievement. *American Educational Research Journal, 42*(2), 371–406.

Hodgen, J. (2011). Knowing and identity: A situated theory of mathematics knowledge in teaching. In T. Rowland & K. Ruthven (Eds.), *Mathematical knowledge in teaching* (pp. 27–42). New York, NY: Springer.

Horn, I. S., & Little, J. W. (2010). Attending to problems of practice: Routines and resources for professional learning in teachers' workplace interactions. *American Educational Research Journal, 47*(1), 181–217.

Hutchins, E. (1996). Learning to navigate. In S. Chaiklin & J. Lave (Eds.), *Understanding practice: Perspectives on activity and context* (pp. 35–63). Cambridge, MA: Cambridge University Press.

Kazemi, E., & Hintz, A. (2014). *Intentional talk: How to lead productive mathematical discussions*. Portland, ME: Stenhouse.

Kennedy, M. (2015). Parsing the practice of teaching. *Journal of Teacher Education, 67*(1), 6–17.

Lampert, M., Beasley, H., Ghousseini, H., Kazemi, E., & Franke, M. (2010). Designed instructional activities to enable novices to manage ambitious mathematics teaching. In M. K. Stein & L. Kucan (Eds.), *Instructional explanations in the disciplines* (pp. 129–141). New York, NY: Springer.

Lampert, M., Franke, M., Kazemi, E., Ghousseini, H., Turrou, A. C., Beasley, H., Cunard, A., & Crowe, K. (2013). Keeping it complex: Using rehearsals to support novice teacher learning of ambitious teaching. Journal of Teacher Education, *64*(3), 226–243.

Lloyd, G. M. (2005). Beliefs about the teacher's role in the mathematics classroom: One student teacher's explorations in fiction and in practice. *Journal of Mathematics Teacher Education, 8*, 441–467.

Saldaña, J. (2013). *The coding manual for qualitative researchers* (2nd ed.). Thousand Oaks, CA: Sage.

Shulman, L. S. (1986). Those who understand: Knowledge growth in teaching. *Educational Researcher, 15*(2), 4–14.

Silver, E. A., Clark, L. M., Ghousseini, H., Charalambous, C. Y., & Sealy, J. T. (2007). Where is the mathematics? Examining teachers' mathematical learning opportunities in practice-based professional learning tasks. *Journal of Mathematics Teacher Education, 10*, 261–277.

Steele, M. D., & Hillen, A. F. (2012). The content-focused methods course: A model for integrating pedagogy and M=mathematics content. *Mathematics Teacher Educator, 1*(1), 53–70.

Stein, M. K., Engle, R. A., Smith, P. S., & Hughes, E. K. (2008). Orchestrating productive mathematical discussions: Five practices for helping teachers move beyond show and tell. *Mathematical Thinking and Learning, 10*(4), 313–340.

Thanheiser, E. (2015). Developing prospective teachers' conceptions with well-designed tasks: Explaining successes and analyzing conceptual difficulties. *Journal of Mathematics Teacher Education, 18*(2), 141–172.

Wilson, S. M., Floden, R., & Ferrini-Mundy, J. (2001). Teacher preparation research: Current knowledge, gaps, and recommendations. In *A research report prepared for the U.S. Department of Education*. Seattle: Center for the Study of Teaching and Policy, University of Washington.

Chapter 8
How Do I Know I Learned Something? Reflecting on Learning by Using Video-Recorded Interviews to Battle Hindsight ("I-Knew-It-All-Along") Bias

Eva Thanheiser

8.1 Introduction

When prospective elementary school teachers (PTs) enter our mathematics content courses, they are often proficient with the procedures for operating on numbers (Harkness & Thomas, 2008; Khoury & Zazkis, 1994; Menon, 2009; Thanheiser et al., 2014; Zazkis & Khoury, 1993) and approach mathematics assuming that knowing how to apply procedures is synonymous with understanding (Graeber, 1999). In addition, PTs often enter their content courses thinking they already know all the mathematics they need to know (Thanheiser, Philipp, Fasteen, Strand, & Mills, 2013) and that elementary mathematics is easy or simple. For example, Carol, a PT entering the course, stated, "I came into the class thinking that I knew all of the math used in an elementary school and therefore could definitely teach it."

After their content course, PTs better understand why correctly applying the procedures for mathematical operations produces correct answers (Thanheiser, 2015) but often do not realize that they did not hold this understanding at the beginning of the course. It is hard for them to imagine that there was a time when they were not able to explain regrouping in addition and subtraction. Thus, they leave their content courses stating that the courses were merely nice "refreshers." Because they lack recognition of their own learning, the PTs fail to appreciate (a) that they actually learned important content in the course as well as (b) that elementary mathematics is complex. Consequently, PTs may undervalue the difficulty of their students' learning elementary mathematics content. Recognizing the complexity of elementary mathematics is, therefore, an essential aspect of mathematical knowledge for teaching (MKT) (Ball, Charalambous, Thames, &

E. Thanheiser (✉)
Portland State University, Portland, OR, USA
e-mail: evat@pdx.edu

Y. Li et al. (eds.), *Developing Mathematical Proficiency for Elementary Instruction*, Advances in STEM Education, https://doi.org/10.1007/978-3-030-68956-8_8

Lewis, 2009), which PTs need to appreciate their elementary school students' difficulties in learning elementary mathematics concepts. PTs who have experienced conceptual change (Chi, Slotta, & De Leeuw, 1994) and remember the challenge of understanding elementary mathematics concepts will better understand and relate to their future students than those who lack these experiences.

In this chapter, I argue that by participating in a videotaped interview at the beginning of the course allows PTs recognize that they have mathematics content to learn, and viewing and reflecting on that same interview at the end of the course allows PTs to recognize and value their own learning and the complexity of elementary mathematics.

I add to the current literature by focusing on the metacognitive rather than the cognitive (concepts) or affective (beliefs) as well as by focusing on the idea that elementary mathematics is complex (not elementary/simple/easy). PTs often think that they know all they need to know to teach K–3 mathematics. Through reflecting on their own knowledge at the beginning and end of the course, the PTs begin to understand and value the difficulty of elementary mathematics. At the beginning of the course, the PTs reflect on their inabilities to explain regrouping, recognizing that their mathematical knowledge is lacking. At the end of the term, PTs reflect on their newfound abilities to explain regrouping, remembering that they could not do so at the beginning of the term. This change focuses their reflection on their own learning and the complexity of elementary mathematics.

The intervention described consists of three parts: (a) a brief one-on-one video-recorded content interview with each PT at the beginning of the course, (b) three online reflections, completed throughout the course, on the experience of being interviewed, and (c) viewing and reflecting on the initial content interview at the end of the course (as part of the third reflection).

In the initial interview, I established a baseline for each PT's content knowledge at the beginning of the course and allowed the PTs to reflect on the notion that they knew all the mathematics content they need for teaching K–3. While reviewing the interview at the end of the course, the PTs could compare and contrast their own understanding of the content at two points in time and recognize the conceptual change. In this chapter I draw on the conceptual-change literature (Chi et al., 1994; Von Glasersfeld, 1995) and define *learning* in the context of multidigit whole numbers as the change in conceptions from a less sophisticated conception of number, based on algorithmic reasoning, to a more sophisticated conception of number, a basis to conceptually explain number operations. I also situate the conceptual change within the larger construct of understanding the MKT, which includes understanding of both mathematics and the challenge of learning mathematics.

The research questions guiding this work are as follows:

1. What is the effect of a brief one-on-one videotaped interview on the PTs' assessment of their own mathematical knowledge at the beginning of the mathematics content course? (Answering this research question replicates and extends the findings of Thanheiser et al. (2013) by scaling up.)

2. What is the effect of the PTs' watching, at the end of the course, their own brief one-on-one videotaped interviews from the beginning of the course? (Answering this research question extends prior work.)

8.2 Conceptual Framework

In this study, I draw on the concepts of (a) mathematical knowledge for teachers (MKT) and (b) conceptual change to situate the learning in the content courses for prospective elementary school mathematics teachers, (c) hindsight bias, to explain the PTs' undervaluing both their own learning in the course and the difficulty of elementary mathematics (the content of their learning), and (d) metacognition and awareness to highlight the need for PTs to have an accurate sense of their knowledge and understanding at various points in time. I discuss each of these in the following.

8.2.1 *Mathematical Knowledge for Teaching (MKT)*

MKT is a form of mathematical knowledge related to the work of teaching mathematics to students (Adler & Davis, 2006; Ball & Bass, 2000; Ball, Thames, & Phelps, 2008; Silverman & Thompson, 2008) or, more precisely, "the knowledge demanded by the work of teaching" (Ball et al., 2009, p. 122). Teachers' MKT has been linked to the level of their students' achievement (Hill, Rowan, & Ball, 2005). MKT includes a sophisticated understanding of the mathematics but goes beyond, for example, to include ability to explain the mathematics and core activities such as "figuring out what students know; selecting, and modifying textbooks; deciding among alternative courses of action; steering a productive discussion" (Ball & Bass, 2000, p. 89). MKT also comprises awareness of which topics are generally difficult for children, why they are difficult for children, and what can be done to mediate such difficulties. It is this latter aspect, in particular, understanding and valuing the difficulty of certain elementary mathematics topics, that I address in this study. By coming to recognize both the effort they expended in learning and understanding number to develop a profound understanding of fundamental mathematics (PUFM) (Ma, 1999) and operation themselves and the time this learning required, PTs may be poised to appreciate the difficulty children may encounter when learning the same topics and the time they will need to accomplish this learning.

8.2.2 *Conceptual Change*

Conceptual change is broadly characterized as "learning that changes some existing conception" (Chi et al., 1994, p. 27) or "the kind of learning required when the new information to be learned comes in conflict with the learners' prior knowledge" (Vosniadou & Verschaffel, 2004, p. 3). Thus, what will change is already known and identified as the prior conception or the conflicting conception. In this chapter, I address dual conceptual changes. First, PTs come to recognize the limitations of their own understandings in relying on procedural explanations of standard algorithms. This recognition conflicts with the notion that they already know everything they need to know to teach mathematics at the elementary school level. Second, the PTs build on their currently held conceptions to develop more sophisticated number conceptions that enable them to then conceptually explain the mathematics underlying these standard algorithms. At the end of the course, the PTs reflect on their knowledge at the beginning and discuss how their understandings have changed. In the next section, I address remembering what one knew or did not know before learning something.

8.2.3 Hindsight Bias

Hindsight bias, or the "I-knew-it-all-along effect" (Fischhoff, 1975; Fischhoff & Beyth, 1975; Kahneman, 2011; Roese & Vohs, 2012), highlights people's tendencies to forget what they knew or believed before they learned something. Being unable to ignore what they have recently learned (Bukszar & Connolly, 1988), people reinterpret their prior knowledge as being equivalent to this new learning, particularly if replacing the old knowledge (for example procedures without rationale) with the new knowledge (procedures based on rationales) is efficient and if the old knowledge is inconsistent with the new knowledge (Loftus & Loftus, 1980). After people have learned something, they are likely to overestimate their knowledge before the learning (Hawkins & Hastie, 1990). Studies in various contexts have shown that students did not recall their original estimates and were "less surprised by what did or did not happen than they would have been, had they recalled their original estimates" (Loftus & Loftus, 1980). If PTs do not recall what they did *not* know at the beginning of the course, they cannot be aware of (Mason, 1998) or value what they have learned. The effort they devoted to learning the material must be measured in conjunction with the learning outcome (i.e., comparing their current knowledge to their prior knowledge), a measurement that can be accurate only if the PTs remember their prior knowledge for comparison with their current knowledge. By correctly assessing the change, they become aware of and appreciate their own efforts in conjunction with the results, which can lead to a growth mindset (Dweck, 2006).

Dweck coined the terms *growth* and *fixed mindset* to describe beliefs about learning and intelligence. A growth mindset is a view of learning in which effort, improvement, and mastery are central; a fixed mindset is a view focused on innate and fixed ability (Dweck, 2000). Students with growth mindsets are more likely to improve their mathematical understanding (Dweck 2007). By reflecting on the fact that one has learned something, a person can move from a fixed to a growth mindset.

8.2.4 Metacognition and Awareness

Typically, the focus of research with PTs is on the cognitive (what PTs know) or affective (what PTs believe) perspectives. In this study, I turn my attention to the metacognitive and awareness. I follow the National Research Council's (2001) definition of *metacognition* as knowledge of and reflection on one's knowledge and learning, more precisely, "knowledge about one's own thinking and ability to monitor one's own understanding and problem-solving activity" (p. 118). To monitor (and recognize) their understandings, PTs must remember what they knew before they learned something and thus recognize a shift in their conceptions. One essential element of an expert mathematics teacher is awareness of one's own expertise and ability to re-enter "the state of not having that expertise" (Mason, 1998, p. 243), thus remembering what not knowing was like.

Metacognition is a powerful predictor of learning (Veenman, Van Hout-Wolters, & Afflerbach, 2006; Wang, Haertel, & Walberg, 1990). PTs with metacognitive capacity have the "ability to plan and monitor their success and to correct errors when necessary" (Bransford, Brown, & Cocking, 1999, p. 234). Teachers (in our case mathematics teacher educators) must help students (in our case PTs) accurately assess their current knowledge and learning.

> If students do not realize they do not know some aspect of factual, conceptual, or procedural knowledge, it is unlikely they will make any effort to acquire or construct new knowledge. Accordingly, I stress the need for teachers to help students make accurate assessments of their self-knowledge, not inflate their self-esteem. (Pintrich, 2002, p. 222)

Students who recognize that they do not know much about a topic might be more likely to pay attention when that topic is discussed in class (Pintrich, 2002). "Lack of self-knowledge can be a constraint on learning" (Pintrich, 2002, p. 223).

Inclusion of metacognitive instruction at all levels, including teacher education, (Bransford et al., 1999) has been called for. Tanner (2012) recommended specific activities to promote student metacognition, suggesting (a) preassessments, to allow students to establish what they do and do not yet know about a topic; (b) retrospective postassessments, to push students to recognize conceptual change; and (c) reflective journals, to provide students a forum to monitor their own learning. Monitoring one's own understanding requires the abilities to accurately represent and compare of one's own knowledge at two points in time. Chapman (2007) used such an approach in having students reflect, inquire, and reflect again in the context of arithmetic word problems. She stated that the initial reflection helped raise the PTs' awareness of their own knowledge, and the second reflection helped raise their awareness of their learning.

In prior work (Thanheiser et al., 2013), I have shown three results stemming from a brief initial content interview with PTs (see Table 8.1). Thus, that work has shown that PTs can become aware of their own (lack of) understanding/knowledge through a brief interview, which might set them on a trajectory to engage more with the class activities when they have become aware of their own lack of understanding/knowledge.

Table 8.1 Thanheiser et al. (2013) themes for interview reflections

Thanheiser's themes for interview reflections
Theme 1. The PTs are not typically asked to think about or explain beyond procedures.
Theme 2. The PTs recognized (a) the limitations of their own understanding, (b) the value of knowing more than the procedures, or (c) both.
Theme 3. The PTs stated that they were motivated to learn the mathematics taught in the class.

In this study, I examined PTs' conceptual changes within the context of MKT. PTs need to solve elementary mathematics problems and explain the mathematics underlying their solutions. Furthermore, they need to know various ways to solve problems and explain solutions. My goals, in addition to fostering conceptual change in the PTs, was for them to understand, first, the limitations of their own understandings (at the beginning of the course) and, second, the value of knowing more than procedures. I anticipated that with these two understandings, the PTs would, in line with my third goal, be motivated to learn the mathematics in the course. After the course, I had the PTs reflect on their own learning and connect that to how knowledge of number concepts and operations develops to achieve my fourth goal, that the PTs recognize that elementary mathematics is complex and that learning elementary mathematics concepts takes time.

I draw on the constructs of metacognition and awareness to discuss helping the PTs recognize the level of their own understanding at the time of the interview (beginning of the term) as well as at the end of the term. In addition, I use the construct of hindsight bias to explain the phenomenon that PTs often forget what they did not know before so that by the end of the term they may not remember not having known something at the beginning. This forgetting can result in PTs' undervaluing both their own learning and the complexity of elementary mathematics. Being aware of the complexity of elementary mathematics is an essential aspect of MKT and, as such, essential for PTs to understand.

8.3 Literature and Background

8.3.1 PTs' Incoming Conceptions of Number

Studies have shown that PTs struggle to understand the mathematics content of elementary school (Ball, 1988; Ma, 1999; Simon, 1993; Simon & Blume, 1992; Thanheiser, Browning, et al., 2014; Thanheiser, Whitacre, & Roy, 2014; Zazkis & Campbell, 1996), including understanding of whole number and operation (Chapman, 2007; Crespo & Nicol, 2006; Kaasila, Pehkonen, & Hellinen, 2010; Thanheiser, 2009, 2010; Thanheiser, Whitacre, & Roy, 2014). PTs have had years of experience with the standard algorithms and are computationally proficient with

Table 8.2 Thanheiser's (2009) framework for PTs' conceptions of multidigit whole numbers

Conception of multidigit whole numbers
Reference units. (Sufficient to explain regrouping) PTs with this conception reliably conceive of the reference units for each digit and relate reference units to one another, seeing the 3 in 389 as 3 hundreds or 30 tens or 300 ones, the 8 as 8 tens or 80 ones, and the 9 as 9 ones. They can reconceive of 1 hundred as 10 tens, and so on.
Groups of ones. (Sufficient to explain regrouping) PTs with this conception reliably conceive of all digits in terms of groups of the correct number of ones (389 as 300 ones, 80 ones, and 9 ones) but not in terms of reference units; they do not relate reference units (e.g., 10 tens to 1 hundred).
Concatenated-digits plus. (Insufficient to explain regrouping) PTs with this conception conceive of at least one digit as an incorrect unit type at least sometimes. They struggle when relating values of the digits to one another (e.g., in 389, 3 is 300 ones but the 8 is only 8 ones).
Concatenated-digits only. (Insufficient to explain regrouping) PTs holding this conception conceive of all digits in terms of ones (e.g., 548 as 5 ones, 4 ones, and 8 ones).

whole number operations but struggle when asked to explain the mathematics underlying those algorithms (Ball, 1988/1989; Ma, 1999; Thanheiser, 2009, 2010; Thanheiser, Whitacre, & Roy, 2014).

In prior work (Thanheiser, 2009, 2010), I developed a framework for the multidigit-number conceptions PTs hold when entering mathematics content courses. These conceptions were categorized into four categories: Two conceptions, thinking in terms of (a) reference units and (b) groups of ones, are sufficient to explain regrouping; the other two, (c) concatenated-digits plus and (d) concatenated-digits only, are insufficient to explain regrouping. Although differing in levels of sufficiency to explain regrouping, all four conceptions enable one to perform the standard vertical US algorithms for addition and subtraction. For a description of each conception, see Table 8.2.

Typically, about 80% of the PTs entering a teacher education course hold one of the concatenated-digits conceptions (Thanheiser, 2018). However, the PTs are often unaware of their own lack of understanding because they equate procedural proficiency with conceptual understanding (Graeber, 1999). PTs have had years of experience with applying the algorithms and are often surprised that a rationale for regrouping exists (Thanheiser et al., 2013). Toward the end of the course, when PTs have realized that mathematical rationales underlie algorithms and have developed conceptions of numbers that enable them to understand how to justify the algorithms mathematically, they often fail to remember that recently they lacked this understanding, thinking that they always knew, but simply forgot, the rationales. This misremembering leads PTs to reflect on the course as a "nice refresher" rather to value their learning and the effort they expended (hindsight bias), misshaping the PTs' MKT with respect to the level of difficulty in understanding multidigit whole numbers.

8.3.2 Preparation for Future Learning

When PTs enter our courses, we mathematics educators do not expect them to already know the things we include in the course. However, we do expect them to be prepared to learn the course materials. Bransford and Schwartz (1999) showed that providing students experiences with contrasting cases can set the stage for future learning. In this study, the two contrasting cases are Case 1, the PTs think that they already know the mathematics they need to know to teach, and Case 2, to prepare them for future learning in the course, an interview with the PTs revealed that they did not yet know all the mathematics they need to know. As a result of the interview, PTs can recognize the need to pay attention in class rather than assume that they already know the materials being taught (Thanheiser et al., 2013).

An essential element of the PTs' preparation for future learning is for them to recognize that being successful or "knowing" mathematics entails more than being able to apply procedures and provide numerically correct answers to questions. Successful PTs also need to be poised to explain the "why."

Bransford and Schwartz (1999) highlighted the need to allow PTs to first generate their own ideas about a phenomenon so that they can later compare their own thinking to that of others. In my work, I also focus on allowing PTs to first generate their own ideas about their own knowledge (in our case their understanding of number and operation). Throughout the course, they compare their own ideas to those of others, including experts; however, at the end of the course, they compare their own ideas (about number and operation) at that point to their own ideas at the beginning of the course.

8.4 Methods

8.4.1 Interviews as Tools in Teacher Education

In this study, I utilize a brief 10–15 min one-on-one interview with each PT with a dual purpose: (a) for PTs to understand the limits of their current understanding and thus to help motivate them to learn (Thanheiser et al., 2013) and (b) to create a record of the PTs' knowledge at the beginning of the course for the PTs' own reflection.

Interviews provide a format that allows for immediate and personal feedback and assessment. The instructor can ask probing follow-up questions to assess the depth of a PT's understanding. Both the instructor and PT are aware of what the PT knows and does not know. In a whole-class or small-group discussion, a PT can melt into the crowd and, intentionally or not, allow others to carry the intellectual load. In an interview setting, the PTs cannot hide their knowledge gaps from either themselves or their instructor. Interviews (if video-taped) also create a record of the PTs' knowledge at the time of the interview. Viewing these videos at a later date, the

PTs can (a) compare their knowledge at the end of the course with their knowledge at the beginning (the time of the interview) and (b) reflect on their knowledge prior to the course, thus accurately assessing their own learning experiences.

8.4.2 Participants and Data Collection

The participants in the study were 23 PTs who met for two 110-min class sessions per week (equivalent to 4 credits in one quarter) in a mathematics content course designed for PTs at a large state university in the Northwestern United States. The course was focused on number and operation. Among the 23 PTs were 4 sophomores, 5 juniors, 6 seniors, 2 fifth-year, and 6 post-baccalaureate students. The PTs reported on the last mathematics class they had taken: four PTs took their last mathematics course in high school, three PTs took a remedial mathematics course in college, nine PTs took an introductory (100 level) mathematics course in college, three took a higher level (200+) mathematics course in college, one had taken the mathematics content course for teachers before but had not received a grade adequate for admission to the teacher education program, and three did not respond to this question. At this university, PTs take three mathematics content courses (equivalent to 1 academic year) as a prerequisite for admission to the Graduate School of Education for a master's degree in education. The described activities take place in the first course in the sequence.

Interviews As part of their regular content course, all PTs participated in two videotaped one-on-one interviews with the instructor (or a TA), one at the beginning and one at the end of the course. The initial interview was designed to establish the PTs' displayed conceptions at that time and to pique the PTs' interest in the content of the course (Thanheiser et al., 2013); the final interview was designed to establish the PTs' displayed conceptions at the end of the course.

During Interview 1, PTs were asked to work three problems: two subtractions and one addition (for the problems and sample student solutions, see Fig. 8.1). After PTs solved each problem, they responded to *press* questions to clarify their thinking (for sample press questions, see Table 8.3). During Interview 2, PTs were asked to solve equivalent problems.

Surveys In addition to participating in the interviews, all 23 PTs responded, throughout the course, to three surveys in which they reflected on their initial-

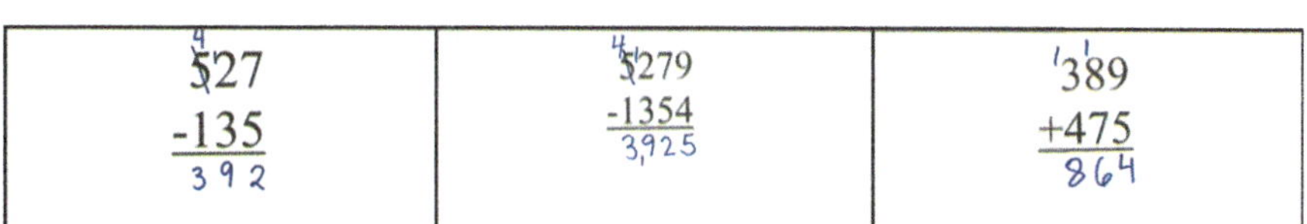

Fig. 8.1 Interview problems with sample PT solutions

Table 8.3 Press questions for 527 − 135

Press questions
• "Can you talk a little bit about what you did here [pointing to 527 − 135]?"
• "What are you doing here [pointing to the regrouping within 527 − 135]? Why does it work?"
• "What exactly is going on here—in the regrouping part?"
• "Did you change the value of the 527?"
• "What do the *small* [regrouped] numbers mean?"
• "Why can we simply cross out a number? How and why does this work?"
• "What about this problem 5279 − 1357?".

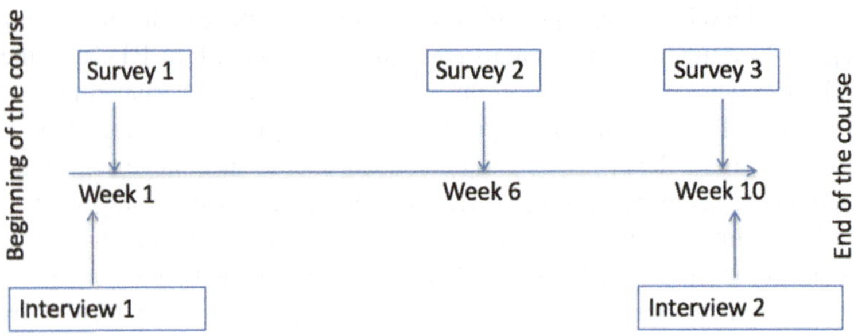

Fig. 8.2 Timeline of data collection

interview experiences. The first survey was collected at the beginning of the course (immediately after the first interview), the second in the middle of the course (Week 6), and the third at the end of the course (before the last interview) (see Fig. 8.2 for the data-collection timeline). As part of the last reflection (Survey 3), the PTs viewed their initial interviews (provided online) and responded to reflection questions about that experience. The first and last reflections were collected via web surveys and the middle survey, via email. See Table 8.4 for sample reflection questions.

8.4.3 Data Analysis

I discuss the various data sets and their analyses below. I describe the analysis for each interview and the surveys.

Interviews Thanheiser's (2009) framework (see Table 8.2) was used to categorize the PTs' conceptions of multidigit whole numbers at the beginning and the end of the course on the bases of the two interviews. To categorize the PTs' conceptions, two coders watched each interview and coded independently, with agreement of 86%. All disagreements were resolved through discussion.

Table 8.4 Sample reflection questions from each survey

Survey #	Reflection questions
1	PTs completed the initial interview and were asked to immediately respond to the first survey questions (collected via web survey). • I would imagine that the interview with me was kind of an unusual experience for you. Would you agree or not? Explain. • Did you learn something from this experience? If so, what?
2	PTs e-mailed the reflection during Week 6 of class. • Thinking back to the interview at the beginning of class, how do you think that interview affected your learning in this class? Please try to be as specific as you can in this response. • Do you think that your experience would have been different without having participated in the interviews? If so, how? • Do you have any additional comments for us?
3	PTs viewed their intake interviews and answered survey questions (collected via web survey). • Watch your first interview and give your general reactions to the interview; we are very interested in your reactions, so feel free to share. • How do you think having this interview at the beginning of the term affected your learning in this class? • Given that it takes a LOT of the instructor's time to conduct interviews with all students (and thus the teacher may rather not do interviews), could you please share with us how you feel about the interviews?

Surveys Coders had an initial list of themes for categorization of PTs' perceptions about the effects of an initial interview; this list was based on a prior, smaller study (Thanheiser et al., 2013) of 13 PTs enrolled in an earlier course. My goal for that study was to examine the effects of an initial content interview on the PTs' reflection of their understanding at the beginning of the course and on the PTs' motivations to engage in the course. The analysis was based on two surveys similar to Surveys 1 and 2 in this study. The three themes identified in that study are shown in Table 8.1. I used that list of themes in conjunction with a *thematic approach* (Braun & Clarke, 2006), "identifying, analyzing and reporting patterns (themes)" (p. 82) to both possibly replicate the results of the prior study and identify additional themes. In contrast to the prior study (in which only Surveys 1 and 2 were analyzed), in this study I analyzed all three surveys, with emphasis on Survey 3. A colleague and myself followed Braun and Clarke's process by first familiarizing ourselves with the survey data. I (as the instructor) read the survey responses as soon as they came in (during the course), and we both read all responses after the course ended and data analysis began. I read through all survey responses to each question several times. During initial reading of all the data, we independently identified the prior established themes and generated new codes. We then met to compare both our tagging of the previously identified themes and our identification of new codes for each survey question. After discussion, we identified three new codes: (a) The PTs reflected on their own experiences as learners; (b) the PTs reflected on their own knowledge; and (c) the PTs reflected on their learning in the course. From these codes, we established a fourth theme, focused on reflection.

After establishing the new coding scheme with its four themes, we both coded all the data using this scheme. We met to discuss all codes, and if we disagreed, we discussed the code until we reached agreement, leading to a final coding scheme and a final set of agreed upon coded survey responses. Because we used this data set to identify new themes, no interrater reliability was established for this part of the data analysis.

In addition to analyzing the survey responses to all three surveys, we analyzed one Survey 3 question ("Watch your first interview and give your general reactions to the interview; we are very interested in your reactions so feel free to share,") in isolation to examine PTs' responses to viewing their Interview 1 videos at the end of the course. We used the same process as above to analyze the responses to this question.

8.5 Results and Discussion

I present the results in four parts. First, I discuss the PTs' conceptual changes discussing the PTs' displayed conceptions of multidigit whole numbers at the beginning and the end of the course (i.e., as coded through the interview analysis). I first show the changes in the PTs' number conceptions during the course because these data relate to the results used to address my research questions.

Second, I relate the PTs' self-estimated depths of understanding (i.e., as stated by the PTs) at the beginning of the course to their displayed conceptions at the beginning of the course, showing the misalignment of the two.

Third, I show how, through the initial interview (and PTs' reflecting on their interview), PTs become aware of (a) the limitations of their own understanding (at the beginning of the course) and (b) the value of knowing more than procedures so that they realize that their self-estimated depths of understanding at the beginning of the course were inaccurate. Through this realization, PTs recognize that they need to learn the mathematics in the course. In addition, the PTs reflect on their own experiences as learners, their own knowledge, and their learning in the course.

Fourth, I show how, by reviewing their initial interviews at the end of the course, the PTs can accurately reflect on their beginning knowledge, compare that to their current understanding, and thus (a) reflect on their own learning and connect that to how knowledge of number concepts and operations develops and (b) understand that these aspects of elementary mathematics are complex and that learning elementary mathematics concepts takes time and effort.

Table 8.5 PTs' multidigit number conceptions before and after the content course

Conception	Pre	Post
Reference units	2	19
Groups of ones	4	2
Concatenated-digits plus	9	2
Concatenated-digits only	8	0

8.5.1 *Part 1: PTs' Displayed Initial Conceptions and Self-Estimated Initial Depths of Understanding*

The first results concern the PTs' conceptual change. To examine this change, I first compared the PTs' displayed conceptions at the beginning and end of the course.

PTs' displayed conceptions at the beginning and the end of the content course Consistent with previous research (Thanheiser, 2018), most PTs exhibited conceptual change from the beginning of the course to the end. At the beginning of the course, 17 of the 23 PTs displayed one of the concatenated-digits conceptions, 4 displayed a groups-of-ones conception, and only 2 PTs displayed a reference-units conception. Although the groups-of-ones conception is sufficient to explain regrouping one way, the reference-units conception is essential to explain all aspects of regrouping (i.e., regrouping 1 hundred as 10 tens).

At the end of the course, 19 PTs displayed a reference-units conception and 2 PTs displayed a groups-of-ones conception; only 2 PTs displayed a concatenated-digits conception. See Table 8.5 for the distribution of conceptions pre- and post course.

Thus, whereas 17 PTs displayed one of the concatenated-digits conceptions at the beginning of the course and only 2 displayed the most sophisticated conception, at the end of the course 19 PTs displayed a reference-units conception, being able to flexibly explain regrouping in various ways, thus exhibiting a level of understanding mathematics essential for MKT; 2 PTs displayed a groups-of-ones conception. Only 2 PTs displayed a concatenated-digits conception at the end of the course.

8.5.2 *Part 2: PTs' Self-Estimated Depths of Understanding at the Beginning of the Content Course*

When asked, at the beginning of the course, whether they thought that their mathematical knowledge was adequate to teach K–3, 16 of the 23 PTs thought theirs was, 2 were ambivalent, and 5 thought that they needed to learn more mathematics. Thus, although most PTs displayed one of the concatenated-digits conceptions coming into the course, conceptions inadequate to explain regrouping, most thought they knew enough mathematics to teach (see Table 8.6).

Of the 16 PTs who considered their mathematics incoming knowledge sufficient to teach K–3, only 2 displayed a conception sufficient to explain regrouping and only

Table 8.6 PTs' self-estimated depths of understanding in relation to their number conceptions

	Pre	Do you think that your current mathematics knowledge is sufficient to teach K–3?		
Conception	Number holding the conception	Yes	No	Ambivalent
Reference units	2	1	1	0
Groups of ones	4	1	3	0
Concatenated-digits plus	9	8	0	1
Concatenated-digits only	8	6	1	1
Total	23	16	5	2

1 held a conception that would enable them to explain all aspects of regrouping. Of the 5 PTs who thought that they did not know enough mathematics, 4 displayed a conception sufficient to explain regrouping, but only 1 displayed a conception sufficient to explain all aspects of regrouping. Of the 2 who were ambivalent, neither displayed a conception sufficient to explain regrouping.

Thus, a mismatch existed between the PTs' displayed conceptions of multidigit whole numbers (most of them did not display a conception of number sufficient to explain regrouping) and their self-estimated depth of understanding (most of them thought that they knew enough mathematics to teach K–3). This mismatch explains the PTs' lack of awareness, at this point, of the need to learn more mathematics before teaching, thinking that their content knowledge was already sufficient.

8.5.3 Part 3: PTs Reflect on Their Self-Estimated Depths of Understanding, Experiences as Learners, Own Knowledge, and Learning in the Course

I first discuss the use of the interviews (and the PTs' reflection on the interviews) to help PTs reflect on their own knowledge and gain awareness of what they do and do not yet know. This information can, in turn, motivate the PSTs to engage in the activities in class. This first part is a replication of prior work (Thanheiser et al., 2013). In the second part of this section, I discuss a new metacognitive theme: The interviews enabled PTs to reflect on their own experiences as learners, their own knowledge, and their learning in the course. This part is an extension of the prior work.

Interviews to allow PTs to reflect on their own knowledge at the beginning of the class [replication] To facilitate PTs' learning and developing the MKT needed to teach, we mathematics educators need to help them align their actual knowledge and their self-estimated depths of understanding at the beginning of class. To be motivated to engage in the classroom activities, PTs must recognize that they need to learn more mathematics in the course to proficiently teach K–3. By reflecting on their individual content interviews at the beginning of class, PTs can reach this alignment.

Thanheiser et al. (2013) showed that such an interview can lead to PTs' changing their beliefs about mathematics and about their understanding of mathematics, leading to the recognition that (a) they need to learn mathematics that extends beyond procedures, (b) their own knowledge is limited and they need to know more to teach, and (c) engaging in the mathematical activities in their content courses will lead to their learning important content. Thanheiser et al. (2013) reported that three themes emerged in the PTs' reflections on the interview experience (see Themes 1–3 in Table 8.1). These themes also emerged in the current study (see Table 8.7) through the analyses of Surveys 1, 2, and 3. However, Theme 1 was only weakly

Table 8.7 Emergence of the themes in Thanheiser et al. (2013) study and the current study

Themes	# of PTs $N = 13$ (2013)	# of PTs $N = 23$ (Current)
1. The PTs are not typically asked to think about or explain beyond procedures.	7	4
2. The PTs recognized (a) the limitations of their own understanding, (b) the value of knowing more than the procedures, or (c) both.	10	20
3. The PTs stated that they were motivated to learn the mathematics taught in the class.	11	21
4. The PTs reflected on their own experiences as learners, their own knowledge, and their learning in the course.	N/A	21

represented (4 of 23 PTs mentioned it). Themes 2 and 3 were strongly represented (by 20 and 21 of the 23 PTs, respectively) and thus could be seen as robust themes. The emergence of the themes in Thanheiser et al.'s (2013) study and the current study can be seen in Table 8.7.

In addition to the replication of the three themes of the prior study, a new metacognitive theme emerged through this data analysis: 21 of the 23 PTs reported reflecting on their own experiences as learners, their own knowledge, and their learning in the course (Theme 4 in Table 8.7). I discuss this new theme in the next section.

New Theme: During interviews PTs reflect on their own experience as learners, their own knowledge, and their learning in the course [extension of prior work] Reflection is an essential part of learning, and reflecting on their own knowledge and learning can help PTs better understand themselves and align their self-perceptions of understanding with a realistic image. Reflection is essential for metacognition and awareness of one's knowledge and learning, which enable PTs to build their MKT with respect to understanding the depth and challenge of their mathematics learning. I discuss this additional theme further.

Some PTs, like Amanda, noted that the interview helped them align their expectations with the content to be learned in this class, different from their earlier experiences as learners:

> [If] I hadn't participated in the interview, my experience [in the class] probably would have been different. After the interview I realized that our class was not going to be us learning how to teach math the way [we] were taught in grade school. [Amanda, Survey 2]

Thus, the interview helped Amanda adjust her expectations, to recognize that the mathematics learning in the course would differ from her previous experiences. PTs must experience a sense-making learning environment in their university content courses—perhaps their first time experiencing such an environment—to acquire an image of such a mathematics classroom and develop their MKT.

When reflecting on the interviews, PTs stated that the interviews helped them realize that they did not know the mathematics they thought they knew in the ways they need for teaching. Carol, for example, expressed this view:

> I think the interview at the beginning of the class definitely gave me a sense of humility. Math has humbled me a lot since elementary school, but I think that I came into the class thinking that I knew all of the math used in an elementary school and therefore could definitely teach it. [Carol, Survey 2]

For Carol, reflecting on the interview strengthened her MKT in helping her recognize that her incoming understanding was insufficient for teaching elementary school mathematics as a sense-making experience for students. Understanding the type of understanding necessary for teaching is one essential component of MKT.

Lena was more specific, adding that her realization that she did not yet know how to explain addition, subtraction, and place value caused her to adjust her approach to the course:

> I think that [the interview] really affected my learning, because I wasn't really aware that I didn't know how to explain simple addition and subtraction and place value. This interview showed me that I needed to learn how to explain these, in order to teach. [Lena, Survey 3]

Through the interview, Lena became aware of aspects of MKT (explaining addition, subtraction, and place value) she was missing so that she became open to learning them in the course.

PTs must align their self-estimated depths of understanding with their displayed conceptions so that they can be realistic about what they do and do not know; this alignment can lead them to "buy-into" the content of the course and develop better understanding of the various aspects of MKT (e.g., being able to explain the mathematics vs. merely being able to "do" mathematics). If a student is confident that she already knows the course content, her effort in the course may be minimal. Kendra, for example, reflected on her own knowledge and on how that affected her approach to the class:

> I think that during [the interview] ... I know I did not know how to explain what the carrying of 1 meant. This made me think to myself if I even knew math as well as I thought. .... Without the interview, I do not think I would have learned much about myself as a math learner. I personally do not think that I would have taken more information and thought to myself more about the 'why' in math. As well, I might have not known where I was lacking in math or how to even teach it in the first place. [Kendra, Survey 2]

Thus, Kendra explained that realizing that her knowledge was less robust than she thought changed her approach to the class and her view of herself as a learner. Several PTs shared the sentiment that without becoming aware of their own lack of understanding they might have been less open to learning in the class and thus less likely to develop their MKT. Saskia was one such student:

> If we hadn't done the interviews prior to class, I think that I wouldn't have realized the things I didn't know [during class]. I would have just thought, "Oh that's right," rather than admitting that I had totally forgotten. [Saskia, Survey 2]

And Simon elaborated further:

> I think the interview affected me by introducing me to the idea that there were things in elementary math I did not know and that I had/have much yet to learn. It was humbling yet motivating. I think without the interview, I might have gone into class a bit cockier (if possible) and possibly have taken the class less seriously in terms of the amount of work I would have to put forth. As it was, I knew from the relative start I had work to do. [Simon, Survey 2]

Both Saskia and Simon credit the interview with helping them realize what they did and did not yet know. Being aware of what one does and does not know is an essential aspect of learning. Alexander reflected on his knowledge of mathematics and recognized the tendency for a correct answer to interfere with a search for understanding the mathematics underlying a strategy:

> Without the interviews, I wouldn't have noticed that the strategies I learned as a child, although they gave me the correct answer, were lacking in helping me explain the "why." Overall, having the interviews at the beginning, before we were exposed to the underlying concepts, was a great way for us to gauge where we were at as teachers and learners. [Alexander, Survey 2]

Alexander referred to contrasting aspects of understanding (how and why) and reflected on the interview as helping him recognize that MKT is more than knowing how.

And finally, in addition to reflecting on their own knowledge and themselves as learners, PTs reflected on their learning in the course. The interview caused PTs like Aubrie to reflect on their knowledge at the beginning of the course and on what they learned during the course:

> I feel the interview affected my learning in this class by making myself think and realize how I have changed and progressed. .... I can see myself understanding what I didn't understand during that interview, and I can see myself progressing with my own knowledge. [Aubrie, Survey 2]

Kelsey expressed the enjoyment she got from looking back and recognizing growth: "I think it is fun to be able to look back and see how much I have grown and how much I have learned" [Kelsey, Survey 2]. Melinda explained that through reflecting, she came to value her learning: "[The interview and reflecting on it] showed how much I have learned. Makes me appreciate it more" [Melinda, Survey 3].

Thus, to address Research Question 1 (*What is the effect of a brief one-on-one videotaped interview on the PTs' assessment of their own mathematical knowledge at the beginning of the mathematics content course?*), I replicated prior work by scaling up (studying more PTs) to show that reflecting on an interview can help PTs realign their initial self-estimated depths of understanding. PTs' responses showed areas of MKT the PTs lacked. Becoming aware that elementary mathematics is not elementary/simple/easy is an essential part of MKT. Extending prior work, I showed that in reflecting on such an interview, PTs reflect on themselves as learners and on their learning in the class.

Table 8.8 Themes emerging from PTs' reflections on viewing their initial interviews

Prompt Watch your first interview and give your general reactions to the interview; we are very interested in your reactions, so feel free to share.	# of PTs$N = 23$
A. Shock/surprise/embarrassment of limited mathematics knowledge at the beginning (akin to Theme 2a above but with much stronger language)	19
B. Pride/notation/recognition of what was learned in the course. (Theme 4)	12
Both above themes	11

Although in reflecting on the interviews, PTs realigned their self-estimated depths of understanding at the beginning of the course and thus recognized their learning during the course, I show next that their memories of their initial mathematics knowledge were not quite accurate at the end of the course.

8.5.4 Part 4: Reviewing Initial Interview Videos, PTs Reflect on Incoming Knowledge and Value Learning

In reviewing their own initial interviews at the end of the term, PTs reflected on their knowledge at the beginning of the term, their own learning of number, and the complexity of elementary mathematics. We identified two themes in these reflections (see Table 8.8), discussed in the next two sections.

Shock/surprise/embarrassment by initial limited mathematics knowledge (akin to Theme 2a above but with stronger language) When the PTs viewed their initial interviews at the end of the course, they seemed to approach this task realizing that they had not known the mathematics they needed to know to teach K–3. Despite that knowledge most PTs used fairly extreme language, such as "shocked," "surprised," and "embarrassed," to describe their feelings in viewing their videos. This result indicates that although they had a general sense that they had not known enough to respond to the interview questions at the beginning of the course, they were surprised/shocked/embarrassed by the extent of their lack of knowledge. Kendra, for example, reflected on viewing her interview, indicating her surprise at her knowledge at the time [Theme A]:

> I was shocked at the level of understanding of math I had. I am surprised that it did not occur to me to think about the ones that were being carried over were tens or hundreds, not just one. ... Overall, I am surprised by my thinking at the time. [Kendra, Survey 3]

Kendra did not seem to remember her level of displayed knowledge, understanding, or thinking before the course. According to Mason (1998), one essential aspect of the expert/teacher is the ability to re-enter the state of expertise they had at various points. To understand how knowledge develops is an essential aspect of MKT, and, thus, remembering one's own learning can help one develop MKT. Other PTs' statements were similar to Kendra's. Josephine also expressed surprise [Theme A], especially at her inability to solve a problem more than one way: "Watching the video now, I am surprised I was only able to see the one way to solve the problem and was unable to even explain it" [Josephine, Survey 3]. Amanda expressed shock [Theme A] at her understanding at the time: "After watching my first interview, I was really shocked to say the least. I couldn't help but laugh at myself at how little I actually knew" [Amanda, Survey 3]. And Melinda could not believe what she saw in her interview [Theme A]: "I can't believe that I couldn't really explain myself, like what I was borrowing" [Melinda, Survey 3].

The notion of experiencing shock, disbelief, embarrassment, or surprise at their own understanding at the beginning of the course was prevalent in 19 of the 23 reflections. One characteristic of elementary mathematics is that after it has been learned, to imagine not knowing it (even if it was a mere 10 weeks ago that one did not know it) is difficult. This "hindsight bias" or "I-knew-it-all-along" effect may be more likely to be present for concepts that, once learned, seem obvious. Consider, for example, the fact that $345 = 300 + 40 + 5$. This fact, once understood, seems so obvious that not knowing it seems unimaginable; it is, however, a fact with which many people struggle. A result of this phenomenon (not remembering what they did not know at the beginning) is that PTs often leave their content courses thinking that nothing or little was learned because, now that they do understand the course content, not knowing it seems shocking, surprising, and unimaginable. This failure to remember may lead to the PTs' undervaluing both their own learning and the difficulty of elementary mathematics. The interview (and the fact that it is recorded so PTs can watch it later) enables PTs to remember what they knew and did not know at the beginning of the course and thus to appreciate what they have learned in the course. I discuss this latter theme, the pride in or recognition and expression of what was learned in the course, in the next section.

Theme B: Pride/expression/recognition of learning in the course Twelve PTs explained that in addition to being shocked and embarrassed by their initial knowledge, they recognized what they had learned [Theme B]. Many noted the differences in their knowledge from the beginning to the end of the course. Amanda, for example, stated (in addition to the quote above), "And now, realizing how far I have come. If I had only knew [*sic*] then what I know now." Melinda elaborated:

> It shocks me to see myself now after I have taken the course. Taking this course really opened my mind, and I see now how much I have learned. It seems silly now that I couldn't explain myself very well. I couldn't really explain that the one was a ten I just called it a one. [Melinda, Survey 3]

Kristin expressed appreciation for the depth of her learning: "I cannot believe how much I have learned in 10 short weeks!" This sentiment was shared by others, including Lena:

> I know I was a little uneasy about the initial interview, but now watching it at the end of this class makes me realize how easily I could explain all of that now. It feels good to see that I have learned so much over the term. [Kristin, Survey 3]

Melinda, expressed shock at how much she had learned: "It shocks me to see myself now after I have taken the course. Taking this course really opened my mind, and I see now how much I have learned" [Melinda, Survey 3].

In viewing their initial interviews, PTs became aware of what they had learned, expressed pride in their learning/knowledge at the end of the course, and recognized the value of the course. Through the videotaped reflection, they re-entered the state of their knowledge before the course to reflect on the knowledge they developed during the course.

Reviewing their videos at the end of the course to realign their initial self-estimated depths of understanding, value their learning, and recognize the complexity of elementary mathematics The prompt for the PTs to view and reflect on their initial interviews was intentionally nonspecific to allow PTs latitude in their comments (see Table 8.8). Despite the openness of the prompt, 19 of the 23 PTs recognized the limitations of their mathematics knowledge at the beginning of the course and expressed shock, surprise, and embarrassment in their confusing explanations (akin to Theme 2), and 12 PTs recognized what they had learned and expressed pride in that learning (Theme 4).

Thus, in response to Research Question 2 (*What is the effect of the PTs' watching their own brief one-on-one videotaped interview from the beginning of the course at the end of the course? [This research question extends prior work]),* I found that as a result of participating in an initial content interviews and viewing those interviews at the end of the course, PTs were able to realign their self-estimates of depth of understanding at the beginning of the course with their displayed incoming conceptions. Although the initial interview helped PTs recognize their need to learn (and thus realigned their self-estimated depths of understanding somewhat), by reviewing the videotaped interview at the end of the course, the PTs completely realigned their self-estimated depths of understanding and came to value the entirety of their own learning and appreciate the complexity of elementary mathematics. This realignment, in turn, informed their MKT so that they became aware of the difficulties their own students may encounter later; without it, PTs might enjoy the course and see it as beneficial for teaching, but they might recognize neither their mathematical learning nor its complexity. Thus, they could discount their own learning as well as the complexity of the elementary mathematics content they will teach.

8.6 Summary and Conclusions

Helping PTs learn content that they think is common knowledge, simple, or easy can be challenging, especially if the PTs are convinced that they already know all the mathematics they need to know to teach elementary school. To address this difficulty, I initially interviewed all PTs regarding what they thought they knew about whole number addition and subtraction and their abilities to explain the place-value ideas underlying the subtraction algorithms. This unexpected challenge to their knowledge alerted them to the idea that MKT comprises more than being able to perform an algorithm to obtain the correct answer. In realizing that they could not explain the mathematics inherent in the algorithms and that explanations are an essential part of MKT, the PTs recognized, at the beginning of the course, their need to learn. However, after learning the course content, the PTs often thought that they had always known this material, forgetting what they did not know at the beginning of the course.

Helping PTs realize that they had actually learned mathematics they think of as common knowledge, simple, or easy can be as difficult as motivating them to learn it When reviewing their videotaped interviews 10 weeks after starting the course, the PTs were surprised by what they had not known at the beginning of the course. By reviewing the interviews, the PTs re-entered "the state of not having that expertise" (Mason, 1998, p. 243) needed to explain the place-value concepts related to multidigit-whole-number algorithms, so they could compare their knowledge at the beginning of the course to the expertise acquired by the end of the course. This comparison and reflection augment their MKT in two ways: First, they recognize *that* they learned (by comparing their current knowledge to their initial knowledge); second, they become aware of *what* specific content they learned, thus recognizing the complexity of the subject and the time required to learn it. In turn, the PTs come to appreciate what they have learned in the course.

Recognizing that elementary mathematics is not elementary may help PTs approach teaching it differently. They may be poised to recognize how difficult these concepts are and that students may need time to understand them. In addition, recognizing that elementary mathematics is not elementary may help PTs value their own learning in the course, knowing that they have learned significant mathematics.

I posit that such an experience (reviewing one's own video at the end of the course) enhances PTs' abilities to reflect on their knowledge at various points (especially at the beginning and end of the course) and thus allows for metacognition and awareness. During such an experience, PTs become aware of and track their own progress throughout the course. They realize what they did and did not know at various times. Most important, I believe that recognizing that elementary mathematics is complex (not simple or easy or something that everyone knows) and that learning it requires time and effort is an essential aspect of MKT. The interview and reflection on it at the end of the term led to this recognition.

I end by contemplating the practicality of my suggestion. Conducting one-on-one interviews with all PTs is a time-consuming endeavor and becomes impractical at a certain number. I have, to date, been able to conduct individual 10–15-min interviews with classes of up to 40 PTs (consuming about 10 h) and have found the effort worthwhile. In Thanheiser (2018), I showed that an interview designed to help PTs assess their own knowledge accurately at the beginning of the course correlates with their developing more sophisticated conceptions by the end of the course. I typically assign the class readings and an activity in lieu of one 2-h class period and schedule interviews with PTs who are available only at class time. Canceling one class period, with the associated preparation and grading, and possibly office hours, the first week could compensate for part of the 10 h consumed by interviews. For numbers, much larger than 40 or instructors teaching many sections each term, another form of assessment may be required. Possibilities include written assessments, explanations the PTs submit on video, or peer interviews. A written assessment, an explanation, or a self-recorded interview may lack the in-depth look at a PT's ability to explain seen in the videotaped interview but could be more feasible as an assessment. The effects of such experiences would need to be explored in further research.

Acknowledgement I would like to thank Krista Strand and Briana Mills who accompanied me through the beginnings of this work and Jodi Fasteen who tirelessly coded this data with me.

References

Adler, J., & Davis, Z. (2006). Opening another black box: Researching mathematics for teaching in mathematics teacher education. *Journal for Research in Mathematics Education, 37*(4), 270–296.

Ball, D. (1988). *Knowledge and reasoning in mathematical pedagogy: Examining what prospective teachers bring to teacher education*. Doctoral dissertation (50 doctoral dissertation), Michigan State University, Ann Arbor.

Ball, D., & Bass, H. (2000). Interweaving content and pedagogy in teaching and learning to teach: Knowing and using mathematics. In J. Boaler (Ed.), *Multiple perspectives on mathematics teaching and learning* (pp. 83–104). Westport, CT: Greenwood.

Ball, D., Charalambous, C., Thames, M., & Lewis, J. (2009). *Teacher knowledge and teaching: Viewing a complex relationship from three perspectives*. Paper presented at the Proceedings of the 33rd Conference of the International Group for the Psychology of Mathematics Education.

Ball, D., Thames, M. H., & Phelps, G. (2008). Content knowledge for teaching: What makes it special? *Journal of Teacher Education, 59*(5), 389–407.

Bransford, J. D., Brown, A. L., & Cocking, R. R. (Eds.). (1999). *How people learn*. Washington, DC: National Academy Press.

Bransford, J. D., & Schwartz, D. L. (1999). Rethinking transfer: A simple proposal with multiple implications. *Review of Research in Education, 24*, 61–100.

Braun, V., & Clarke, V. (2006). Using thematic analysis in psychology. *Qualitative Research in Psychology, 3*(2), 77–101.

Bukszar, E., & Connolly, T. (1988). Hindsight bias and strategic choice: Some problems in learning from experience. *Academy of Management Journal, 31*(3), 628–641.

Chapman, O. (2007). Facilitating preservice teachers' development of mathematics knowledge for teaching arithmetic operations. *Journal of Mathematics Teacher Education, 10*, 341–349.

Chi, M. T., Slotta, J. D., & De Leeuw, N. (1994). From things to processes: A theory of conceptual change for learning science concepts. *Learning and Instruction, 4*(1), 27–43.

Crespo, S., & Nicol, C. (2006). Challenging preservice teachers' mathematical understanding: The case of division by zero. *School Science and Mathematics, 106*(2), 84–97.

Dweck, C. S. (2000). *Self-theories: Their role in motivation, personality, and development*. Philadelphia, PA: Psychology Press.

Dweck, C. S. (2006). *Mindset: The new psychology of success*. New York, NY: Random House.

Dweck, C. S. (2007). Boosting achievement with messages that motivate. *Education Canada, 47*(2), 6–10.

Fischhoff, B. (1975). Hindsight is not equal to foresight: The effect of outcome knowledge on judgment under uncertainty. *Journal of Experimental Psychology: Human Perception and Performance, 1*(3), 288–299.

Fischhoff, B., & Beyth, R. (1975). I knew it would happen: Remembered probabilities of once—Future things. *Organizational Behavior and Human Performance, 13*(1), 1–16.

Graeber, A. O. (1999). Forms of knowing mathematics: What preservice teachers should learn. *Educational Studies in Mathematics, 38*(1–3), 189–208.

Harkness, S. S., & Thomas, J. (2008). Reflections on "multiplication as original sin": The implications of using a case to help preservice teachers understand invented algorithms. *Journal of Mathematical Behavior, 27*(2), 128–137.

Hawkins, S. A., & Hastie, R. (1990). Hindsight: Biased judgments of past events after the outcomes are known. *Psychological Bulletin, 107*(3), 311.

Hill, H., Rowan, B., & Ball, D. (2005). Effects of teachers' mathematical knowledge for teaching on student achievement. *American Educational Research Journal, 42*(2), 371–406.

Kaasila, R., Pehkonen, E., & Hellinen, A. (2010). Finnish pre-service teachers' and upper secondary students' understanding of division and reasoning strategies used. *Educational Studies in Mathematics, 73*(3), 247–261.

Kahneman, D. (2011). *Thinking, fast and slow*. New York, NY: Farrar, Straus and Giroux.

Khoury, H. A., & Zazkis, R. (1994). On fractions and non-standard representations: Pre-service teachers' concepts. *Educational Studies in Mathematics, 27*(2), 191–204.

Loftus, E. F., & Loftus, G. R. (1980). On the permanence of stored information in the human brain. *American Psychologist, 35*(5), 409–420.

Ma, L. (1999). *Knowing and teaching elementary mathematics: Teachers' understanding of fundamental mathematics in China and the United States*. Mahwah, NJ: Erlbaum.

Mason, J. (1998). Enabling teachers to be real teachers: Necessary levels of awareness and structure of attention. *Journal of Mathematics Teacher Education, 1*(3), 243–267. https://doi.org/10.1023/A:1009973717476

Menon, R. (2009). Preservice teachers' subject matter knowledge of mathematics. *The International Journal for Mathematics Teaching and Learning*. Retrieved from www.cimtplymouth.ac.uk/jjournal/menon.pdf.

National Research Council (Ed.). (2001). *Adding it up: Helping children learn mathematics*. Washington, DC: National Academy Press.

Pintrich, P. R. (2002). The role of metacognitive knowledge in learning, teaching, and assessing. *Theory into Practice, 41*(4), 219–225. https://doi.org/10.1207/s15430421tip4104_3

Roese, N. J., & Vohs, K. D. (2012). Hindsight bias. *Perspectives on Psychological Science, 7*(5), 411–426.

Silverman, J., & Thompson, P. (2008). Toward a framework for the development of mathematical knowledge for teaching. *Journal of Mathematics Teacher Education, 11*(6), 499–511. https://doi.org/10.1007/s10857-008-9089-5

Simon, M. (1993). Prospective elementary teachers' knowledge of division. *Journal for Research in Mathematics Education, 24*(3), 233–254.

Simon, M., & Blume, G. W. (1992). *Mathematization as a component of the concept of ratio-as-measure: A study of prospective elementary teachers*. Paper presented at the Annual Meeting of the American Educational Research Association, San Francisco, CA. Retrieved from http://stats.lib.pdx.edu/proxy.php?url=http://search.ebscohost.com/login.aspx?direct=true&db=eric&AN=ED349175&site=ehost-live.

Tanner, K. D. (2012). Promoting student metacognition. *CBE-Life Sciences Education, 11*(2), 113–120.

Thanheiser, E. (2009). Preservice elementary school teachers' conceptions of multidigit whole numbers. *Journal for Research in Mathematics Education, 40*(3), 251–281.

Thanheiser, E. (2010). Investigating further preservice teachers' conceptions of multidigit whole numbers: Refining a framework. *Educational Studies in Mathematics, 75*(3), 241–251. https://doi.org/10.1007/s10649-010-9252-7

Thanheiser, E. (2015). Developing prospective teachers' conceptions with well-designed tasks: Explaining successes and analyzing conceptual difficulties. *Journal of Mathematics Teacher Education, 18*(2), 141–172. https://doi.org/10.1007/s10857-014-9272-9

Thanheiser, E. (2018). Brief report: The effects of preservice elementary school teachers' accurate self-assessments in the context of whole number. *Journal for Research in Mathematics Education, 49*(1), 39–56.

Thanheiser, E., Browning, C., Edson, A. J., Lo, J., Whitacre, I., Olanoff, D., & Morton, C. (2014). Prospective elementary mathematics teacher content knowledge: What do we know, what do we not know, and where do we go? *The Mathematics Enthusiast, 11*(2), 433–448.

Thanheiser, E., Philipp, R., Fasteen, J., Strand, K., & Mills, B. (2013). Preservice-teacher interviews: A tool for motivating mathematics learning. *Mathematics Teacher Educator, 1*(2), 137–147. https://doi.org/10.5951/mathteaceduc.1.2.0137

Thanheiser, E., Whitacre, I., & Roy, G. (2014). Mathematical content knowledge for teaching elementary mathematics: A focus on whole-number concepts and operations. *The Mathematics Enthusiast, 11*(2), Article 4.

Veenman, M. V., Van Hout-Wolters, B. H., & Afflerbach, P. (2006). Metacognition and learning: Conceptual and methodological considerations. *Metacognition and Learning, 1*(1), 3–14. https://doi.org/10.1007/s11409-006-6893-0

Von Glasersfeld, E. (1995). A constructivist approach to teaching. In L. P. Steffe & J. E. Gale (Eds.), *Constructivism in education*. New York, NY: Lawrence Erlbaum.

Vosniadou, S., & Verschaffel, L. (2004). Extending the conceptual change approach to mathematics learning and teaching. *Learning and Instruction, 14*(5), 445–451.

Wang, M. C., Haertel, G. D., & Walberg, H. J. (1990). What influences learning? A content analysis of review literature. *The Journal of Educational Research, 84*(1), 30–43. https://doi.org/10.1080/00220671.1990.10885988

Zazkis, R., & Campbell, S. (1996). Divisibility and multiplicative structure of natural numbers: Preservice teachers' understanding. *Journal for Research in Mathematics Education, 27*, 540–563.

Zazkis, R., & Khoury, H. A. (1993). Place value and rational number representations: Problem solving in the unfamiliar domain of non-decimals. *Focus on Learning Problems in Mathematics, 15*(1), 38–51.

Part III
Professional Development in Mathematics for In-Service Elementary School Teachers

Chapter 9
Mathematics in the Yale Teachers Institute

Roger E. Howe

The Yale Teachers Institute offers, and in fact is defined by, a distinctive approach to professional development for in-service teachers. This approach involves several features. It is organized around a core of seminars on a variety of topics, from the humanities through social sciences to the physical sciences and mathematics, each led by a university faculty member (referred to as a Seminar Leader) expert in the topic. The seminars are heavily content based, and do not attempt to deal with pedagogy. Instead, reliance is placed on the participating experienced teachers (referred to as Fellows) to translate the content into effective lessons. This is the main job of the Fellows: rather than show mastery of the content by undergoing some sort of formal examination, they incorporate it into a curriculum unit of their own design. This arrangement completely obviates the question of whether the PD content gets used in the classroom: it will. Teachers often choose seminars to take in order to solve a classroom need they have sensed: a topic, or set of topics about which they feel a need for better understanding in order to teach more effectively. By approaching the topic from a new point of view, and learning more deeply about it, they are able to create curriculum units in which they take pride of ownership, and substantial satisfaction. For several of the Fellows in my seminars on mathematics, the value of their units has been affirmed by their school districts, which have asked them explain their units locally to their colleagues, thus spreading the new approach from a single classroom to multiple classrooms in multiple schools. The units are also published online (YNIpub, n.d.), and over the years, thousands have been downloaded.

The Yale Teachers Institute began in the late 1970s as an outreach program from Yale to New Haven, and was originally the Yale-New Haven Teachers Institute (YNHTI). Around the turn of the century, it was decided to spread the model of

R. E. Howe (✉)
Department of Teaching, Learning & Culture, Texas A&M University, College Station, TX, USA
e-mail: roger.howe@yale.edu; rogerhowe@tamu.edu

Y. Li et al. (eds.), *Developing Mathematical Proficiency for Elementary Instruction*, Advances in STEM Education, https://doi.org/10.1007/978-3-030-68956-8_9

YNHTI more broadly. Thus was born the Yale National Initiative (YNI), which offers seminars during the summer for teachers from other districts, and guidance for local educational administrators and university faculty who want to support formation of a Teachers Institute in another location. It is targeted at high-need school districts. Over the past two decades, districts from many places across the US have participated, including Charlotte (NC), Chicago (IL), Diné Nation (NM and AZ), Houston (TX), Newcastle County (DE), Philadelphia (PA), Pittsburgh (PA), Richmond (VA), San Jose and San Francisco Bay (CA), and Tulsa (OK).

Several of these places ((the whole state of) Delaware, the Diné Nation, Philadelphia, Tulsa) currently have operating Teachers Institutes. Several other places, including Houston and Pittsburgh, have had successful institutes for a number of years, but it has been found that, when the leader of a Teachers Institute retires, continuation can be a delicate matter. So also is getting one started. This is probably partly due to the somewhat complex nature of the YNI model of a Teachers Institute. It requires cooperation between a university and a school district, including strong buy-in from the university faculty and administration, ongoing teacher leadership, support from district administration, and not least, a reliable source of funding. A Teachers Institute is not intended to be a volunteer activity on the part of the participants. Each Seminar leader is paid a month's salary for leading a seminar, and each seminar Fellow receives a stipend, in the range of $1000–2000, for completing their curriculum unit. Besides the direct seminar costs, there is the salary of the Institute leader, and a certain amount of other administrative cost. Coordinating all the different constituents, and keeping the multiple activities running smoothly requires judgment and flexibility. The typical Fellow in the National Seminars is a highly motivated and highly dedicated teacher. In this author's view, if the only positive benefit of a Teachers Institute were to keep such teachers in the profession, it more than justifies its cost. But there are other benefits. Units developed by Fellows increase their individual teaching effectiveness; and since Fellows' units are often propagated locally, the influence of a Teachers Institute spreads well beyond the teachers who formally participate.

The YNI model emphasizes teacher leadership. The main decisions are made by a committee of district teachers. The committee solicits proposals for seminars from interested faculty of the participating university. This committee then canvases their colleagues to see which of the proposals draw the most interest from teachers. These seminars are then selected to run. District teachers are invited to apply to the seminar that appeals to them most strongly. They may also be allowed a second choice. Fellows are then selected from the applicants by the same committee of teachers, with input from the prospective Seminar Leaders.

A Teachers Institute seminar consists of twelve two hour meetings. In the local Institutes, the meetings are scheduled once a week, beginning in the spring and extending into the summer. The National Seminars, for purposes of introducing the idea to other districts, have two introductory meetings over a weekend in April/May, ten meetings in a 2-week period in July, aptly named the "Intensive Session," with a final reunion meeting in October. Seminars are limited to at most 12 Fellows. Although this might seem overly restrictive, the small size is important for fostering collegiality, which is an important value of the program. Also, it makes feasible for the seminar leader to work closely with each Fellow, and to carefully edit the multiple drafts of each curriculum unit, to ensure the quality of the final products. Writing a curriculum unit is a considerable challenge. The first curriculum unit that a Fellow writes may well be the most ambitious writing project s/he has ever attempted. Yet many Fellows find inspiration in that hard work, and come back for more.

For Seminar Leaders, running a Teachers Institute seminar is likely to be a considerably different experience from their normal university teaching duties. Topics must be chosen carefully, and especially, must be educationally robust, in the sense that they should be relevant to a wide range of grade levels in the K–12 curriculum since the Fellows in a given seminar collectively may represent elementary, middle, and high school. In this context, humanities faculty may be in a somewhat better situation than social science, laboratory science, or mathematics faculty since they sometimes get to teach small seminars on broad topics, e.g., poetry of a certain era, and their methods of analysis may enable more refined treatment of standard material across a range of grades. Nevertheless, a wide range of subjects is desirable for the seminars. The YNI seminars for 2019 were as follows:

Reading for Writing: Modeling the Modern Essay
The Problem of Mass Incarceration
American Democracy and the Promise of Justice
Energy Sciences
Perimeter, Area, Volume, and All That: A Study of Measurement

For me, a university mathematician with a longstanding interest in K–12 mathematics education, running YNI seminars has given me a chance to get to know practicing classroom teachers, to see how they think, and how they react to various topics and approaches. This has been very helpful for guiding my thinking about mathematics education, and about teacher preparation in particular. The seminars I have given in recent years are as follows:

2019: Perimeter, Area, Volume, and All That: A Study of Measurement
2018: Big Numbers, Small Numbers
2017: From Arithmetic to Algebra: Variables, Word Problems, and the Rules
2016: The Number Line in the Common Core
2015: Problem-Solving and the Common Core
2014: Place Value, Fractions, and Algebra: Improving Content Learning through the Practice Standards

In these seminars, I have tried to emphasize the principles underlying key ideas in the curriculum. Below are some examples of things I have discussed with and/or learned from my teachers. Because of their central nature, I have written about some of these topics elsewhere (Epp & Howe, 2008; Howe, 2012, 2018, 2019; Howe & Reiter, 2012). I will try to minimize repetition here.

every fourth century. (This is why we had a leap day in 2000.) This produces an approximation of the year as 3.652425×10^2 days, two extra decimal places of accuracy. We will be lucky if we last long enough to need another adjustment.

9.2 The Number Line

Another topic I have emphasized in my seminars is the number line. The number line seems to get used in the early grades, but then fades out until algebra, where it is submerged into Cartesian coordinates. In my 2016 seminar, the number line was the focal topic. I emphasized its metric aspect, a feature that, in my observation, may not be appreciated by elementary students. The relevant quantity is length or distance. A number labeling a point on the number line tells you how far the point is from the origin, as a multiple of the unit length.

There is a natural operation of addition of lengths: put two bars together to make a bar whose length is the sum of the two constituents. This is actually a pre-numerical operation—it does not depend on knowing the lengths of the two constituent bars, so it makes sense across all types of numbers—whole numbers, fractions, and decimals, even irrational numbers.

In the seminar on the number line, two of my Fellows from Chicago had observed that their students tended to think of these several types of numbers as distinct from each other, and dealt with them according to separate rules, each adapted to the differing symbolic representations of the various types. These two Fellows had the idea of using the number line to give their sixth/seventh grade students a unified picture of all these numbers, by showing them how they all fit on the number line, and that addition and subtraction can be viewed in a unified fashion, based on the geometric interpretation of the operations in terms of length measurement.

A very interesting development, from the point of view of YNI practice, was that they wrote coordinated curriculum units. One of them (Rossiter, 2017) focused on placing the various types of numbers on the number line, and the other (Bingea, 2017) built on that, and discussed how to interpret addition and subtraction using the number line. In their classrooms the following year, both Fellows taught both units.

An added benefit of the number line approach was that the sign of a number was interpreted as a direction—to the right of zero, or to the left—and this turned the segment corresponding to a number into a vector, that is, a directed segment. Then the addition of a negative number was naturally the same as what had previously been done to subtract the corresponding positive number, so addition and subtraction were merged naturally into a single operation in the geometric setup.

This approach proved so successful that it was adopted by the local group of schools associated with the school where these Fellows taught.

9.3 Word Problems

One of the recurring themes in my YNI seminar has been word problems. Word problems are a particularly weak point in an overall weak (PISA, 2018) system of math education in the USA, but they play a crucial role in mathematics education. Word problems are what connects standard mathematical topics with the real world. They show how mathematics actually can get used. At the primary level, one-step word problems help students understand the contexts where the basic arithmetic operations are relevant. In this way, they help students form a concept of what the operations are and how they are used.

My experience with YNI has led to what seems to be a productive approach to enabling teachers to teach word problems. The Common Core State Standards for Mathematics (CCSSM) has two appendices with taxonomies of one-step addition/subtraction problems, and one-step multiplication/division problems. These taxonomies are based in part on the work of the CGI project at the University of Wisconsin (Carpenter, Fennema, Franke, Levi, & Empson, 2014; Carpenter, Franke, & Levi, 2003).

The taxonomy for addition/subtraction problems has three main categories of problem: problems that involve change of a quantity over time; problems that involve comparison of two quantities of the same type; and part-part whole problems. Change problems can be further articulated into six subtypes, as can comparison problems, and part-part-whole problems can be subdivided into two types. Thus, a list of 14 different problem types emerges. If you suppose that a two-step problem can have any of the 14 types for its first step, and any of the 14 for its second step, this would make $14 \times 14 = 196$ kinds of two-step addition/subtraction problems. Under the same supposition, there would be 2744 kinds of three-step problems. Even if some triples of types cannot be combined to make a three-step problem, it is clear that the literature of multistep word problems is formidable, which partly explains the difficulties in dealing with it, and shows the need for an explicit strategy.

There are also nine types of one-step multiplication/division problems in the CCSSM taxonomy. Thus, the potential for three-step mixed operation problems would be $(14 + 9)^3 = 23^3 = 12{,}167$ possibilities! No one will have encountered them all.

It is important for primary grades teachers to be familiar with the taxonomies, and to make sure that their students see a representative sample of word problems, in order that they form a robust conception of what the operations are used for, as described above. Fortunately, learning the taxonomies, and analyzing and creating one-step problems seems to be quite feasible.

Familiarity with the taxonomies may also provide a basis for analysis of more complicated problems. At a higher level, word problems can be a vehicle for teaching children to reason. A multistep word problem is a short passage of text that requires careful analysis to determine what needs to be done to answer the question posed. Consider the following two problems.

Rhonda and Beth both love toy trucks. They have collections of red and green toy trucks. Rhonda has 6 in all, and 4 of them are red. Beth has 2 more toy trucks than Rhonda, including 3 green ones. How many red toy trucks does Beth have?

Rhonda and Beth both love toy trucks. They have collections of red and green toy trucks. Rhonda has 6 in all, and 4 of them are red. Beth has 2 more toy trucks than Rhonda, including 3 green ones. How many green toy trucks does Rhonda have?

Both of these questions give the same information, and the questions they ask may seem similar.

However, to answer the first one, you have first to calculate the number of toy trucks that Beth has. (This is given as 2 more than what Rhonda has, which is given as 6, so it is $6 + 2 = 8$.) Then to find the number of red trucks Beth has, you subtract the number of green trucks (3) from the total of 8, to get 5. This is a two-step problem with the first being a comparison, with the larger quantity unknown, and the second being a part-part-whole, with a part being unknown.

The second question, however, is a one-step problem. Since you are asked about Rhonda's green trucks, you can use her total number of trucks (6), and subtract the number of red trucks (4) to find the number of green trucks Rhonda has (2). The information about Beth's trucks is irrelevant to answering the question.

These examples show that there are problems that are similar in many ways, but require quite different calculations to produce the answers. Presenting such problems requires students to read carefully and to analyze the relationships in a given problem. In short, it requires them to think. A teacher who is adept at problem analysis and construction, who can quickly analyze and compose problems that students can relate to, could lead class discussions about such problems and enhance the analytical skills of the class. In short, in the classes of skilled teachers, arithmetic word problems could become a vehicle for improving students' ability to engage with and think about complex passages of writing, enhancing their potential as learners.

This is the kind of result one might hope for from enabling teachers to deal flexibly and incisively with word problems. I cannot claim to have taken this work quite this far in the seminars I have led up to now. However, this is the vision that has slowly come into focus through the several seminars I have given about solving word problems, and it will be my goal in the next such seminar.

References

Bingea, A. (2017). Retrieved from http://teachers.yale.edu/curriculum/viewer/initiative_16.05.01_u.

Carpenter, T. P., Fennema, E., Franke, M. L., Levi, L., & Empson, S. B. (2014). *Children's mathematics* (2nd ed.). Portsmouth, NH: Heinemann.

Carpenter, T. P., Franke, M. L., & Levi, L. (2003). *Thinking mathematically*. Portsmouth, NH: Heinemann.

Epp, S., & Howe, R. (2008). *Taking place value seriously*. Retrieved January 27, 2018, from http://www.maa.org/sites/default/files/pdf/pmet/resources/PVHoweEpp-Nov2008.pdf.
Howe, R. (2012). From arithmetic to algebra. *Mathematics Bulletin: A Journal for Educators, 49*(Special issue), 13–21.
Howe, R. (2018). Cultural knowledge for teaching mathematics. In Y. Li, W. J. Lewis, & J. J. Madden (Eds.), *Mathematics matters in education* (Advances in STEM Education) (Vol. 1). Cham, Switzerland: Springer Nature.
Howe, R. (2019). Learning and using our base ten place value number system: Theoretical perspectives and twenty-first century uses. *ZDM, 51*(1), 57–68.
Howe, R., & Reiter, H. (2012). Retrieved May 24, 2018, from http://www.teachersofindia.org/en/article/five-stages-place-value.
PISA. (2018). Retrieved December 26, 2019, from https://www.oecd.org/pisa/.
Rossiter, J. (2017). Retrieved from http://teachers.yale.edu/curriculum/viewer/initiative_16.05.07_u.
Thanheiser, E. (2009). Preservice elementary school teachers' conceptions of multidigit whole numbers. *Journal for Research in Mathematics Education, 40*, 251–281.
Thanheiser, E. (2015). Leveraging historical number systems to build an understanding of the base 10 place value system. In X. Sun, B. Kaur, & J. Novotná (Eds.), *Proceedings ICMI Study 23 'Primary Mathematics Study on Whole Numbers'* (pp. 149–156). Macao: University of Macau. Retrieved January 27, 2018, from http://www.umac.mo/fed/ICMI23/proceedings.html.
YNIpub. (n.d.). Retrieved from http://teachers.yale.edu/curriculum/index.php?skin=h.

Chapter 10
Sustainable, Scalable Professional Development in Early Mathematics: Strategies, Evaluation, and Tools

Douglas H. Clements and Julie Sarama

10.1 Background

Young children can learn substantial mathematics (National Research Council, 2001; Sarama & Clements, 2009), but many children in the USA have limited opportunities to do so. Unfortunately, many early childhood teachers are not well-prepared to engage children in rich experiences in domains other than literacy (Institute of Medicine (IOM) and National Research Council (NRC), 2015; National Research Council, 2009). The meager training most early childhood teachers receive in mathematics especially increases the importance of designing effective programs for preservice education and in-service professional development (PD), especially for scale-up efforts. Fortunately, research provides some guidance regarding the essential features of high-quality PD (Björklund & Barendregt, 2016; Brendefur, Strother, Thiede, Lane, & Surges-Prokop, 2013; Carpenter, Fennema, Franke, Levi, & Empson, 2014; MacDonald, Davies, Dockett, & Perry, 2012; Perry, 2010;

This research was supported by the Institute of Education Sciences, U.S. Department of Education, through grants R305K05157 and R305A110188 and also by the National Science Foundation, through grants ESI-9730804 and REC-0228440. The opinions expressed are those of the authors and do not represent views of the IES or NSF. Although the research is concerned with the scale-up model, not particular curricula, a minor component of the intervention used in this research has been published by the authors, who thus could have a vested interest in the results. An external auditor oversaw the research design, data collection, and analysis and other researchers independently confirmed findings and procedures. The authors wish to express appreciation to the school districts, teachers, and students who participated in this research.

D. H. Clements (✉) · J. Sarama
Marsico Institute for Early Learning and Literacy, Morgridge College of Education, University of Denver, Denver, CO, USA
e-mail: Douglas.Clements@du.edu; Julie.Sarama@du.edu

Y. Li et al. (eds.), *Developing Mathematical Proficiency for Elementary Instruction*, Advances in STEM Education, https://doi.org/10.1007/978-3-030-68956-8_10

Piasta, Logan, Pelatti, Capps, & Petrill, 2015; Sarama & Clements, 2018; Sarama, Clements, Wolfe, & Spitler, 2016). In developing our approach, we followed this guidance, placing a unique focus on the use of *learning trajectories*.

10.1.1 Learning Trajectories-Based Professional Development (LTPD)

Many researchers concerned with preservice teachers and in-service professional development emphasize a need for teachers to understand children's mathematical thinking (Baroody & Ginsburg, 1990; Franke, Carpenter, Levi, & Fennema, 2001), and some have documented the connection between teachers' understanding of children's mathematical thinking and achievement (e.g., Carpenter, Franke, Jacobs, Fennema, & Empson, 1998). A growing body of evidence shows that children's thinking does not always follow historical or logical developments of mathematics as a discipline, nor adult ways of reasoning (Björklund & Barendregt, 2016; Sztajn, Confrey, Wilson, & Edgington, 2012), but does follow predictable sequences of levels, or patterns, of thinking (Battista, 2013; Bobis et al., 2005; B. A. Clarke, 2008; Confrey, Maloney, Nguyen, & Rupp, 2014; Sarama & Clements, 2009).

The construct of learning trajectories (LTs) (Simon, 1995) further stimulated focus on the natural sequences of children's mathematical growth (Clements & Sarama, 2014a; Confrey et al., 2014; Sarama & Clements, 2009) and the implications for designing curricula and pedagogical approaches. Importantly, learning trajectories are more complex than sequencing of topics or tasks; we define them as "descriptions of children's thinking and learning in a specific mathematical domain, and a related, conjectured route through a set of instructional tasks designed to engender those mental processes or actions hypothesized to move children through a developmental progression of levels of thinking, created with the intent of supporting children's achievement of specific goals in that mathematical domain" (Clements & Sarama, 2004, p. 83).

In other words, each learning trajectory has three parts: (a) a goal, (b) a developmental progression, and (c) instructional activities. To attain a certain mathematical competence in a given topic or domain (the goal), students learn each successive level (the developmental progression), aided by tasks (instructional activities) designed to build the mental actions-on-objects that enable thinking at each higher level. The inclusion of these three components separates this definition of learning trajectories from "progressions" or other research that only documents natural sequences of children's thinking. For us the mathematical goal is essential: Beyond an "educational objective," our goal includes conceptual understanding of the mathematics we wish children to learn (and wider understanding of its place in mathematics for educators). The third component is also important: Instruction goes beyond general teaching approaches—each level of the developmental progression is achieved through instructional tasks and strategies specific to that level. All aspects of the present study, especially the professional development, were based on this tripartite definition of LTs: goal, development progression, and instruction.

Support for basing PD on LTs comes from studies that show learning trajectories-based instruction served as a tool for teacher education by providing a framework for organizing theory on teaching and learning (Sztajn et al., 2012). For example, an LT on equi-partitioning was shown to support teachers' understanding of mathematics, as well as their students' reasoning (Wilson, Mojica, & Confrey, 2013). This approach to PD was supported by the AMTE Standards for Mathematics Teacher Preparation (Association of Mathematics Teacher Educators, 2017) that emphasized the central role learning trajectories should play, especially for early childhood.

10.1.2 A Scale-Up Study Using LTPD

We conducted a large cluster randomized trial designed to evaluate the effectiveness of a research-based model for scaling up educational intervention that places learning trajectories at the core of all aspects—including the PD. Effects on children were positive in the short and long terms (e.g., Clements, Sarama, Spitler, Lange, & Wolfe, 2011; Clements, Sarama, Wolfe, & Spitler, 2013). Here we focus on how the intervention affected teachers.

10.2 Methods

10.2.1 The TRIAD Model

As the name implies, the Technology-enhanced, Research-based, Instruction, Assessment, and professional Development, or TRIAD, uses technology and research to implement the triad of critical components for scale up: (1) instruction, including curriculum, (2) formative assessment, and (3) extensive professional development. The model emphasizes teaching early mathematics for understanding via learning trajectories and technology and includes collaboration of key groups to establish and maintain the following:

- A pre-K mathematics curriculum, with all components of the curriculum—a teacher's manual, demonstration videotapes, manipulatives, software, teaching strategies, assessments, and professional development—based on a common core of understanding the learning trajectories through which children develop mathematically.
- Professional development via teacher training.
- Professional development via coaching during the school year both also built upon learning trajectories.

These are comprehensive components necessary for improving teaching (Association of Mathematics Teacher Educators, 2017; Hiebert & Stigler, 2017). The TRIAD model includes ten research-based guidelines (for a complete description with research citations, see Sarama & Clements, 2013; Sarama, Clements, Starkey, Klein, & Wakeley, 2008). We present here only the TRIAD guideline for PD:

Provide professional development that is ongoing, intentional, reflective, goal oriented, focused on content knowledge and children's thinking, grounded in particular curriculum materials, and situated in the classroom and the school. A focus on content includes accurate and adequate subject-matter knowledge both for teachers and for children. A focus on children's thinking emphasizes the learning trajectories' developmental progressions and their pedagogical application in formative assessment. Grounding in particular curriculum materials should include all three aspects of learning trajectories, especially their connections. This also provides a common language for teachers in working with each other and other groups. Situated in the classroom does not imply that all training occurs within classrooms. However, off-site intensive training remains focused on and connected to classroom practice and is completed by classroom-based enactment with coaching. In addition, this professional development should encourage sharing, risk-taking, and learning from and with peers. The aim is preparing professional to teach a specific curriculum and develop their knowledge and positive beliefs that the curriculum is appropriate and its goals are valued and attainable. To do so, the LTPD work is grounded in the classroom, formatively evaluating teachers' fidelity of implementation and providing feedback and support from coaches in real time.

10.2.2 The TRIAD Evaluation

The TRIAD evaluation tested for generalizability of the model over two diverse settings in two states in the USA. We employed hierarchical linear modeling (HLM) to measure the effects of the intervention on children's mathematics performance (Clements et al., 2011, 2013; Sarama, Clements, Wolfe, & Spitler, 2012). These studies revealed that the implementation increased the mathematical competencies (effect size = 0.72 standard deviations) of the 1305 children, as well as showing transfer to other domains, such as oral language (Sarama, Lange, Clements, & Wolfe, 2012). Here we review studies of the impact of the intervention on teachers.

10.2.3 Teacher Participants

Original adult participants were 106 preschool teachers from 42 schools. Schools were randomly assigned, by site, to one of three treatment groups: two were experimental (72 teachers in 26 schools), with no differences relevant to this study (differences in the experimental treatments began after the preschool year), and one was the control (34 teachers in 16 schools). Schools served low-resource, ethnically diverse children. Every teacher was required by their administration to participate. Teachers in the TRIAD group were required to attend all PD sessions (control teachers received PD after the end of the study).

10.2.4 Professional Development

Teachers actively participated in LTPD for 8 days during Year 1 and 5 days during Year 2 (control teachers for approximately 8 days during Year 3). The LTPD sessions, held during the school day approximately monthly, were paced and organized to follow the sequence of a 30-week early mathematics curriculum.

As they were for all aspects of TRIAD, learning trajectories were the core of the professional development program. During the first half of each session, teachers actively learned about the three components of the learning trajectories (mathematical goals, children's developmental progressions, and instructional tasks). Presentations on mathematical concepts and ideas, followed by extensive work on mathematical explorations and problems (at the pre-K level and above), facilitated teachers' understanding of mathematical goals and the mathematical concepts, practices, and skills for each main topic (Polly et al., 2015; Stosich, 2016). For example, work in geometry including the pre-K activities, of course, but far beyond these, such as placing a diverse set of shapes in Venn diagrams, challenging each other with "guess my rule" games, and exploring tilings of pattern blocks and tangrams, with extension discussions about mathematical justifications for ideas and hypotheses these activities engendered. As an example in measurement, teachers were given an 8.5 by 11 cm picture and asked to figure out what number would be used to scale it (e.g., on a copy machine) to fit an 8.5 by 11-in. sheet of paper.

Teachers learned about the second element of the trajectories, the development progressions, using the Learning and Teaching with Learning Trajectories (LTLT) website (see LearningTrajectories.org, a recent update of this application). LTLT was a critical tool, both for our own PD and for the scalability of the TRIAD model. Here teachers viewed videos of children at specific levels of the developmental progression and read descriptions and commentaries. The multiple videos illustrated particular characteristics of thinking of children at each level, serving as landmarks (aka "growth points," D. M. Clarke et al., 2001) for teachers to identify levels of their own children's mathematical thinking, as well as children's growth through the developmental progression. As a brief example, children learn to subitize, or quickly quantify sets without counting ("subitize" comes from the Latin word for "suddenly"), first very small numbers (1 to 2 or 3) and then 4 and 5, two levels of *perceptual subitizing*. They then progress to levels of *conceptual subitizing* when they recognize, for example, a dice pattern for 6 by quickly combining 3 dots with 3 dots.

Each level of the developmental progression is linked to specific instructional activities, which are similarly illustrated by videos and pictures. This provided opportunities to see connections between the learning trajectory's level of thinking and the specific instructional activities that were designed to promote children's learning of those mathematical competencies. For example, discussions would be prompted by the question, "*Why* would this activity help children achieve the next level of thinking?" The instructional tasks, teaching strategies, and resources that facilitate movement along the progression from informal ideas to more sophisticated understandings were thus demonstrated, explored, and discussed, setting the stage for the afternoon sessions.

During the second half of the LTPD day, teachers practiced. Teachers gathered for whole-group activities in a simulated setting, enacting the whole group lessons by taking turns leading the lesson, and playing the part of preschoolers. They similarly practiced small-group activities by gathering in groups of five to simulate the small-group sessions they would lead in their classrooms. Early in the year, project staff demonstrated instruction. Gradually, teachers took turns leading the lessons without demonstrations, but only by reading the curriculum. A summary of the learning trajectories was available for teachers for reference.

Formative assessment of children's place on the continuum of the learning progressions was an expected part of the small-group implementation. Teachers became familiar with the record sheets they would use to formatively assess every student during every small-group lesson they taught in their classrooms. This became routine for the teachers over time. They became skilled at using the names of the LT levels (e.g., for counting beyond verbal counting, "corresponder," "counter (small numbers)," and "producer"). These terms facilitated quick and accurate formative assessments. Teachers also discussed how to observe, listen to, and interpret children's thinking. They learned that formative assessment is a powerful tool for decision-making, for knowing what to teach to whom and when. In particular, developmental levels provided teachers with a useful range of levels to utilize when working with children who were gifted and/or had special needs.

At the computer, teachers also independently practiced the curricular computer activities they would introduce to the children in the upcoming weeks. Additionally, teachers could interact with the data management system that tracked the progress of each of their children on every software activity. Tracking children's progress allowed teachers to tailor assigned software activities to the level of both most children in the class as well as individuals (i.e., if any children needed either earlier or more challenging activities) at the beginning of each week.

During Year 2, teachers focused less on the curricular activities, and more on the mathematical learning trajectories for preschoolers, presenting case studies to their peers. Teachers discussed the cases, assessed where the child was on a developmental progression, and discussed what activity, or modification of an activity, might be appropriate for the particular child. They also shared their ideas for productive adaptations of the activities.

LTPD also took place in the teachers' own classrooms via the LTLT website and visits from project coaches and peer coaches. Project coaches were TRIAD staff who received 2 days of training on TRIAD coaching. They participated in all LTPD sessions and provided individualized support during these sessions.

Project coaches interacted with their assigned teachers (four to five teachers per coach) approximately twice per month. At times the teacher would observe as the project coach demonstrated small or whole group, or software activities, but usually the project coach would observe and support while the teacher implemented the curricular activities. Project coaches completed fidelity ratings 3–4 times per year, provided teachers with feedback, and offered teachers a fidelity rating sheet teachers could voluntarily use to self-evaluate.

School-based peer coaches were teacher participants who volunteered to lead all participating teachers at their school. They served as a liaison to the project staff and provided in-school/in-classroom support for the teachers as they implemented the intervention. Technology mentors (project staff) rotated into classrooms to troubleshoot software, hardware, or network problems, demonstrate software, and answer teachers' technology questions. Website activities allowed teachers to test themselves by observing a video of a child and naming the child's location on a progression, using the descriptors for each level.

10.2.5 Data Sources

To determine the effect of the LTPD on teachers, we analyzed data from two classroom observation instruments, a classroom administered instrument used to measure the mathematical environment and teacher/student interactions and a classroom administered fidelity instrument. We also administered a written teacher questionnaire, an extended version of one teachers had completed at the beginning of the project.

Classroom Observation "Classroom Observation of Early Mathematics—Environment and Teaching" (COEMET) was created based on a body of research on the characteristics and teaching strategies of effective teachers of preschool mathematics (e.g., D. M. Clarke & Clarke, 2004; Clements, Sarama, & DiBiase, 2004; Fraivillig, Murphy, & Fuson, 1999). The COEMET measures the quality of the mathematics environment and activities with an observation of three or more hours and is not connected to any specific curriculum. Thus, it allows for intervention-control group contrasts, no matter what the source of the enacted curriculum. There are 31 items, all but 4 of which are 5-point Likert scales. An example of one of the three items in the section "Personal Attributes of the Teacher," with potential codes varying from "Strongly Disagree" to "Strongly Agree" is, "the teacher appeared to be knowledgeable and confident about mathematics (e.g., demonstrated accurate knowledge of mathematical ideas and procedures, demonstrated knowledge of connections between, or sequences of, mathematical ideas)."

Assessors spend no less than a half-day in the classroom, for example, from before the students arrive until the end of the half-day (e.g., until lunch). All mathematics activities are observed and evaluated, without reference to any printed curriculum. The COEMET has three main sections: classroom elements, classroom culture, and specific mathematics activities (SMA). Assessors complete the first two sections once to reflect their entire observation. They complete an SMA form for each observed mathematics activity, defined as one conducted intentionally by the teacher, involving several interactions with one or more students, or set up to develop mathematics knowledge (this would not include, for instance, a single, informal comment).

Fidelity Observation The main data source for studying sustainability of the TRIAD intervention was a classroom observation measure, rather than teacher self-reports, dubious instruments that unfortunately make up the main body of evidence of sustainability (Timperley, Wilson, Barrar, & Fung, 2007). Based on the same research base as the COEMET, the fidelity instrument was designed to evaluate whether or not, and to what degree, teachers were faithfully implementing the specific components of the implemented curriculum—thus this measure could only be reasonably applied to intervention classrooms (Clements et al., 2011).

An example of the Likert scale questions within each subscale is as follows: "The teacher facilitated students' responding: e.g., elicited many solution methods for one problem, encouraged elaboration of students' responses, waited for and listened attentively to individual students, responded to errors as learning opportunities." Project coaches collected data using this instrument, after participating in several training meetings and conducting inter-rater reliability visits over 2 years.

Questionnaire To further understand teacher change as a result of participation in learning trajectories-based professional development (LTPD) and curriculum implementation, we invited all 106 original teachers to fill out a paper and pen Teacher Questionnaire 7 years after they joined the project.

10.3 Results

10.3.1 Classroom Observations: The COEMET

The main objective of the PD was to increase the quality and quantity of teachers' mathematics instruction. This was achieved: TRIAD teachers had higher scores than the control teachers on the Classroom Culture subscale (Effect Size = 1.23 standard deviations), SMA subscale (ES = 0.78), total number of math activities observed in SMAs (ES = 1.02), and the number of computers on and working for students to use (ES = 0.90) (Clements et al., 2011).

Another set of analyses sought to ascertain whether these increases were responsible for children's achievement gains. That is, the mediational hypothesis was that the COEMET components are influenced by the TRIAD intervention and they *in turn* cause changes in the child outcomes.

Three subscores derived from the COEMET mediated the effects of TRIAD (Clements et al., 2011). The classroom culture component, the total number of computers on and working for students, and the total number of math activities significantly, partially mediated the impact of treatment group on math and language student outcomes (Clements et al., 2011). We also examined individual items (Sarama et al., 2016). The largest differences were in two types of items. First, TRIAD teachers used computers in mathematics far more. Second, TRIAD teachers interacted with students more. They more frequently were observed to "support listener understanding" and to "observe, listen, and take notes."

10.3.2 Sustainability of Fidelity

Longitudinal evaluation is particularly important for large-scale interventions because a full concept of scale requires not only consequential implementation and diffusion but also endurance over long periods of time and a transfer of responsibility from any external organization to the internal resources of a school district (Coburn, 2003; Dearing & Meyer, 2006). Perhaps most important is *sustainability*, teachers' continued use of program components in the achievement of desirable student outcomes and professional practice over time, with a focus on the maintenance of core beliefs and values, and the use of these core beliefs to guide adaptations (Century, Rudnick, & Freeman, 2012; Scheirer & Dearing, 2011; Timperley, 2011).

We found that teachers exposed to the TRIAD intervention continued to demonstrate high levels of fidelity to the underlying curriculum 2 years past the end of the external intervention, without continued project support (Clements, Sarama, Wolfe, & Spitler, 2015). Teachers appeared to first learn to implement the instructional components (whole group, small group, centers, computer center) and made continual improvements in such implementation, but they made even more substantial growth in the General Curriculum subscale only after they had some skill and confidence in the separate instructional components. That is, they improved on specific implementation of the instructional activities first, and only later on such general items as use of everyday activities to teach mathematics and the *extension* of the curriculum's activities to enhance the quality of children's learning (e.g., making up math problems using felt board characters from a story book or having all staff join in dramatizing a math problem). This finding, triangulated by both fidelity implementation over several years and teachers' responses on the questionnaire, suggested that only when skilled in the several components did teachers appear to be able to synthesize them into a coherent, positive classroom culture for early mathematics learning. Further, the main factor that supported such sustainability was perceptions of students' learning. Qualitative and quantitative data supported the position that observations of such learning along LTs motivated teachers to productively face challenges and improve their teaching (Clements et al., 2015).

We then returned 6 years after the external intervention, three times the duration of this original sustainability analyses (Sarama et al., 2016). At this point, data were available only for 28 teachers (who were not significantly different from the others in demographics, years of experience, or any other variable we checked). Across the items, the pattern of sustained fidelity to the curriculum was sustained with only one exception. Computer usage declined, probably because the initial licenses for the computer program lapsed. Results on other items support the original sustainability findings. For example, teachers scored almost identically as in earlier timepoints on featuring mathematics materials in their classrooms, using everyday math activities, extending and enhancing the written activities, and teaching those activities with fidelity. They encouraged students to actively think, reason, solve problems, or reflect and involved and supported students in discussions of mathematics about the same or even more than at earlier timepoints. They maintained or increased their ability to use formative assessment (Sarama et al., 2016).

10.3.3 Final Questionnaire

The preschool teachers in this study were experienced and well-educated; yet, when asked to describe self-changes after participating in Learning Trajectories-based Professional Development (LTPD), their most salient and dominant response was surprise at their children's mathematical capabilities. Two-thirds of the teachers focused on their newfound understanding of preschoolers' mathematical capabilities; teachers' increased expectations for preschoolers' mathematical achievement; their fresh recognition of preschoolers' ability to learn complex mathematical concepts; and their new awareness of the variety of concepts/topics preschoolers could understand. Many of these teachers explicitly described their changed understanding of what their children could do mathematically, including the ability to reason, explicate, justify, and problem solve; lead the way; and understand complex ideas, make connections, and understand relationship. When asked how they changed, teachers kept the focus on the children, explaining the change in their perceptions of children, and thus maintaining the child-centered focus of the LTPD approach. A related theme addressed changes in how they taught various groups (gender, ethnic/race, ability), always in the direction of equity and appreciation of all children's ability to learn mathematics.

In addition to the theme of children's mathematical capabilities, teachers, when asked about how they changed, described the effects of learning about (a) learning trajectories, (b) how to teach mathematics to young children, (c) how important small-group instruction to implementing LT instruction, (d) how to use formative assessment in early mathematics, (e) how to augment small-group instruction, and (f) how early mathematics could constitute activities suitable for young children, incorporating mathematical activities through play.

Finally, teachers also explicitly mentioned the learning trajectories, describing self-change as they began to understand the components of learning trajectories.

They also described the process of becoming familiar with the probable levels of thinking of preschoolers along developmental progressions, as well as instructional activities. Impressively, they also mentioned learning about the goals—the mathematical *content*.

10.4 Discussion and Implications

The *T*echnology-enhanced, *R*esearch-based, *I*nstruction, *A*ssessment, and professional *D*evelopment (TRIAD) model was designed to address the challenges of scaling up interventions, especially interventions in mathematics situated in the diverse settings of preschool education. Because research suggests that knowledgeable and responsive adults are the most important feature of successful educational interventions (National Research Council, 2009), we evaluated TRIAD's effects on teachers' practices, whether teachers' practices mediated the effects of the intervention on students' outcomes, and whether these practices were maintained 2–6 years after the cessation of project support. We also asked teachers about their opinions in a final questionnaire.

Findings showed that the TRIAD intervention positively affected teachers' practices in mathematics education compared to control teachers (who also received mathematics curricula and professional development, but different from the TRIAD learning trajectories approach). The intervention based on children's mathematical learning trajectories provided a coherent program of teaching and learning, which may have promoted the significant levels of high-quality practice found in this study (Wilson et al., 2013).

Mediational analysis confirmed that increasing specific aspects of the environment (e.g., the number of activities, quality of the math activities, the number of computers, and the overall quality of mathematics in the classroom) accounted for significant change in student outcome scores over and above the direct influence of exposure to high-quality mathematics curriculum. This suggests that these core processes of change serve to support the growth of mathematics learning as well as transfer to expressive oral language. The differences on two individual items were particularly striking, relating to teachers' willingness and ability to listen to students and support students' understanding as they listen to others.

The classroom culture subscore assesses teachers' general approach to mathematics education, indicated by "environment and interaction" variables such as responsiveness to students, use of "teachable moments," and environmental signs of mathematics, as well as "personal attributes of the teacher" variables, including appearing knowledgeable and confident about mathematics as well as showing enjoyment in, curiosity about, as and enthusiasm for, teaching mathematics. Teachers within the experimental condition demonstrated increases in the target practices measured within this subscale as compared to control.

The finding involving number of computers suggests that increased use of the *Building Blocks* math software improves math scores. Finally, the mediational impact of the total number of classroom mathematics activities appears to be a simple "more time on task is better" result. However, a separate COEMET measure of total time on task did not mediate the impact, where the number of specific mathematics activities did. Thus, the number of distinct mathematics activities in which students engaged was more important than total time on task in supporting their learning of mathematics (cf. Sylva, Melhuish, Sammons, Siraj-Blatchford, & Taggart, 2005).

In sum, these results specify which practices contribute to student outcomes. Those practices relating to teachers' willingness and ability to listen to students and support students' understanding as they listen to others confirm the work of others regarding the importance of dialog (Borko, 2004; Fennema et al., 1996; Knapp, Marder, Zucker, Adelman, & Needels, 1995) as well as the role of the research-based learning trajectories at the core of the TRIAD model. In a similar vein, the mediation of classroom culture on math scores is consistent with the literature supporting the connection between academic performance and general features of the classroom, including signs of mathematical activity and teachers who display both knowledge of and enthusiasm about mathematics and who interact with and respond to students frequently (D. M. Clarke & Clarke, 2004; Clements & Sarama, 2007; Fraivillig et al., 1999).

The number of mathematics activities also contributed to the mediation. We hypothesize that students of this age learn more from a variety of activities emphasizing the same level of thinking as they may learn concepts more readily from generalizing mathematics structures from different problematic situations that require the same mathematical concepts and processes for their solution. Further, such multiple situations may create a greater number of cognitive paths for retrieval. Finally, the mediational role of the computer software confirms separate evaluations of the software as the solitary (unconfounded) component of the intervention (Foster, Anthony, Clements, & Sarama, 2016).

Finally, teachers taught the curriculum with increasing fidelity over the following 6 years (fidelity was collected only in intervention classrooms, so we compared within the same classrooms over time), even though research project staff was no longer able to provide support. The participating teachers seemed to have internalized the program (Timperley et al., 2007) and to have made sense of the curricular activities involved with whole group, small group, and other components, within an overall structure of LTs that progressed toward a known mathematical goal. By engaging in the initial professional development, and then, becoming empowered by their own knowledge of the trajectories and the practices to support learners through the trajectories, they became progressively more faithful to the intended program, instead of drifting from it as time elapsed and support disappeared, a contrasting negative trend found in other studies (Datnow, 2005; Hargreaves, 2002).

One implication is that a coherent model of professional development, curriculum, instruction, and assessment based on LTs may provide the conditions for promoting high-quality instruction as well as sustainability in such practices (Clements & Sarama, 2014b). This may be particularly beneficial in addressing the climate of low expectations (Johnson & Fargo, 2010), as teachers increase their understanding of the capacities of all students to learn mathematics.

As teachers come to understand students' probable developmental paths and become adept at anticipating students' strategies and misconceptions, their teaching practices may become more grounded and solidified; this is one way that practices and beliefs may interact. As they notice students' multiple strategies, and probe for the ways in which students' mathematical thinking fits the structure of the trajectory, their teaching practices may become reinforced as student reactions provide positive feedback for their practices (Guskey, 2002). Teachers who demonstrate sustained fidelity of implementation to a program that has demonstrated improved student achievement will have a positive impact on many more students than teachers who implement with fidelity only during treatment. Thus, another general implication is that helping teachers develop the skills and practices necessary to perceive and document their students' learning may be an effective way to maintain and even increase fidelity of implementation. These positive perceptions of learning may be especially important in motivating teachers to productively face the challenges inherent in fully implementing all aspects of the curriculum. For example, educational technology challenges include limited hardware, hardware and software problems and limited troubleshooting competencies, difficulty scheduling computer use for all students, and inconsistency between computer use and customary practice including contextually constrained choices. Solving problems successfully, such as engaging students productively in technology activities, may engender confidence and risk-taking in future work. Simply, success breeds success, and such changes in practice may lead to positive changes in beliefs (e.g., Showers, Joyce, & Bennett, 1987), again, resulting in co-mutually reinforcing changes in beliefs and practices (Caudle & Moran, 2012) rather than conflicts between them (cf. Einarsdottir, 2003), supporting bidirectional path between teachers' knowledge of practices and their beliefs.

Our data suggest that certain factors are sine qua non for an effective scale-up effort. *Our evaluations (including mediators of effects, not discussed here due to space constraints) indicate that effectiveness depends on the development of teachers' skills and knowledge.* A total of 50–70 h of professional development—we provide about 75 h—is consistent with previous research documenting what is necessary to achieve measurable effectiveness (Yoon, Duncan, Lee, Scarloss, & Shapley, 2007). Situating the materials not just in the classroom but also each school is important for related but additional reasons. Because it establishes a cultural practice and provides peer support, school-based implementation supports both fidelity and sustainability of a new curriculum (Zaslow, Tout, Halle, Vick, & Lavelle, 2010) and reaches over 90% of students in the school (not just one classroom, hoping for "spillover").

Several findings indicate that learning trajectories played a critical role in both teachers' and children's learning. Finally, respect of teachers by project staff and the resultant acceptance of in-class coaches and technology support staff appeared to be as or more important than whole-group professional development sessions. Note that this was especially striking in another project, just being completed, in which we used coaches employed by the school district. They were not able to schedule meetings regularly, and our initial impression is that they were not as effective. If valid, this has serious ramifications for scaling up coaching-intensive models.

The TRIAD model is not simply about a new curriculum or training teachers to use it. Success required complex changes, including a change in instructional structures, pedagogical strategies, and classroom communication and culture (Grubb, 2008). Given the importance of early competence in mathematics (e.g., Duncan et al., 2007; Paris, Morrison, & Miller, 2006), the TRIAD implementation described here has implications for practice and policy, as well as research. TRIAD's guidelines should be considered when planning to increase the quality and quantity of preschool mathematics education.

In summary, experience with the LTPD generated fundamental change in teachers, as evidenced by teachers' responses (100% reported change) and by observation. They were more aware and appreciative of young children's mathematical capabilities and of the pedagogical use of all three components of learning trajectories: mathematical goals and content, developmental progressions, and instructional tasks and strategies (cf. Jacobs & Empson, 2016).

References

Association of Mathematics Teacher Educators. (2017). *AMTE standards for mathematics teacher preparation*. Raleigh, NC: AMTE.

Baroody, A. J., & Ginsburg, H. P. (1990). Children's mathematics learning: A cognitive view. In R. B. Davis, C. A. Maher, & N. Noddings (Eds.), *Constructivist views on the teaching and learning of mathematics* (Journal for Research in Mathematics Education Monograph Number 4) (pp. 79–90). Reston, VA: National Council of Teaching of Mathematics.

Battista, M. T. (2013). *Elementary teachers' learning, understanding, and classroom use of learning progressions*. Paper presented at the NCTM, Denver.

Björklund, C., & Barendregt, W. (2016). Teachers' pedagogical mathematical awareness in diverse child-age-groups. *Nordic Studies in Mathematics Education, 21*(4), 115–133.

Bobis, J., Clarke, B. A., Clarke, D. M., Gill, T., Wright, R. J., Young-Loveridge, J. M., & Gould, P. (2005). Supporting teachers in the development of young children's mathematical thinking: Three large scale cases. *Mathematics Education Research Journal, 16*(3), 27–57. https://doi.org/10.1007/BF03217400

Borko, H. (2004). Professional development and teacher learning: Mapping the terrain. *Educational Researcher, 33*(8), 3–15.

Brendefur, J. L., Strother, S., Thiede, K., Lane, C., & Surges-Prokop, M. (2013). A professional development program to improve math skills among preschool children in Head Start. *Early Childhood Education Journal, 41*(3), 187–195. https://doi.org/10.1007/s10643-012-0543-8

Carpenter, T. P., Fennema, E. H., Franke, M. L., Levi, L., & Empson, S. B. (2014). *Children's mathematics: Cognitively guided instruction* (2nd ed.). Portsmouth, NH: Heinemann.

Carpenter, T. P., Franke, M. L., Jacobs, V. R., Fennema, E. H., & Empson, S. B. (1998). A longitudinal study of invention and understanding in children's multidigit addition and subtraction. *Journal for Research in Mathematics Education, 29*(1), 3–20. https://doi.org/10.2307/749715

Caudle, L., & Moran, M. J. (2012). Changes in understandings of three teachers' beliefs and practice across time: Moving from teacher preparation to in-service teaching. *Journal of Early Childhood Teacher Education, 33*, 38–53.

Century, J., Rudnick, M., & Freeman, C. (2012). *Defining and measuring sustainability of reform: Factors that affect our abilities to generate enduring change*. Paper presented at the Annual Meeting of the American Educational Research Association, Vancouver, British Columbia, Canada.

Clarke, B. A. (2008). A framework of growth points as a powerful teacher development tool. In D. Tirosh & T. Wood (Eds.), *Tools and processes in mathematics teacher education* (pp. 235–256). Rotterdam, The Netherlands: Sense Publishers.

Clarke, D. M., Cheeseman, J., Clarke, B., Gervasoni, A., Gronn, D., Horne, M., ..., Sullivan, P. (2001). Understanding, assessing and developing young children's mathematical thinking: Research as a powerful tool for professional growth. In J. Bobis, B. Perry, & M. Mitchelmore (Eds.), *Numeracy and beyond (Proceedings of the 24th Annual Conference of the Mathematics Education Research Group of Australasia)* (Vol. 1, pp. 9–26). Reston, Australia: MERGA.

Clarke, D. M., & Clarke, B. A. (2004). Mathematics teaching in grades K-2: Painting a picture of challenging, supportive, and effective classrooms. In R. N. Rubenstein & G. W. Bright (Eds.), *Perspectives on the teaching of mathematics (66th yearbook)* (pp. 67–81). Reston, VA: National Council of Teachers of Mathematics.

Clements, D. H., & Sarama, J. (2004). Hypothetical learning trajectories. *Mathematical Thinking and Learning, 6*, 81.

Clements, D. H., & Sarama, J. (2007). Early childhood mathematics learning. In F. K. Lester Jr. (Ed.), *Second handbook of research on mathematics teaching and learning* (Vol. 1, pp. 461–555). New York, NY: Information Age Publishing.

Clements, D. H., & Sarama, J. (2014a). *Learning and teaching early math: The learning trajectories approach* (2nd ed.). New York, NY: Routledge.

Clements, D. H., & Sarama, J. (2014b). Learning trajectories: Foundations for effective, research-based education. In A. P. Maloney, J. Confrey, & K. H. Nguyen (Eds.), *Learning over time: Learning trajectories in mathematics education* (pp. 1–30). New York, NY: Information Age Publishing.

Clements, D. H., Sarama, J., & DiBiase, A.-M. (2004). *Engaging young children in mathematics: Standards for early childhood mathematics education*. Mahwah, NJ: Erlbaum.

Clements, D. H., Sarama, J., Spitler, M. E., Lange, A. A., & Wolfe, C. B. (2011). Mathematics learned by young children in an intervention based on learning trajectories: A large-scale cluster randomized trial. *Journal for Research in Mathematics Education, 42*(2), 127–166. https://doi.org/10.5951/jresematheduc.42.2.0127

Clements, D. H., Sarama, J., Wolfe, C. B., & Spitler, M. E. (2013). Longitudinal evaluation of a scale-up model for teaching mathematics with trajectories and technologies: Persistence of effects in the third year. *American Educational Research Journal, 50*(4), 812–850. https://doi.org/10.3102/0002831212469270

Clements, D. H., Sarama, J., Wolfe, C. B., & Spitler, M. E. (2015). Sustainability of a scale-up intervention in early mathematics: Longitudinal evaluation of implementation fidelity. *Early Education and Development, 26*(3), 427–449. https://doi.org/10.1080/10409289.2015.968242

Coburn, C. E. (2003). Rethinking scale: Moving beyond numbers to deep and lasting change. *Educational Researcher, 32*(6), 3–12.

Confrey, J., Maloney, A. P., Nguyen, K. H., & Rupp, A. A. (2014). Equipartitioning: A foundation for rational number reasoning. Elucidation of a learning trajectory. In A. P. Maloney, J. Confrey, & K. H. Nguyen (Eds.), *Learning over time: Learning trajectories in mathematics education* (pp. 61–96). New York, NY: Information Age Publishing.

Datnow, A. (2005). The sustainability of Comprehensive School Reform models in changing district and state contexts. *Educational Administrative Quarterly, 41*(1), 121–153.

Dearing, J. W., & Meyer, G. (2006). Revisiting diffusion theory. In A. Singhal & J. W. Dearing (Eds.), *Communication of innovations: A journey with Ev Rogers* (pp. 29–60). New Delhi, India: Sage.

Duncan, G. J., Dowsett, C. J., Claessens, A., Magnuson, K., Huston, A. C., Klebanov, P., …, Japel, C. (2007). School readiness and later achievement. *Developmental Psychology, 43*(6), 1428–1446. https://doi.org/10.1037/0012-1649.43.6.1428

Einarsdottir, J. (2003). Beliefs of early childhood teachers. In O. N. Saracho & B. Spodek (Eds.), *Studying teachers in early childhood settings* (pp. 113–134). Greenwich, CT: Information Age Publishing.

Fennema, E. H., Carpenter, T. P., Frank, M. L., Levi, L., Jacobs, V. R., & Empson, S. B. (1996). A longitudinal study of learning to use children's thinking in mathematics instruction. *Journal for Research in Mathematics Education, 27*, 403–434.

Foster, M. E., Anthony, J. L., Clements, D. H., & Sarama, J. (2016). Improving mathematics learning of kindergarten students through computer assisted instruction. *Journal for Research in Mathematics Education, 47*(3), 206–232. https://doi.org/10.5951/jresematheduc.47.3.0206

Fraivillig, J. L., Murphy, L. A., & Fuson, K. C. (1999). Advancing children's mathematical thinking in *Everyday Mathematics* classrooms. *Journal for Research in Mathematics Education, 30*(2), 148–170.

Franke, M. L., Carpenter, T. P., Levi, L., & Fennema, E. H. (2001). Capturing teachers' generative change: A follow-up study of professional development in mathematics. *American Educational Research Journal, 38*, 653–689.

Grubb, W. N. (2008). Multiple resource, multiple outcomes: Testing the "improved" school finance with NELS88. *American Educational Research Journal, 45*, 104–144.

Guskey, T. R. (2002). Professional development and teacher change. *Teachers and Teaching: Theory and Practice, 8*(3/4), 381–391.

Hargreaves, A. (2002). Sustainability of educational change: The role of social geographies. *Journal of Educational Change, 3*(3–4), 189–214.

Hiebert, J., & Stigler, J. W. (2017). Teaching versus teachers as a lever for change: Comparing a Japanese and a U.S. perspective on improving instruction. *Educational Researcher, 46*(4), 169–176. https://doi.org/10.3102/0013189X17711899

Institute of Medicine (IOM) and National Research Council (NRC). (2015). *Transforming the workforce for children birth through age 8: A unifying foundation*. Washington, DC: National Academy Press.

Jacobs, V. R., & Empson, S. B. (2016). Responding to children's mathematical thinking in the moment: An emerging framework of teaching moves. *ZDM Mathematics Education, 48*, 185–197. https://doi.org/10.1007/s11858-015-0717-0

Johnson, C. C., & Fargo, J. D. (2010). Urban school reform enabled by transformative professional development: Impact on teacher change and student learning of science. *Urban Education, 45*(1), 4–29.

Knapp, M. S., Marder, C., Zucker, A. A., Adelman, N. E., & Needels, M. C. (1995). The outcomes of teaching for meaning in high-poverty classrooms. In M. S. Knapp (Ed.), *Teaching for meaning in high-poverty classrooms* (pp. 124–144). New York, NY: Teachers College Press.

MacDonald, A., Davies, N., Dockett, S., & Perry, B. (2012). Early childhood mathematics education. In B. Perry, T. Lowrie, T. Logan, A. MacDonald, & J. Greenlees (Eds.), *Research in mathematics education in Australasia: 2008–2011* (pp. 169–192). Rotterdam, The Netherlands: Sense Publishers.

National Research Council. (2001). *Eager to learn: Educating our preschoolers*. Washington, DC: National Academy Press.

National Research Council. (2009). *Mathematics learning in early childhood: Paths toward excellence and equity*. Washington, DC: National Academy Press.

Paris, S. G., Morrison, F. J., & Miller, K. F. (2006). Academic pathways from preschool through elementary school. In P. Alexander & P. Winne (Eds.), *Handbook of research in educational psychology* (pp. 61–85). Mahwah, NJ: Erlbaum.

Perry, B. (2010). Mathematical thinking of preschool children in rural and regional Australia: An overview. *Journal of Australian Research in Early Childhood Education, 16*(2), 1–12.

Piasta, S. B., Logan, J. A. R., Pelatti, C. Y., Capps, J. L., & Petrill, S. A. (2015). Professional development for early childhood educators: Efforts to improve math and science learning opportunities in early childhood classrooms. *Journal of Educational Psychology, 107*(2), 407–422. https://doi.org/10.1037/a0037621

Polly, D., McGee, J., Wang, C., Martin, C., Lambert, R., & Pugalee, D. K. (2015). Linking professional development, teacher outcomes, and student achievement: The case of a learner-centered mathematics program for elementary school teachers. *International Journal of Educational Research, 72*, 26–37. https://doi.org/10.1016/j.ijer.2015.04.002

Sarama, J., & Clements, D. H. (2009). *Early childhood mathematics education research: Learning trajectories for young children*. New York, NY: Routledge.

Sarama, J., & Clements, D. H. (2013). Lessons learned in the implementation of the TRIAD scale-up model: Teaching early mathematics with trajectories and technologies. In T. G. Halle, A. J. Metz, & I. Martinez-Beck (Eds.), *Applying implementation science in early childhood programs and systems* (pp. 173–191). Baltimore, MD: Brookes.

Sarama, J., & Clements, D. H. (2018). Designing, scaling up, and evaluating comprehensive professional development based on learning trajectories. In P. Sztajn, P. H. Wilson, & C. Edgington (Eds.), *Learning trajectories-based professional development*. New York, NY: Teachers College Press.

Sarama, J., Clements, D. H., Starkey, P., Klein, A., & Wakeley, A. (2008). Scaling up the implementation of a pre-kindergarten mathematics curriculum: Teaching for understanding with trajectories and technologies. *Journal of Research on Educational Effectiveness, 1*(1), 89–119. https://doi.org/10.1080/19345740801941332

Sarama, J., Clements, D. H., Wolfe, C. B., & Spitler, M. E. (2012). Longitudinal evaluation of a scale-up model for teaching mathematics with trajectories and technologies. *Journal of Research on Educational Effectiveness, 5*(2), 105–135. https://doi.org/10.1080/19345747.2011.627980

Sarama, J., Clements, D. H., Wolfe, C. B., & Spitler, M. E. (2016). Professional development in early mathematics: Effects of an intervention based on learning trajectories on teachers' practices. *Nordic Studies in Mathematics Education, 21*(4), 29–55.

Sarama, J., Lange, A., Clements, D. H., & Wolfe, C. B. (2012). The impacts of an early mathematics curriculum on emerging literacy and language. *Early Childhood Research Quarterly, 27*(3), 489–502. https://doi.org/10.1016/j.ecresq.2011.12.002

Scheirer, M. A., & Dearing, J. W. (2011). An agenda for research on the sustainability of public health programs. *American Journal of Public Health, 101*(11), 2059–2067.

Showers, B., Joyce, B., & Bennett, B. (1987). Synthesis of research on staff development: A framework for future study and a state-of-the-art analysis. *Educational Leadership, 45*(3), 77–87.

Simon, M. A. (1995). Reconstructing mathematics pedagogy from a constructivist perspective. *Journal for Research in Mathematics Education, 26*(2), 114–145. https://doi.org/10.2307/749205

Stosich, E. L. (2016). Joint inquiry: Teachers' collective learning about the Common Core in high-poverty urban schools. *American Educational Research Journal, 53*(6), 1698–1731. https://doi.org/10.3102/0002831216675403

Sylva, K., Melhuish, E., Sammons, P., Siraj-Blatchford, I., & Taggart, B. (2005). *The effective provision of pre-school education [EPPE] project: A longitudinal study funded by the DfEE (1997–2003)*. London, UK: DCSF/Institute of Education, University of London.

Sztajn, P., Confrey, J., Wilson, P. H., & Edgington, C. (2012). Learning trajectory based instruction: Toward a theory of teaching. *Educational Researcher, 41*, 147–156. https://doi.org/10.3102/0013189X12442801

Timperley, H. (2011). *Realizing the power of professional learning*. London, UK: Open University Press.

Timperley, H., Wilson, A., Barrar, H., & Fung, I. (2007). *Teacher professional development: Best Evidence Synthesis iteration*. Wellington, New Zealand: Ministry of Education.

Wilson, P. H., Mojica, G. F., & Confrey, J. (2013). Learning trajectories in teacher education: Supporting teachers' understandings of students' mathematical thinking. *The Journal of Mathematical Behavior, 32*(2), 103–121. https://doi.org/10.1016/j.jmathb.2012.12.003

Yoon, K. S., Duncan, T., Lee, S. W.-Y., Scarloss, B., & Shapley, K. L. (2007). *Reviewing the evidence on how teacher professional development affects student achievement* (Issues & Answers Report, REL 2007–No. 033). Retrieved from: http://ies.ed.gov/ncee/edlabs.
Zaslow, M. J., Tout, K., Halle, T. G., Vick, J., & Lavelle, B. (2010). *Towards the identification of features of effective professional development for early childhood educators: A review of the literature*. Retrieved from Washington, DC: http://www2.ed.gov/rschstat/eval/professional-development/literature-review.pdf.

Chapter 11
Inservice Mathematics Professional Development: The Hard Work of Learning Mathematics

Hung-Hsi Wu

11.1 Introduction

We all want to improve school mathematics education, but before making any recommendations on how to take it to the next level, we would do well to first find out where we stand. The answer: not in a good place. For the past five decades or so, the mathematics we teach in school has been mostly flawed and unlearnable.[1] For example, the fractions that students have to compute extensively with from grade 5 to grade 12 are supposed to be thought of as pieces of pizza. The resulting fraction phobia has been something of a national pastime for decades (see, e.g., https://www.gocomics.com/peanuts/1966/04/21). Another example: we do not make any effort to teach students proofs (reasoning) in the K-12 curriculum outside the high school geometry course, and yet in that one geometry course alone, students are suddenly called upon to *prove everything*—no matter how trivial or boring—on the basis of a collection of new objects called "axioms." The situation would not be as bleak if we had educated our mathematics teachers properly so that they could help smooth students' learning path along such a rugged obstacle course, but we haven't. Since teachers are only equipped with this body of flawed and unlearnable mathematical knowledge, they inevitably inflict the same flawed and unlearnable mathematical knowledge on their students. So the vicious cycle continues to this day.

[1] We use "unlearnable" in this article to mean "unlearnable by a majority of students." We note that, in this case, learning *mathematics* includes learning how to reason; see Sect. 11.3 below for the fundamental principles of mathematics.

H.-H. Wu (✉)
Department of Mathematics, University of California at Berkeley, Berkeley, CA, USA
e-mail: wu@berkeley.edu

Y. Li et al. (eds.), *Developing Mathematical Proficiency for Elementary Instruction*, Advances in STEM Education, https://doi.org/10.1007/978-3-030-68956-8_11

Obviously, two things must be in place before there can be any improvement in school mathematics education: a correct and learnable version of mathematics is made available in textbooks to students and a corps of teachers who are capable of teaching the same. These are the tasks before us, and they have recently acquired increased urgency because of the advent of the CCSSM (Common Core State Standards for Mathematics, [CCSSM]). The mathematics advocated by the CCSSM represents the first—but major—step towards meeting the goal of being correct and learnable, so the need for better school textbooks and teachers who are mathematically more knowledgeable can no longer be put off to the distant future.

To achieve the first goal of getting better school textbooks, some recent developments have given us hope, but in any case, there is now a detailed mathematical guide on what constitutes correct and learnable school mathematics. To achieve the second goal about a corps of mathematically competent teachers, we need a serious commitment to content-based PD (professional development) to meet this problem head-on. However, two large-scale impact studies of PD for teachers carried out in the past decade by IES (the Institute of Education Sciences) have raised serious doubts about the ability of content-intensive PD for inservice *mathematics* teachers to improve student learning (see Garet et al. 2011 and Garet et al. 2016). We are therefore forced to take a close look at this claim by the two IES studies. In our view, the claim is not supported by the available evidence, and we will make some effort to explain why not. Along the way, our explanation will also suggest the kind of PD that may be more likely to produce mathematically knowledgeable teachers who can improve student learning.

This article will expand on the preceding rather cryptic statements. A brief outline follows. Section 11.2 gives a description of the flawed and unlearnable body of knowledge—what we call TSM (*Textbook School Mathematics*)—that has dominated school mathematics education for the past half century. Section 11.3 introduces the *Fundamental Principles of Mathematics*, which are the *sine qua non* of mathematics. School mathematics that respects these fundamental principles will be called PBM (*principle-based mathematics*, see Poon 2014), and we will explain why PBM, because of its transparency, is *learnable*. In Sect. 11.4, we briefly discuss the situation regarding school textbooks that respect PBM. In Sect. 11.5, we give a fairly detailed discussion of the kind of PD needed to produce inservice teachers who can teach PBM *and* of the obstacles that stand in the way of implementing such PD. Section 11.6 presents an in-depth analysis of the aforementioned 2011 IES impact study and explains why its PD could not have produced teachers capable of teaching PBM: the PD did not help them overcome the handicap of knowing only TSM but not correct mathematics. The last section offers a variety of comments, including the need for "mathematics teachers" in elementary school and what may be preventing effective *preservice* PD from becoming a reality on university campuses.

11.2 Mathematical Engineering and TSM

To understand the kind of "mathematics" that has dominated school mathematics education for the past 50 years or so after the demise of the New Math around 1970, we need to step back to get some perspective on the nature of school mathematics and the overall state of school math education.

School mathematics is not part of *mathematics* proper—the mathematics we teach in universities and use in science and mathematics research—but is, rather, a particular version of mathematics that has been customized for consumption by K–12 students (see Wu 2006). This is analogous to the case of electrical engineering, which is not part of physics but is a customized version of it for the purpose of creating electrical and electronic products to meet humans' everyday needs. It is in this sense that school mathematics is a product of *mathematical engineering*, and a good part of school mathematics education is just mathematical engineering (Wu, *loc. cit.*). Of the need to customize university mathematics for consumption in K–12, there can be no doubt. After all, we do not introduce fractions to elementary students as the positive elements in the quotient field of the ring of integers. Rather, we directly develop fractions from whole numbers using the number line (Jensen 2003; Wu 1998, 1999a, and 2011a); this will be discussed in Sect. 11.3.2. Similarly, in K–12, a line in the plane is not a linear map from $\mathbf{R}$ to $\mathbf{R}^2$ but the unique curve joining any two of its points as specified by Euclid's first postulate.[2] And so on.

This engineering takes many forms. Sometimes it recasts the whole concept in a different but equivalent setting, as in the case of fractions and rational numbers. Sometimes it makes use of advanced theorems without any proof (so long as there is no circular reasoning), such as the fundamental theorem of algebra, the Jordan curve theorem for polygons, or the existence of the exponential function e^x. At other times it simply leaves out topics that are too conceptually sophisticated, such as the structure of the real numbers[3] or the concept of continuity. But regardless of the engineering decisions, there will always be good and bad engineering. In the same way that bad engineering in electrical engineering produces electronic gadgets that are hazardous to users, bad mathematical engineering produces a body of mathematical knowledge for K-12 that is unlearnable, basically because it is often *wrong as mathematics*. The mathematical knowledge that has dominated school mathematics education for the past five decades is unfortunately one example of what bad mathematical engineering has wrought. We call it **TSM**, *Textbook School Mathematics*, because its most complete realization resides in all the standard school mathematics textbooks and almost all the textbooks for mathematics teachers'

[2] Nevertheless, some recent publications have done just that: defining a line in the plane as the graph of an equation $y = mx + b$. See, e.g., page 711 of Billstein-Liebeskind-Lott 2007.

[3] This particular engineering decision makes a tremendous impact on the school curriculum because, as a result, the principal number system in K–12 is actually the rational numbers $\mathbf{Q}$ and *not* the real numbers $\mathbf{R}$. This is the reason why fractions are so important in K–12.

professional development (see Askey 2018; Baldridge 2013; Douglas 2015; Cuoco-McCallum 2018; Wu 2011c and 2018).

Although TSM looks superficially like mathematics, it differs from mathematics in important ways, especially in its lack of precise definitions and reasoning. TSM is not concerned with students understanding concepts or developing a capacity for reasoning, but instead focuses on *getting right answers to problems that TSM sees fit to pose.* To this end, TSM offers students a set of procedures, which, when followed conscientiously, lead to the right answers to these problems. To make the procedures more attractive to students, TSM uses only intuitive language to describe the concepts to make students believe that they "get it." The absence of precise definitions—and the attendant absence of reasoning—in TSM is therefore part of the design.

We must confront TSM directly because of its tenacious and pervasive hold on school mathematics education. It is the mathematics used by teachers and education researchers in their work, and its omnipresence can be easily explained. Teachers and educators[4] learned TSM in their K-12 years, and when they were students in institutions of higher learning, they learned mainly about the *pedagogical* issues of the K–12 curriculum. On the rare occasion that they got to take a course on school mathematics, almost all the textbooks for such courses—as mentioned above—consisted of little more than polished presentations of TSM. Once teachers and educators begin their professional lives, the mathematics they deal with is once again TSM. This is especially true for teachers because textbooks are "the authority on knowledge and the guide to learning . . . many teachers see their job as just 'covering the text' " (Romberg and Carpenter 1985). We therefore have a vicious cycle that reinforces the dominance of TSM in American school mathematics education, including education research. Thanks to this well-established recycling program, it would be fair to say that TSM *is now part and parcel of the mathematics education literature.* The article of Armstrong and Bezuk (1995) illustrates this point very well. These authors discuss the difficulty teachers have trying to teach the multiplication and division of fractions in middle school. They observe that teachers teach these concepts procedurally (without reasoning) not because they intentionally want to "withhold conceptual understanding from their students," but because

> It is quite possible that the teachers do not know that a conceptual base for multiplication and division of fractions even exists. Nothing in their mathematics learning experiences would have provided a hint of that existence. (*loc. cit.*, page 91)

From our perspective, what this says is that most teachers have been denied the opportunity to learn a correct approach to the multiplication and division of fractions simply because all they have access to is TSM.

---

[4]We use the term "educators" to refer to university faculty in schools of education.

11.2.1 Some Examples of TSM

Some examples will clarify why we object to TSM.

Example 11.1 TSM explains equivalent fractions by using what is often called the *Giant One*. For example, to show $\frac{3}{2} = \frac{12}{8}$, TSM reasons as follows:

$$\frac{3}{2} = \frac{3}{2} \times \mathbf{1} = \frac{3}{2} \times \boxed{\frac{4}{4}} = \frac{3 \times 4}{2 \times 4} = \frac{12}{8} \tag{11.1}$$

This "reasoning" is probably too well-known to require any comments. Formally, the starting point of this "reasoning" is that $\frac{3}{2}$ and $\frac{12}{8}$ *are fractions*, and the conclusion is that *the two fractions are equal.*

From the outset, this "reasoning" faces two insurmountable obstacles: *TSM has no precise definition of a fraction* and, therefore, *it is unclear what it means for two fractions to be equal.* So TSM begins with a vague hypothesis and arrives at a conclusion that is equally vague. Hardly an ideal setting for doing mathematics. Yet, the greater obstacle is the use of fraction multiplication in this attempted "proof." Since the concept of equivalent fractions appears almost as soon as fractions are introduced, *before* students get to know how to add or multiply them, fractions are not ready to be multiplied in this argument. In this light, the transgression implicit in the first step, $\frac{3}{2} = \frac{3}{2} \times 1$, seems relatively harmless because "1 times anything is the thing itself." The key step in Eq. (11.1) that

$$\frac{3}{2} \times \boxed{\frac{4}{4}} = \frac{3 \times 4}{2 \times 4}$$

is, however, totally out of place because the validity of the product formula that says $\frac{a}{b} \times \frac{c}{d} = \frac{ac}{bd}$ (for any positive integers a, b, c, and d) itself depends on the use of equivalent fractions (see pp. 62-63 of Wu 2016a). Therefore, this "proof" in TSM is guilty of circular reasoning at the very least.

TSM's inability to define what it means for two fractions to be equal also plays a role in the next example.

Example 11.2 The article of Otten et al. (2010) tries to give a demonstration of the *cross-multiplication algorithm* (**CMA**): If two fractions $\frac{a}{b}$ and $\frac{c}{d}$ are equal, then $ad = bc$. Because the authors were working within TSM, they had no precise definition of a fraction at their disposal, so they made up an *ad hoc* definition of *equality for fractions* by saying that $\frac{a}{b} = \frac{c}{d}$ means that there is a nonzero whole number k so that $c = ka$ and $d = kb$. Then they used this definition of "equal fractions" to prove the theorem. The fact that this definition of *equality* is incorrect (e.g., $\frac{6}{9} = \frac{14}{21}$, but there is no whole number k so that $14 = k \times 6$ and $21 = k \times 9$) and that such a hypothesis trivializes the theorem is almost beside the point here. What is striking is that we get to witness the struggle the authors were going through in trying to break free from TSM, and how TSM ultimately defeated them.

Incidentally, CMA should be taught in grade 5, not long after the theorem on equivalent fractions has been proved, and it (together with its various extensions) belongs in the survival kit of every student and every teacher in K–12.

Example 11.3 What does it mean to add two fractions such as $\frac{3}{8}+\frac{5}{6}$, and what is the sum? TSM provides no answer to the first question; for the second, it prescribes the following procedure: Get the least common denominator (**LCD**) 24 of 8 and 6 and observe that $24 = 3 \times 8,\ 24 = 4 \times 6$. Then add as follows:

$$\frac{3}{8}+\frac{5}{6} = \frac{3\times 3}{3\times 8}+\frac{4\times 5}{4\times 6} = \frac{9+20}{24}$$

In terms of students' mathematics learning, one has to take note of the fact that when elementary students encounter the addition of fractions for the first time, they expect that it will be more or less the same as the addition of whole numbers, i.e., addition is "putting things together." However, not only is there no indication in TSM that adding fractions has anything to do with "putting things together," but there is also nothing in the preceding procedure—LCD and all—to suggest any connection with "putting things together." TSM makes learning how to add any two fractions more complicated and difficult than it needs to be.

Example 11.4 What does it mean to multiply two fractions such as $\frac{2}{3}\times\frac{5}{8}$ and what is the product? Again TSM has nothing to say about the first question, and it answers the second by declaring that fractions are multiplied by the following rule: $\frac{a}{b}\times\frac{c}{d}=\frac{ac}{bd}$ for any whole numbers a, b, c, and d (with the understanding that $bd \neq 0$). It then follows easily that the preceding product is equal to $\frac{10}{24}$. No explanation is given for this rule, but there is usually some effort to make this rule seem reasonable by discussing the special case where $b = 1$ (i.e., a whole number multiplies a fraction) and also the special case where $d = 1$ and b divides c (i.e., a fraction multiplies a whole number which is a multiple of the denominator of the fraction). How then do we use multiplication in word problems? Again, do it by rote: when the word "of" appears, it means "multiply" (see Moynahan 1996).

Example 11.5 TSM introduces the concept of a *mixed number* right after the definition of a fraction—but before the addition of fractions is discussed. Thus, $2\frac{3}{4}$ is, by definition, "2 *and* $\frac{3}{4}$." TSM also explains the conversion of mixed numbers to improper fractions by rote, e.g.,

$$2\frac{3}{4} = \frac{(2\times 4)+3}{4} = \frac{11}{4}$$

This procedure has to be done by rote because of TSM's refusal to *define* a mixed number as the sum of a positive integer and a *proper fraction*, and that, for example, $2\frac{3}{4}$ is the shorthand notation for $2+\frac{3}{4}$. Notice that the word "and" has been purposely used to hide the fact that the addition of fractions is involved, an inexcusably bad piece of mathematical engineering. If mixed numbers were introduced *after* the addition of fractions, they would be a perfectly simple topic to learn.

Example 11.6 TSM considers finite decimals to be a different kind of number from fractions and it teaches finite decimals on a parallel track, independent of fractions. For example, a finite decimal such as 2.307 is defined to be "2 and 3 tenths and 7 thousandths." Once again, the word "and" is purposely used to hide the fact that the addition of fractions is involved, so that 2.307 is actually the following sum of fractions,

$$2+\frac{3}{10}+\frac{0}{100}+\frac{7}{1000}$$

So TSM knows that a finite decimal *is* a fraction, but nonetheless tries to hide it. Bad mathematical engineering again. Such an approach to the teaching of finite decimals has produced misconceptions that are legendary (see, for example, https://tinyurl.com/y6k59uqp).

These examples are cited for their relevance to our discussion, but we must emphasize that they do not come close to exhausting the sins of TSM. A few other examples are the obsession in TSM with the so-called *order of operations*, which elevates a notational convention to a major topic in middle school mathematics, or the use of FOIL in TSM to expand the product of two linear polynomials, or the convention in TSM geometry that precludes a square from being a rectangle, an equilateral triangle from being an isosceles triangle, a parallelogram from being a trapezoid, etc. There is another glaring defect that should not be overlooked: the cavalier way TSM handles real numbers. In middle school, irrational numbers begin to encroach on many mathematical discussions because numbers such as π and square roots of whole numbers can no longer be avoided. Real numbers are not the province of K–12 mathematics, granted, but when students are asked to believe—*without a word of explanation*—that $\frac{\sqrt{2}\cdot\sqrt{3}}{\sqrt{2}}=\sqrt{3}$ because of the "usual" cancellation law for fractions (whose numerators and denominators are *whole numbers*), things clearly have gotten out of hand.

In summary, TSM represents a major transgression against what is acceptable in mathematics.

11.2.2 The Neglect of Definitions in TSM

To get an idea of the scope of TSM's devastation of school mathematics, it may be of some interest to see at least a *partial* list of the fundamental mathematical concepts in K-12 that are either not defined or defined incorrectly in TSM:

- the *remainder* in the division-with-remainder of whole numbers;
- fraction; *equality* of fractions; one fraction being *bigger* or *smaller* than another; addition of fractions; multiplication of fractions; division of fractions;
- finite decimal; *equality* of decimals; one decimal being *bigger* or *smaller* than another; addition of decimals; multiplication of decimals; division of decimals;
- ratio; percent; rate; *constant* rate;
- expression, equation; *graph* of an equation; *graph* of an inequality; half-plane;
- *slope* of a line;
- the 0-th power of a number, negative power of a number;
- polygon, regular polygon, parabola;
- *congruent* figures; *similar* figures; scale-drawing;
- length of a curve; area of a plane region; volume of a solid.

Because reasoning is impossible without definitions, TSM has to teach the skills related to all the concepts on this list entirely *by rote*. For example, because there is no definition for either the "graph of a linear inequality in two variables" or a "half-plane," there is no explanation in TSM for the fact that the graph of a linear inequality in two variables is a half-plane. The absence of precise definitions for fraction, decimal, ratio, percent, and rate will be particularly pertinent to the discussion of the PD program of the 2011 IES impact study (mentioned in the Introduction) in Sect. 11.6.2 below.

Beyond its failure to define key concepts, TSM also does great harm to mathematics learning by introducing spurious concepts, notably "variable" and "proportional reasoning." It is not difficult to see that neither can be defined in a way that makes any sense as mathematics, but if a fuller explanation is needed for why these are not *mathematical* concepts, see Section 3.2 of Wu (2018). (One can find a more detailed discussion of "variable" in pp. 2–3, 28–29, 38–39 of Wu 2016b, and of "proportional reasoning" in Section 7.2 of Wu 2016b.)

11.3 Fundamental Principles of Mathematics and PBM

Thus far, we have criticized TSM for its many mathematical flaws, and we have referred vaguely to the need for *correct and learnable* mathematics in the classroom. Now it is time to explain in greater detail what "correct and learnable" school mathematics is.

11.3.1 Fundamental Principles of Mathematics

First, consider the following five **Fundamental Principles of Mathematics**:

(I) Every concept has a precise definition.
(II) Every statement is supported by reasoning.

(III) Precision attends every statement.
(IV) The progression from topics to topics is coherent.
(V) The progression from topics to topics is purposeful.

It will be clear from the following discussion that all five overlap each other and that the first three form a close-knit unit. The examples of the last section illustrate the fact that TSM violates every one of these principles, but we will provide more of such examples below.

These principles form a minimal set of characteristic properties of mathematics, and any mathematical exposition of that violates any one of these principles is not a faithful representation of mathematics. For our present purpose, we call *school mathematics* that respects these five fundamental principles **PBM** (*principle-based mathematics;* this term was coined by Poon 2014). Thus PBM is a body of knowledge that is consonant with both the progression of the K-12 school mathematics curriculum and the fundamental principles of mathematics. Henceforth, we will use PBM as a shorthand for *correct and learnable* mathematics.

We now explain the preceding fundamental principles from the specific vantage point of learning *school mathematics*.

(I) The need for precise definitions stems from the fact that the learning of mathematics involves the learning of many new concepts. A *precise definition* of a concept tells students what it is (a number? a pair of numbers? a function? an equality? a geometric figure? etc.), and what properties it is assumed to possess. From a pedagogical perspective, the purpose of having precise definitions is to lighten students' cognitive load by clearly setting forth—for the purpose of learning—*everything they need to know about the concept in question.* A precise definition of a concept eliminates second-guessing: it assures students that they are already in possession of all they need to know for any reasoning involving this concept, no more and no less. *This is how mathematics works.* There is no need for students to wonder whether the textbook and the teacher have something up their sleeve that is not being shared with them.

To understand what this means, suppose a fraction is defined to be like a piece of pizza. But every student knows that the metaphorical piece of pizza will inevitably turn into something else at a moment's notice. After all, if TSM asks them how long it will take a faucet to fill a tub of $57\frac{1}{2}$ gallons given that the rate of the water flow is a constant $14\frac{2}{3}$ gallons per minute, their common sense would tell them to forget whatever has been taught about fractions-as-pizzas and, instead, concentrate on their rote skills. This illustrates how students in TSM are put in a state of constant distrust. How can real learning take place under the circumstances? Worse, if a concept such as the *division of fractions* is taught without a definition, students are left to cope with problems about fraction division *without knowing what they are doing.* This is why we have "ours is not to reason why, just invert and multiply" and the attendant fraction phobias.

Having a precise definition of a concept—and consistently basing any reasoning involving this concept only on what is in the definition—is therefore a necessary first step to build trust and make it possibly for students to learn about reasoning with the concept. The precise definition eliminates any need for students to constantly look over their shoulders and try to guess what additional information about the concept may be coming their way.

Two further comments about definitions will round out the picture. The first is that, insofar as a definition is supposed to inform students of everything they need to know about a concept, the need for *simplicity* in a definition should be obvious. For example, consider the following "definition" of a *right triangle*: *it is a triangle so that one of its angles is* 90° *and so that if* a, b, *and* c *are the lengths of its sides and* c *is the largest, then* $c^2 = a^2 + b^2$. Such a "definition" is not wrong in a formal sense, but it clearly fails to be informative because students would wonder whether there are any "right triangles" in this world that can meet both requirements. After all, what is the equality $c^2 = a^2 + b^2$ all about? If students' first reaction to this definition is one of disbelief, how to convince them to learn about right triangles? Therefore, we have to pare such a definition down to "*a right triangle is a triangle so that one of its angles is* 90°" and then show how to use reasoning *on the basis of this definition* to prove the equality $c^2 = a^2 + b^2$.

The second general comment about definitions is that the connection between *precise definitions* and *reasoning*—to the effect that any reasoning about a concept must be based *only* on what is contained in the definition—seems to have stayed under the radar in the mathematics education literature for the past few decades. This could be because of the dominance of TSM, which considers "definitions" to be largely superfluous and completely separate from the many rote-learning rules that make up TSM.

(II) We have just seen that having precise definitions is not an end in itself but, rather, the means to an end, the end being to make reasoning possible. Reasoning is the lifeblood of mathematics; there is no difference between reasoning and what is called *problem solving* in the education literature[5] when the latter is correctly interpreted. However, for the purpose of mathematics learning, reasoning plays the pivotal role of serving as the glue that connects concepts and skills. It is well-known that such connections make mathematics more learnable than a collection of concepts and skills that are memorized by rote (see pp. 118–120 of National Research Council 2001 for the large body of research evidence supporting this claim). Another way that reasoning helps to make school mathematics learnable is that it empowers all students to decide *for themselves* whether what they are doing is correct or not without having to submit themselves to the authority of their teacher or textbook. Learning how to reason therefore enhances students' self-confidence and their disposition[6] toward learning, which will in turn generate more learning.

[5]Problem solving is currently the main goal of school mathematics education in certain circles. It is well to note that there is no way to get students ready for problem solving (i.e., reasoning, which is the second fundamental principle of mathematics) without the help of the other four fundamental principles.

[6]Compare the fifth strand of *mathematical proficiency*—productive disposition—in Chapter 4 of National Research Council (2001).

(III) It is a truism that precision minimizes misunderstanding in teaching and learning. In the case of mathematics, however, we can be more specific: *without precision, learning about reasoning becomes well-nigh impossible.* For example, a typical definition of the division of fractions in TSM is the following:[7]

> Division and multiplication are **inverse operations**. Inverse operations are operations that undo each other.

These sentences sound plausible, but ultimately make no sense because multiplication sends *two* numbers, e.g., 2 and 3, to the third number ($2 \times 3 = 6$ in this case). Similarly, division sends *two* numbers, e.g., 6 and 3, to the third number ($6 \div 3 = 2$ in this case). So start with 2 and 3 (let us say), multiplication sends them to 6. Now how to "undo" 6 to send it back to 2 and 3 by division? If the definition does not make sense, how can we teach students to reason about fraction division using the definition?

A little bit more attention to precision would likely have averted this travesty by rephrasing the preceding "definition" as follows: *if a fraction $\frac{a}{b}$ is fixed, then dividing it by a nonzero $\frac{c}{d}$ yields a fraction so that, when the latter is multiplied by $\frac{c}{d}$, we get back $\frac{a}{b}$.*

Perhaps a more telling example of the need for precision is the way CMA (cross-multiplication algorithm) is used in TSM. Let x be a number that satisfies a proportion:

$$\frac{4.6}{13\frac{4}{5}} = \frac{x}{8\frac{1}{2}} \tag{11.2}$$

Then, a standard procedure to solve for x is to use CMA to get $13\frac{4}{5} \cdot x = 4.6 \cdot (8\frac{1}{2})$, thereby obtaining $x = 2\frac{5}{6}$. This solution method has unquestioned authority until we stop to ask: why is CMA applicable to Eq. (11.2)? Here, we need the fact that $\frac{a}{b} = \frac{c}{d}$ implies $ad = bc$. In TSM, CMA is either not proved (see Example 11.2 in Sect. 11.2.1) or proved only for *fractions* $\frac{a}{b}$ and $\frac{c}{d}$, in which case, $a, \ldots, d$ are *whole numbers*. The numerators and denominators in (11.2) are definitely not whole numbers, and it is shocking to realize that TSM never proves the CMA when the numbers involved are not whole numbers! How then can we inspire students to learn how to reason when they consistently bear witness to the fact that TSM plays fast and loose with results obtained by reasoning? What is the point of reasoning?

Such imprecision also has a pernicious side effect: it implicitly invites students not to take what they read literally, because *anything they read is likely to be correct in a wider context.* Consequently, students who are taught that

> if A, B, and C are nonzero fractions, then $A < B$ implies $CA < CB$

[7] This is taken directly from a textbook.

have every right to believe that this must also be true when A, B, and C are *any numbers*. Reports that the author heard consistently from teachers in the field is that many students are dismayed by the fact that

if A, B, and C are rational numbers and $C < 0$, then $A < B$ implies $CA > CB$.

Such imprecision puts students in a difficult position: how to decide when to believe—or not to believe—what they are taught?

Moral For mathematical learning to take place, precision must be the rule so that students know at each step *exactly* what is true and what is false.

(IV) Roughly, the *coherence* of mathematics means that mathematics, far from being a mere random collection of facts, is a tapestry in which all the concepts and skills are logically interwoven to form a single fabric. Mathematics unfolds logically, from basic assumptions (axioms) and definitions to theorems, and from theorems and other definitions to more theorems. Because of this logical progression, different parts of mathematics, even when far apart, often echo each other or are interconnected. It is this interconnectedness that comes from the unfailingly logical development of mathematics that we call *coherence*.[8]

The impact of coherence on learning can be seen in the learning of the arithmetic operations on whole numbers, fractions, rational numbers, and eventually real numbers. These operations are *conceptually the same* across the various number systems. (This fact is a main emphasis in Wu 2011a.) As a consequence of this coherence, if these operations on whole numbers are taught correctly, then the learning of these operations on fractions becomes streamlined and the popular perception in TSM that "fractions are such *different* numbers from whole numbers"[9] will be banished forever from school mathematics education. (Again, see Wu 2011a.)

The impact of coherence on learning can also be seen in a smaller scale in the most mundane of all school mathematics topics: the standard algorithms for whole numbers. When taught as rote skills, these algorithms are the embodiment of mindless tedium. But they are in fact held together by a single leitmotif:

> A knowledge of the addition, subtraction, multiplication, and division of *single-digit numbers* empowers us to perform all arithmetic operations with ease on any whole numbers, no matter how large. (See Chapter 3 of Wu 2011a.)

---

[8]According to Cuoco-McCallum (2018), what we have just defined is the *coherence of content*. The Cuoco-McCallum article is, in their terminology, concerned with the *curricular* coherence of the school mathematics curriculum.

[9]This is a direct quote of what one parent told the author. Such a popular perception is in fact a reflection of not only what transpires in a TSM classroom (compare, e.g., Examples 11.3 and 11.4 in Sect. 11.2.1 above), but also what is in the education literature. For example, "Such difficulty with fractions is often attributed to the fundamental differences between whole numbers and fractions." (Namkung and Fuchs 2016). Or, "Children must adopt new rules for fractions that often conflict with well-established ideas about whole numbers." (Bezuk and Kramer 1989).

The phrase "with ease" refers to the fact that any one of these operations on large numbers can be indescribably tedious (e.g., 2573 × 496 means adding *2573* copies of 496), but when reduced to single-digit computations (which is what the standard algorithms do) it becomes relatively simple. If this leitmotif is made known to elementary students—and of course if the algorithms are explained to them too—they are more likely to learn the algorithms and especially the multiplication table with greater enthusiasm and, more importantly, they will also learn *a substantial amount of valuable mathematics* because these four algorithms bring to light a recurrent theme in all of mathematics: reducing the complex to the simple.

Yet another example of how coherence can impact learning is in the teaching of fractions. Here are the seven most basic topics in fractions:

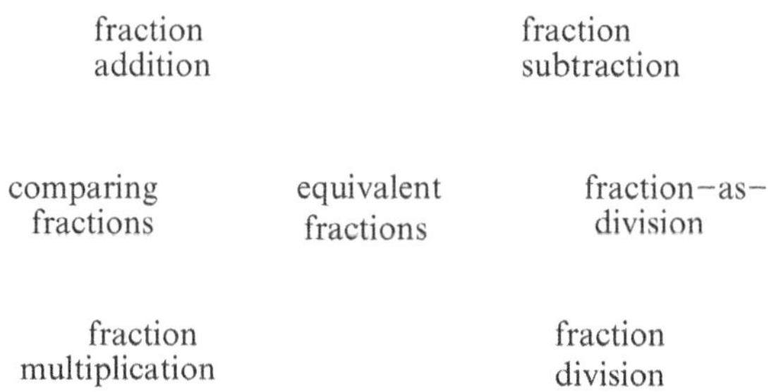

It is difficult to make sense of them when they are presented starkly as seven rote skills. But when reasoning is introduced into the discussion, a clear picture emerges: the other six topics are now seen to follow from the one central fact on equivalent fractions (see, e.g., Chapters 13-18 of Wu 2011a). From this perspective, we can make sense of all seven topics, and *fractions begin to be learnable.*

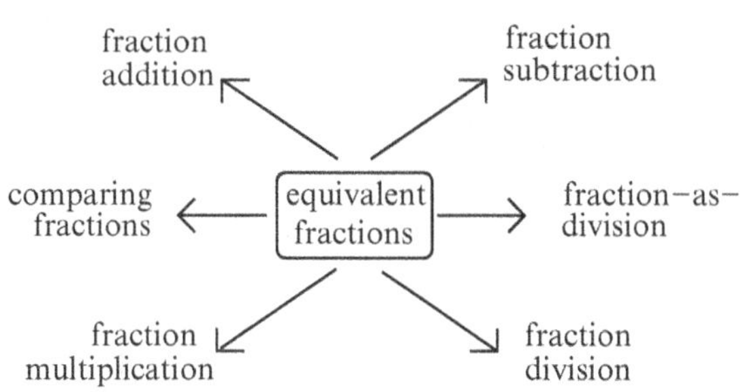

Incidentally, this is analogous to the phenomenon that while it is impossible to commit to memory the contents of even one page from a telephone book,[10] a thousand-page book like *Don Quixote* is quite memorable.

(V) *Purposefulness* refers to the fact that everything in mathematics is done with a purpose; this fact is of vital importance for the purpose of doing and learning mathematics but is unfortunately not something that is brought out in most books in mathematics education, least of all in TSM.

It is easy to explain the important role of purposefulness in school mathematics. Many skills and concepts have competed to stay in the (more or less) universally accepted school curriculum for more than a century, if not longer, and those that have survived to stay in the present day curriculum are the winners of many rounds of elimination. The reason these skills and concepts are still here could only be because *they serve a vital purpose.* If we can bring out this purpose to make students see why these skills and concepts are worth learning, students will be more motivated to learn them and student achievement will improve as a result. For example, we have already alluded to the likelihood that emphasizing the purpose (the leitmotif) of teaching the standard algorithms will increase student learning.

There is probably no better illustration of how *purposefulness* can impact student learning than the topic of *rounding whole numbers.* During my many years of doing inservice PD, I was once asked by a teacher why we bother to teach *rounding*, a skill that she considered to be meaningless. She said her students had no idea why they should learn it. Subsequently, other teachers concurred. Their complaint was entirely justified because TSM never explains that, quite often, one wants to round off a number because *precision is not wanted or is simply unattainable, or both.*

For example, the Census Bureau's estimated population of Houston was 2,303,482 in 2016. If a visitor from afar asks you how many people live in Houston, are you going to say "2,303,482"? *You had better not,* because you would sound ridiculous. Such precision is not the intent of the question. Your visitor probably only wants to know, roughly, how Houston compares with New York (population approx. 8,540,000) or San Francisco (population approx. 870,000). In other words, you are probably only expected to say whether the Houston population is closer to nine million or nine hundred thousand. With this in mind, you would likely *round* 2,303,482 *to the nearest million* to get two million. Then you look at your visitor in the eye and say with great confidence, "about two million."

One can also point to another kind of purpose for rounding: *when precision is not attainable.* Consider the 2016 estimate of Houston's population again. The Census Bureau probably had to release the figure of 2,303,482 for bureaucratic reasons, but such precision clearly makes no sense given the instability of a major city's population due to the unending cycle of births and deaths, the presence of a large transient population, and its ever-changing homeless population. Therefore, the most conservative estimate of Houston's population in 2016 is that the last three

[10]If a reader doesn't know what a "telephone book" is, please ask anyone over 60 or email the author!

digits, 482, are completely meaningless. We can de-emphasize them by *rounding to the nearest thousand* and list 2,303,000 as Houston's population in 2016. But if you round it to the nearest hundred thousand and list Houston's population as 2.3 million, I doubt that eyebrows would be raised.

If TSM would take the trouble to explain the purpose of rounding, many of our teachers probably would cease being exasperated by having to teach it. Students too would likely approach the learning of this skill with greater enthusiasm.

Of course, there is no end of examples to illustrate how the teaching of a concept or skill would be enhanced by bringing out the purpose of introducing said concept or skill. In addition to the four standard algorithms, think of *place value* (no, it is not due to a decree from on high that the 3 in 35 must be 30; see Chapter 1 of Wu 2011a), the introduction of *negative numbers* (see, for example, Chapter 26 of Wu 2011a), the introduction of *absolute value* (see Section 31.3 in Wu 2011a), etc.

11.3.2 PBM vs. TSM

Thus far, we have discussed in general terms some special features of TSM and PBM. Because the overriding theme of this article is to help teachers get rid of their knowledge of TSM and replace it with PBM, we will now revisit the six examples in Sect. 11.2.1 from the perspective of PBM. Because all these examples are about fractions, we will begin with a brief presentation of the definition of a fraction using the number line (first presented in Wu 1998, but see Wu 2002 and Chapter 1 in Wu 2016a). We will try to be brief, except that the discussions of Example 11.3 (adding fractions) and Example 11.4 (multiplying fractions) will be intentionally detailed because we want to illustrate explicitly how to use definitions in reasoning (see the discussion on *learning about definitions* in Sect. 11.5.1 below).

We begin with a (horizontal) number line on which a sequence of equidistant points marching to the right are labeled by the whole numbers. Now proceed as follows to get the **fractions with denominator 3** (and by extension any and all of the other fractions). Partition the **unit segment** [0,1] into three **equal parts** (= three segments of *equal length*). The part adjoining 0 is the third. Denote its right endpoint by $\frac{1}{3}$.

Fix the distance between 0 and $\frac{1}{3}$. Marking off *equidistant* points to the right of $\frac{1}{3}$ as we have done with whole numbers, we obtain a sequence of points, denoted by $\frac{2}{3}$, $\frac{3}{3}$, $\frac{4}{3}$, etc.

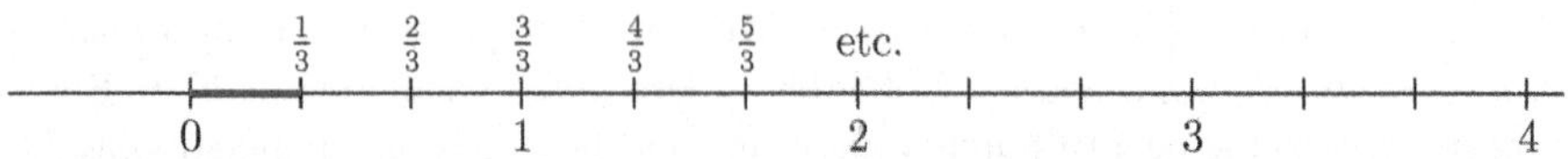

The segment $[0, \frac{1}{3}]$, by convention, is identified with its right endpoint, $\frac{1}{3}$. Similarly, the segment $[0, \frac{2}{3}]$ is identified with its right endpoint $\frac{2}{3}$, the segment $[0, \frac{5}{3}]$ with its right endpoint $\frac{5}{3}$, etc. Call these **the sequence of thirds**. Also call $\frac{n}{3}$ the **length** of the segment $[0, \frac{n}{3}]$ for any nonzero whole number n.[11]

Similarly, the nonzero fractions with denominator 5 are the **sequence of fifths**, determined by the partition of [0, 1] into 5 equal parts and by repeating the construction as in the sequence of thirds. For example, $\frac{8}{5}$ is the last point on the right:

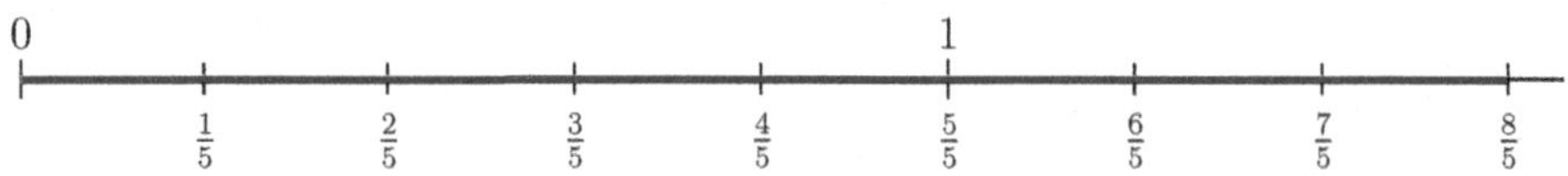

Then we call $\frac{8}{5}$ the **length** of the segment $[0, \frac{8}{5}]$, etc. (See Wu 1998; Chapter 12 of Wu 2011a has more details.)

If n is any nonzero whole number, then we obtain the **sequence of $n$ths** by partitioning the unit segment [0, 1] into n equal parts, denoting the right endpoint of the part adjoining 0 by $\frac{1}{n}$, and marking off equidistant points to the right of $\frac{1}{n}$. The union of 0 and the collection of all the sequences of $n$ths for $n = 1, 2, 3, \ldots$ is what we call the **fractions**.

Now that we know what a fraction is, we may ask if this definition amounts to anything. First of all, a fraction is an abstract concept and there is no point in hiding this fact[12] because introducing students gradually to abstractions is an integral part of school mathematics education. Defining a fraction as a certain point on the number is therefore nothing more than an honest acknowledgement of the abstract nature of the fraction concept. A teacher can mention to elementary students the fact that "$\frac{2}{3}$" is an abstraction the same way "5" is an abstraction.[13] But if fractions are just abstractions, i.e., points on the number line, how do they get involved in describing so many things that seem to have nothing to do with the number line? Furthermore, does the definition shed light on the addition and multiplication of fractions?

Let us answer the first question first. The key is the meaning we assign to the unit 1: it is the meaning of the unit that connects the number line to *every possible real-world situation* involving fractions. Consider, for example, the following problem:

[11]By convention, we also define $\frac{0}{n}$ to be 0 for every nonzero n.

[12]But there is also no need to emphasize it in elementary school either.

[13]It is all too easy to forget that the symbol "5" *is emphatically* an abstraction.

if $\frac{1}{4}$ *of a bucket of water is added to a* $\frac{2}{3}$ *of a bucket of water, how much water is now in the bucket?* To do this problem, we let the unit 1 on the number line be *the volume* of this bucket of water. The length of the unit segment [0, 1] now has to be interpreted as *the volume* of one bucket of water. So $\frac{1}{4}$ of a bucket of water—which is one part when the bucket of water is divided into 4 equal parts *by volume*—will be represented on the number line by one segment when the length of unit segment (= the volume of one bucket of water) is divided into 4 equal parts (= 4 segments of *equal length*). Therefore $\frac{1}{4}$ of a bucket of water is represented on this particular number line by the point $\frac{1}{4}$ (= the first point to the right of 0 in the sequence of fourths). In a similar way, $\frac{2}{3}$ of a bucket of water is represented by the fraction $\frac{2}{3}$ on this number line (= the second point to the right of 0 in the sequence of thirds). The total volume of water obtained by adding $\frac{1}{4}$ buckets of water to $\frac{2}{3}$ buckets of water is therefore what we normally call "$(\frac{1}{4}+\frac{2}{3})$ buckets of water." We will explain this sum in the discussion of Example 11.3.

Notice that the unit segment [0, 1] is what TSM calls "the whole," and that is a blatant error. The unit 1 has to be *the volume* of one bucket of water, but *not* "one bucket of water." The latter would leave open the question of whether we are dividing the bucket of water into "equal parts" by height, weight, or volume, or in fact, by another kind of measurement. Mathematics has no room for such ambiguity. It is sobering to realize that, in TSM, even the meaning of "the whole" is not correct.

There is another way the number line makes contact with other real-world situations, and we should touch on this briefly. Consider the following problem: *if Helena walks* $3\frac{1}{2}$ *miles in* 1 *hour and* 20 *minutes, what is her average speed in this walk?* Here we have to deal with two number lines: one whose unit is 1 mile, and another whose unit is 1 h. Since 1 h and 20 min is $1\frac{1}{3}$ h, we have the following two number lines:

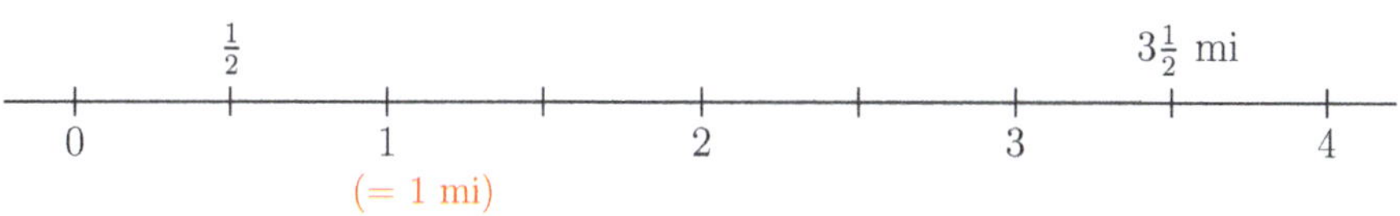

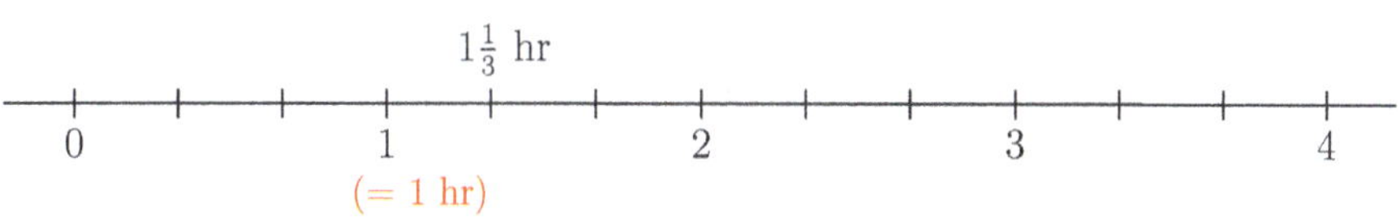

Now the average speed of Helena's walk is, by definition, the division

$$\text{average speed of walk} = \left(\frac{\text{distance traveled}}{\text{time duration}}\right) = \frac{3\frac{1}{2}}{1\frac{1}{3}} \tag{11.3}$$

Since division takes place only between two numbers on the same number line (see Wu 2011a, Chapter 18), Eq. (11.3) does not make sense as it stands. We rectify the situation by *identifying* the two number lines,[14] i.e., by identifying the two units, and obtain this picture:

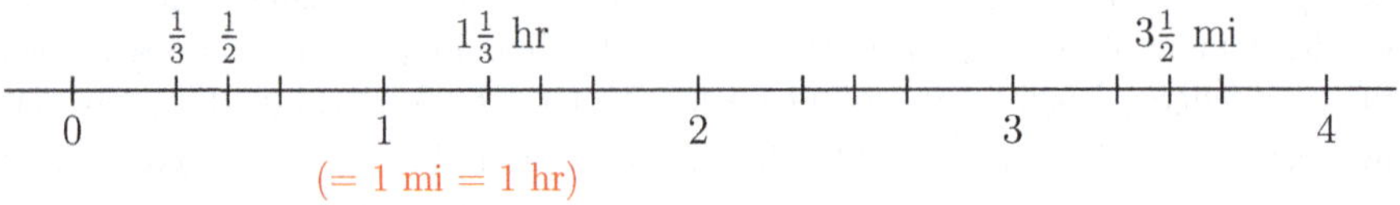

The division in Eq. (11.3) can now take place.

It is time to return to the six examples in Sect. 11.2.1.

Example 11.1 Revisited We can show $\frac{3}{2} = \frac{12}{8}$ as follows. $\frac{3}{2}$ is the third point (to the right of 0) in the sequence of halves.

Now divide each of the segments $[0, \frac{1}{2}]$, $[\frac{1}{2}, 1]$, $[1, \frac{3}{2}]$, etc., into 4 equal parts. Then together with the sequence of halves, these new division points become the sequence of eighths. The point $\frac{3}{2}$ now becomes the 12th point in the sequence of eighths, and it follows from the definition of fractions that $\frac{3}{2} = \frac{12}{8}$.

The reasoning for showing equivalent fractions in general, $\frac{ca}{cb} = \frac{a}{b}$ for all fractions $\frac{a}{b}$ and nonzero whole numbers c, is entirely similar (see page 29 of Wu 2016a).

Example 11.2 Revisited We will prove CMA, i.e.,

$$\frac{a}{b} = \frac{c}{d} \quad \text{implies} \quad ad = bc \tag{11.4}$$

by making use of equivalent fractions (see Example 11.1), but without making use of the multiplication of fractions. We have $\frac{a}{b} = \frac{ad}{bd}$ and $\frac{c}{d} = \frac{bc}{bd}$ by equivalent fractions. Therefore the hypothesis means $\frac{ad}{bd} = \frac{bc}{bd}$. Thus in the sequence of bd-ths, the ad-th point coincides with the bc-th point. This can happen only if $ad = bc$.

It may be mentioned that in the mathematics education literature, CMA is regarded as an algorithm that is "rote and without meaning" (see page 348 of Billstein-Liebeskind-Lott 2007, for example). This is a piece of misinformation

[14]By *combining* the two number lines into one, if one prefers to think of it this way.

that begs to be corrected. As we already remarked at the end of Example 11.2 in Sect. 11.2.1, CMA *is a basic skill in K–12 that should be in the repertoire of every student and every teacher.* In addition, the fact that CMA in the form of (11.4) continues to hold for *rational numbers* a, b, c, and d is given on page 180 of Wu (2016a). The extension to *real numbers* a, b, c, and d is guaranteed by what is called FASM (Fundamental Assumption of School Mathematics); see Wu (2016a), Section 2.7 (the proof of FASM is given in Section 2.1 of the third volume of Wu 2020).

Example 11.3 Revisited To compute $\frac{3}{8} + \frac{5}{6}$, we begin by *defining* the **addition of fractions**. To this end, we look to whole numbers for guidance because, as points on the number line, whole numbers and fractions are on an equal footing. For whole numbers, addition holds no mystery: $4 + 3$, for example, is the total length obtained by combining segments of lengths 4 and 3, respectively. Precisely, consider the **concatenation** of the two segments of lengths 4 and 3, which is the segment obtained by placing these segments end-to-end on the number line:

Now suppose we are given two fractions $\frac{4}{7}$ and $\frac{1}{5}$ (for example). By definition (Wu 1998; also Section 14.1 of Wu 2011a), the **fraction addition** $\frac{4}{7} + \frac{1}{5}$ is the length of the concatenation of the two segments of lengths $\frac{4}{7}$ and $\frac{1}{5}$:

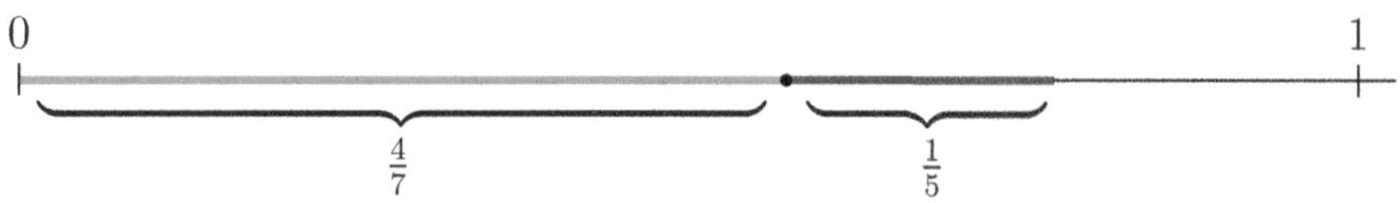

This definition of fraction addition immediately shows that addition—even for fractions—is still just *putting things together*. (See the discussion of the *coherence* of mathematics in (IV) of Sect. 11.3.1.)

Now that we know what we are asked to do regarding $\frac{3}{8} + \frac{5}{6}$, we can try to compute it, i.e., obtain a formula for $\frac{3}{8} + \frac{5}{6}$. First, observe that the addition of fractions with the same denominator becomes very simple. For example,

$$\frac{4}{7} + \frac{6}{7} = \frac{4+6}{7} \tag{11.5}$$

because $\frac{4}{7}$ is the total length of 4 segments of length $\frac{1}{7}$ and $\frac{6}{7}$ is the total length of 6 segments of length $\frac{1}{7}$, so that by the definition of fraction addition, the left side of Eq. (11.5) is the total length of $(4+6)$ segments of length $\frac{1}{7}$ and is therefore equal to the right side of (11.5). Observe that, conceptually, there is no difference between

$\frac{4}{7} + \frac{6}{7}$ and $4 + 6$. In general, $\frac{a}{b} + \frac{c}{b} = \frac{a+c}{b}$ for all fractions $\frac{a}{b}$ and $\frac{c}{b}$, for the same reason.

Although $\frac{3}{8}$ and $\frac{5}{6}$ do not have the same denominator, we can "make them have the same denominator" by appealing to equivalent fractions (see **Example 11.1 Revisited** above). Thus, both fractions $\frac{3}{8}$ and $\frac{5}{6}$ belong to the sequence of 48-ths ($48 = 6 \times 8$) because

$$\frac{3}{8} = \frac{18}{48} \text{ and } \frac{5}{6} = \frac{40}{48} \tag{11.6}$$

Therefore

$$\frac{3}{8} + \frac{5}{6} = \frac{18}{48} + \frac{40}{48} \tag{11.7}$$

By the preceding observation (see (11.5)), we have

$$\frac{18}{48} + \frac{40}{48} = \frac{18 + 40}{48} \tag{11.8}$$

Putting equations (11.7) and (11.8) together, we obtain

$$\frac{3}{8} + \frac{5}{6} = \frac{58}{48} \tag{11.9}$$

More generally, if we retrace our steps and do not multiply out everything, then what this computation shows is actually that

$$\frac{3}{8} + \frac{5}{6} = \frac{(3 \times 6) + (5 \times 8)}{6 \times 8}$$

Now if we introduce symbolic notation, the same reasoning shows in general that for all fractions $\frac{a}{b}$ and $\frac{c}{d}$,

$$\frac{a}{b} + \frac{c}{d} = \frac{ad + bc}{bd}$$

(Whether or not this symbolic formula should be *proved* in a 5-th grade classroom will depend on the teacher's judgment of the quality of the students. In general, a symbolic proof may be too much of a good thing for the average fifth grader.)

Critical Observations The computation in Eq. (11.9) is the culmination of steps (11.6)–(11.8), and each of which is based *strictly* on the definition of what a fraction is, the definition of what fraction addition means, the *prior* established facts on equivalent fractions and (11.5), and standard logical deduction. There is nothing about some abstruse higher-order "conceptual understanding" that students are supposed to "get" but often don't, and nothing that students have never seen before.

So it is learnable. Furthermore, this reasoning only requires that the two fractions $\frac{a}{b}$ and $\frac{c}{d}$ be changed to two fractions with the same denominator (as in (11.6)), and it doesn't matter what that denominator is. Therefore, any thoughts about the *least* common denominator would be extraneous to this reasoning. Apparently, this approach to the addition of fractions (Wu 1999a) has been implemented in school classrooms with some success (Bingea undated).

Example 11.4 Revisited To compute $\frac{2}{3} \times \frac{5}{8}$, once again, we have to first find out what these symbols mean. So we need a *definition* of the multiplication of fractions: $\frac{2}{3} \times \frac{5}{8}$ means the total length of *2* of the parts if we partition the length of the segment $[0, \frac{5}{8}]$ between 0 and $\frac{5}{8}$ into *3* parts of equal length.

Now, how to partition the segment $[0, \frac{5}{8}]$ into *3* parts of equal length? For this purpose, we call on equivalent fractions to rewrite $\frac{5}{8}$ as

$$\frac{5}{8} = \frac{3 \times 5}{3 \times 8} \tag{11.10}$$

The motivation for doing this is that the numerator of the right side, 3×5 now exhibits an obvious partition into 3 equal parts, namely $3 \times 5 = 5 + 5 + 5$. This then leads to the following simple fact (since we are doing fraction multiplication, of course *the addition of fractions is already an established skill*):

$$\frac{5}{8} = \frac{5+5+5}{3 \times 8} = \frac{5}{24} + \frac{5}{24} + \frac{5}{24} \tag{11.11}$$

According to the definition of fraction addition (see the preceding **Example 11.3 Revisited**), the right side of (11.11)—being a concatenation of 3 segments each of length $\frac{5}{25}$ —exhibits a partition of $[0, \frac{5}{8}]$ into three parts of equal length, with each part having length $\frac{5}{24}$. Therefore, using "part" as an abbreviation for "one of the parts when $[0, \frac{5}{8}]$ is partitioned into 3 parts of equal length," we obtain

$$\text{total length of 2 parts} = \frac{5}{24} + \frac{5}{24} = \frac{10}{24} \tag{11.12}$$

In view of the definition of $\frac{2}{3} \times \frac{5}{8}$, (11.12) implies that

$$\frac{2}{3} \times \frac{5}{8} = \frac{10}{24} \tag{11.13}$$

Once again, if we retrace our steps and do not multiply out everything, what this reasoning demonstrates is the fact that

$$\frac{2}{3} \times \frac{5}{8} = \frac{2 \times 5}{3 \times 8}$$

Critical Observations As in the case of adding fractions, the conclusion in Eq. (11.13) is reached via steps (11.10)–(11.12), and each of the latter is based on either a definition (e.g., fraction addition, fraction multiplication) or an established fact (e.g., equivalent fractions, how to add fractions), or both, and the use of logic. This kind of reasoning, i.e., the ability to envision a rough sketch of the intermediate steps (11.10)–(11.12) together with the argument supporting each step, does not come easily to most people, especially beginners. It takes plenty of exposure *and* practice to learn it, and we have to convince students that it is worth learning because *this process of reasoning is the basic methodology of mathematics.*[15] Of course, beginners learn by imitation (as do we all, including professional mathematicians) during their halting first steps towards proficiency, so a classroom teacher can ask students, right after showing this piece of reasoning, to go to the board to explain something like $\frac{2}{3} \times \frac{11}{7} = \frac{22}{21}$ or $\frac{4}{5} \times \frac{5}{8} = \frac{20}{40}$. Then, perhaps, also $\frac{5}{8} \times \frac{2}{3} = \frac{10}{24}$. In due course, the teacher can point out the obvious, namely the fact that if the preceding reasoning is written out in greater detail, then it actually proves that $\frac{2}{3} \times \frac{5}{8} = \frac{2 \times 5}{3 \times 8}$, that $\frac{2}{3} \times \frac{11}{7} = \frac{2 \times 11}{3 \times 7}$, etc., so that in general,

$$\frac{a}{b} \times \frac{c}{d} = \frac{a \times c}{b \times d}$$

for all fractions $\frac{a}{b}$ and $\frac{c}{d}$. (The symbolic statement likely will not be appropriate for all classrooms.) There is no end of variations on this pedagogical theme, and each teacher will find his or her own preferred method of delivery.

In the two preceding examples, the method of logical inference used is standard and therefore available to all, and the established facts (such as equivalent fractions) are also available to all. If precise definitions are also routinely given in textbooks, then the whole process of reasoning will become an open book that is available to all. *This is the necessary first step that will make mathematics learnable to one and all.* Therefore having precise definitions for all concepts is a critical ingredient in making reasoning—and hence mathematics itself—learnable.

At this point, it should be clear that we insist on having precise definitions and reasoning in school mathematics education, not because they are mathematicians' professional fixations, but because, as we said earlier, *school mathematics is not learnable without them.* Concepts and skills not connected by reasoning become isolated factoids that can only be learned by brute force memorization. Therefore one may speculate that, as students go up through the grades, such concepts and skills pile up in TSM and, at some point, they overwhelm students' memory banks[16] by sheer volume and TSM ceases to be learnable even by memorization. This speculation about the effects of TSM on student learning is consistent with the performance of US students on TIMSS in 1995 (TIMSS 1995 Results, 1995). However, when *reasoning* is there to connect the concepts and skills, it includes

[15]To a large extent, this is the basic methodology of science as well.

[16]In the terminology of computers, not enough RAM.

them in a story line that makes sense of them; it renders them learnable (also see pp. 118-120 of National Research Council 2001).

Example 11.5 Revisited As noted in Sect. 11.2.1, if mixed numbers are introduced *after* the addition of fractions, then the mixed number $7\frac{2}{3}$ would be defined as the abbreviation for $7 + \frac{2}{3}$, so that

$$7\frac{2}{3} = 7 + \frac{2}{3} = \frac{7 \times 3}{3} + \frac{2}{3} = \frac{23}{3}$$

and no memorization would be necessary.

Example 11.6 Revisited The correct definition of a finite decimal is that it is a fraction whose denominator is 10^n for some whole number n. For example, 2.307 is the fraction

$$\frac{2307}{1000}$$

Once we know how to add fractions, then the expanded form of 2307 being $2307 = 2000 + 300 + 7$, we get

$$\frac{2307}{1000} = \frac{2000 + 300 + 7}{1000} = \frac{2000}{1000} + \frac{300}{1000} + \frac{7}{1000} = 2 + 0.3 + 0.007$$

Hence, the fact that 2.307 *is* “2 and 3 tenths and 7 thousandths” becomes a *provable* theorem if students are taught about finite decimals *after* fractions.

11.4 Textbooks

We can now return to the first of our two main concerns: how to give students access to PBM rather than TSM.

A main thrust of this article is about how to repair the damage inflicted on teachers and students by TSM, *Textbook* School Mathematics. An obvious question is why we are wasting our time here talking about damage control instead of directly going to the source and writing better school mathematics textbooks. The simple answer is that most of the school textbooks come from major publishers, and there are no ready-made tools to combat the bottom-line mentality of big business (in this regard, the article Keeghan 2012 is very informative). For this reason, most of the nation’s schools are still dependent on TSM textbooks from the major publishers. It is also the case that the publishing industry is not under any kind of federal or state control and is free to produce any textbooks it can afford to put out. From a publisher’s standpoint, so long as its products are welcomed by enough teachers, there is little incentive to change anything, TSM and all. Since there are many teachers out there who were brought up by TSM and are therefore comfortable teaching TSM as of 2019, there is still a ready-made market for the publishing industry to exploit. It therefore seems likely that, until the majority of teachers reject TSM-infested textbooks, TSM will live on in school classrooms. This then adds urgency to our second topic of concern: how to produce inservice teachers who are capable of teaching PBM. Getting better-informed teachers who reject TSM out of hand would seem to be the best hope of breaking the vicious cycle of TSM.

Since the release of CCSSM in 2010, there have been several attempts to write curricula according to the CCSSM by exploiting the internet using online publishing. A few of the new curricula show promise, according to some reports. However, since so few in the world of education seem to be at all concerned with mathematical content or aware of the continued menace of TSM, many of the textbook evaluation agencies should be approached with a great deal of caution. Overall, much remains to be done in the arena of curricular evaluation.

Common Core was quite aware of the inadequacy of existing textbooks for the implementation of CCSSM. It has published two documents for the benefit of publishers: a 24-page document (Common Core 2012) on the K–8 curriculum and a 20-page document (Common Core 2013) on the high school curriculum. They exhort publishers to meet the goals of focus, coherence, and rigor in their textbooks. Neither document mentions the phenomenon of TSM, however.

A more ambitious undertaking is a six-volume, 2500-page project from this author that gives a complete exposition[17] of the K-12 mathematics curriculum according to PBM (Wu 2011a, 2016a,b, and 2020). There are presumably many ways to present the school mathematics curriculum in accordance with the fundamental principles of mathematics, but for now we can make use of what we have got. These six volumes are not student textbooks; they are textbooks for teachers' PD. Given the level of detail in these 2500 pages, however, it should not be difficult to create student texts out of them with the help of some standard pedagogical embellishments. In any case, an eighth-grade student textbook based on these volumes will be offered online (https://math.berkeley.edu/~wu/) in the near future. Since the drafts of some of these six volumes have served as blueprints for a good many standards in CCSSM, there is no fear that any curricular materials based on these volumes will be out-of-date anytime soon.

11.5 Professional Development

Everything we have said so far points to the urgency of replacing our teachers' knowledge of TSM by PBM. This will certainly tax our ability to do effective PD. Let us be clear about what we expect the PD to accomplish. It will not be about

[17]Strictly speaking, these six volumes do not cover geometry in grades K-5 because the file on this topic promised in Wu (2011a), has not yet been made available. However, Chapters 4 and 5 in Wu (2016a), serve to fill this gap to a large extent.

tweaking teachers' content knowledge here and there but, rather, about a revamping of their knowledge of mathematics from the ground up. Because of our systemic negligence, teachers have never been exposed to anything resembling PBM (see, e.g., Wu 2011b), yet we want them to master PBM in short order and turn around to teach it to their students. This is not going to be easy.

In the first subsection, we will go into some detail to explain the kind of hard work that is involved. We will focus on PD for *inservice* teachers[18] because the current implementation of CCSSM (Common Core 2010) requires teachers who can teach PBM. For example, CCSSM asks teachers to teach mathematics in a way that is "coherent," "stresses conceptual understanding of key ideas," helps students to "reason abstractly and quantitatively," encourages students to "construct viable arguments and critique the reasoning of others" and "attend to precision," etc. (pp. 3-7 in Common Core 2010). The long-term neglect of the mathematical education of teachers leads us to believe that most teachers may not be able to rise to this lofty challenge and that their need for content-based inservice PD will be considerable. Although there is apparently no hard data as yet to substantiate this belief, the available anecdotal evidence (cf. Education Week 2014; Loewus 2016, 2017, and Sawchuk 2016) does point in this direction. In addition, what the author has personally learned from teachers and math coaches in several states—including California—is also consistent with this belief. Our proposed PD therefore cannot be the routine variety and its parameters must be carefully prescribed. This is what we will try to do in the second subsection. In the third subsection, we will describe—for the sake of providing a point of reference—one PD program that has been tried with some success to teach teachers PBM.

There is a jarring note hidden behind this optimistic discussion of PD, however. In the last decade, two studies by IES on the impact of content-based PD on student learning have appeared, Garet (2011) and Garet (2016). They seem to shut the door on any hope that PD can help teachers raise student achievement. If there is any validity to the IES studies, the present article on what "good" PD is and how to implement it would simply be a waste of everybody's time. For this reason, we must make an effort to examine these studies *critically*. This will be carried out in the next section.

11.5.1 The Hard Work of Learning PBM

For inservice mathematics teachers trying to learn PBM, a useful analogy may be learning the second language.[19] The immense difference between PBM and TSM

[18] We will also make some comments about PD for preservice teachers in the last section of this article (Sect. 11.7).

[19] One should not push this analogy too far, however. No language has anything like the logical coherence of mathematics.

dwarfs what little they happen to have in common: the topics and the skills, for instance. Since PBM asks teachers to repackage these topics and sometimes even to teach them in a different order (e.g., define mixed numbers only after fraction addition has been discussed; see **Example 11.5 Revisited** in Sect. 11.3.2), the prospect of learning PBM will be daunting to most. There is also a paradoxical aspect to the attempt by inservice teachers to learn PBM, and it is the fact that while we find fault with TSM for oversimplifying school mathematics to a few sound bites, it is actually easier to just "teach" sound bites! Some teachers who have gotten used to "teaching" the sound bites of TSM *may* find teaching PBM with its many attendant cognitive complexities to be a very big stretch. What is good for the learners may not always be easy for the teachers! Let this be a warning. What follows is a more detailed explanation of the hard work involved in learning PBM.

Learning About Reasoning

Learning how to reason is painstaking work under the best of circumstances.[20] Except for the most rudimentary, one-step variety that we inherit from our ancestors on the African savanna tens of thousands years ago, such as "fright → flight," reasoning is not an inborn skill like speech or running. For most teachers who have been immersed in TSM all their lives, learning how to reason about basic tasks that they used to teach by rote with ease is difficult enough. Having to also learn how to *explain* the reasoning process to students makes it doubly difficult, and trying to empower students with the fundamentals reasoning skills is trebly difficult.

Take the case of adding fractions (see Example 11.3 in Sect. 11.2.1). We can complain all we want about the use of LCD and the absence of any explanation of what "addition" means in TSM, but to most inservice teachers, this rote skill has probably become second nature. This LCD skill is simple to teach by rote! The procedure is short, and all a teacher has to do is give students lots of drills. Now PBM changes all that: a teacher has to explain what it *means* to add two fractions, use equivalent fractions to put the two given fractions into the same sequence of $n$ths for some n, and remind students of the *meaning* of adding whole numbers (see **Example 11.3 Revisited** in Sect. 11.3.2). We know how some students hate to be reminded of anything other than what is right in front of them! It is definitely a lot more work than teaching the rote skill using LCD.

The case of multiplying fractions (see Example 11.4 in Sect. 11.2.1) is somewhat similar. Even some mathematicians mistakenly consider fraction multiplication to be a pleasure to teach because it is procedurally so simple (see, e.g., Aharoni 2015): just multiply across the top and the bottom. By contrast, look at **Example 11.4**

---

[20] It has been suggested that there is an apparent contradiction between this statement and the earlier one made in **Example 11.4 Revisited**, to the effect that PBM will make reasoning learnable to one and all. But there isn't. Such a misunderstanding would arise only if one erroneously equates "learning" with "learning without effort." Learning anything worthwhile in life requires effort, e.g., learning how to read require the strenuous effort of memorizing the alphabet and a continuous influx of new vocabulary. What is at issue is whether unnecessary roadblocks are thrown in the learner's path. TSM throws such roadblocks—too many to count—but PBM doesn't.

Revisited in Sect. 11.3.2: the definition of fraction multiplication is among the longest and most complex in elementary mathematics, and the reasoning in the teaching of this so-called simple skill becomes quite delicate according to PBM (see Eqs. (11.10)–(11.12) therein). Real effort is now required for its mastery.

We should mention another telling example about reasoning: the teaching of speed problems and the related so-called *rate* problems. Consider the following:

> Luis usually walks the 1.5 miles to his school in 25 minutes. However, due to road repair, he has to take a 1.7-mile route today. If he walks at his usual speed, how much time will it take him to get to his school? (Siegler et al. 2010, page 38.)

In TSM, the phrase "at his usual speed" (or "at this speed") is code for *setting up a proportion*. In other words, given that Luis walks 1.5 miles in 25 min, if Luis walks 1.7 miles in x minutes "at his usual speed," TSM instructs us to invoke what is known as *proportional reasoning* to set up a proportion:

$$\frac{1.5}{25} = \frac{1.7}{x} \tag{11.14}$$

Now use CMA[21] to get $1.5x = 25 \times 1.7$. So $x = 28\frac{1}{3}$ min.

In the present context of getting teachers to learn about reasoning, something almost leaps off the page: the simple solution involves a rote skill but *no* reasoning! But how can one arrive at (11.14) by the use of reasoning?

The fact is that, as is, the problem cannot be solved because since we have no idea how Luis normally walks to school,[22] we know nothing about how he walks "at his usual speed." Consequently, we have no information about how he walks to school during the road repair either. *We have not been given sufficient information to know how to proceed.*

If we want to base mathematics on reasoning, then we will have to add some precise assumptions about the way he walks. Here is a standard one: let Luis walk a total distance of $f(t)$ miles after t minutes, then we assume that $f(t)$ is a *linear function of t without constant term*, i.e., $f(t) = vt$ for a fixed constant v. This assumption would justify equation (11.14) because all it says is that

$$\frac{f(25)}{25} = \frac{f(x)}{x}$$

Indeed, both are equal to v in this case.

Now, because such problems are usually introduced into the curriculum before students learn about linear functions, we will describe another way—suitable for

---

[21] As noted at the end of **Example 11.2 Revisited** in Sect. 11.3.2, what is being used here is actually not CMA but the extension of CMA to rational numbers.

[22] Does he run the first mile in 10 min and slowly stroll to school in the remaining 15 min?

use in the 6th or 7th grade—to deal with Eq. (11.14). For an object in motion, we introduce the concept of its **average speed over the time interval from t_1 to t_2**, ($t_1 < t_2$), as

$$\frac{\text{total distance traveled from time } t_1 \text{ to } t_2}{t_2 - t_1} \tag{11.15}$$

In terms of *average speed*, the Luis problem can be properly reformulated as follows:

> Luis usually walks the 1.5 miles to his school in 25 minutes. However, due to road repair, he has to take a 1.7-mile route today. If his two trips *have the same average speed,* how much time will it take him to get to his school?

Now equation (11.14) is correct because it is based on the assumption that the two average speeds are the same. We are using the *definition* of "average speed"!

The more common way—and a more nuanced way—of handling the Luis problem is to formulate it in terms of *constant speed*. By definition, a motion **has constant speed v** if its average speed over *any* time interval is always equal to v. Then a correct formulation of the preceding problem in terms of constant speed is the following:[23]

> Luis usually walks the 1.5 miles to his school in 25 minutes. However, due to road repair, he has to take a 1.7-mile route today. If he always walks at the same constant speed, how much time will it take him to get to his school?

Equation (11.14) is now justified by recognizing the fact that its left side is Luis' average speed over the time interval it takes him to walk the normal 1.5 miles, and the right side is the average speed over the time interval it takes him to walk the 1.7 miles. The assumption that he walks at the *same* constant speed then implies that these two average speeds are equal, which is Eq. (11.14).

In all three cases, we get to witness one of the basic characteristics of reasoning: *make explicit use of precise definitions* to draw conclusions. The goal of PBM is to get students used to the habit of analyzing each problem on its own merits by the use of explicit assumptions, explicit definitions, and reasoning.

One may object that the amount of reasoning used to solve the preceding problem formulated in terms of constant speed is too little to be cause for celebration. Granted, but look at the alternative of appealing to "proportional reasoning": the latter is rote learning, plain and simple.

Let us not forget that our teachers were brought up in TSM and are used to setting up proportions. If they want to teach PBM, then they must know this background information about speed problems to be able to answer students' questions, e.g., "why don't we just set up a proportion?" Above all else, teachers who want to

[23] It can be shown that the assumption of constant speed is equivalent to the earlier assumption that the distance function describing Luis' distance from his starting point is a linear function without constant term. See Theorem 7.1 on page 138 of Wu (2016b).

promote PBM have to know why "proportional reasoning" is not *reasoning* at all. For example, let us do the following problem by proportional reasoning:

> A free-falling stone is dropped from 600 ft. It drops 64 ft in 2 seconds. How far does it drop in 3 seconds?

Obviously, proportional reasoning yields an answer of 96 ft, whereas physics gives the correct answer of 144 ft. The reason is that *the falling stone does not move at constant speed*. So unless teachers insist on having precise definitions for all the relevant concepts—so that *constant speed* gets defined—and unless teachers insist on solving problems by reasoning, they cannot even explain to students *why* the free-falling stone problem cannot be done by setting up a proportion.

Altogether, we see that we are imposing a heavy cognitive load on teachers in trying to get them to embrace PBM. Nevertheless, we must do all we can to help teachers acquire the reasoning skill because we have no choice. Students must learn to reason for their survival in year 2019, and if teachers cannot learn to reason mathematically, how can we hope that their students will? So we must try harder.

It remains to make some general remarks on the need for flexibility in teaching reasoning in the school classroom. When we say reasoning should attend every statement in mathematics, we actually mean *grade-appropriate* reasoning. For example, in **Example 11.3 Revisited** of Sect. 11.3.2, we mentioned that although the reasoning suffices to prove the general formula for addition,

$$\frac{a}{b} + \frac{c}{d} = \frac{ad + bc}{bd}$$

it may not be appropriate to give this general reasoning in a typical fifth grade classroom. A more reasonable alternative is to prove the formula only for many specific values of a, b, c, and d. We also made a similar remark about the product formula for arbitrary fractions in **Example 11.4 Revisited** of the same subsection:

$$\frac{a}{b} \times \frac{c}{d} = \frac{a \times c}{b \times d}$$

To firm up this message, we will use an example from middle and high school geometry: the teaching of the theorem that the angle sum of a triangle is 180°. There is the standard, intuitive proof obtained by drawing a line parallel to side AB and passing through the vertex C of the following triangle:

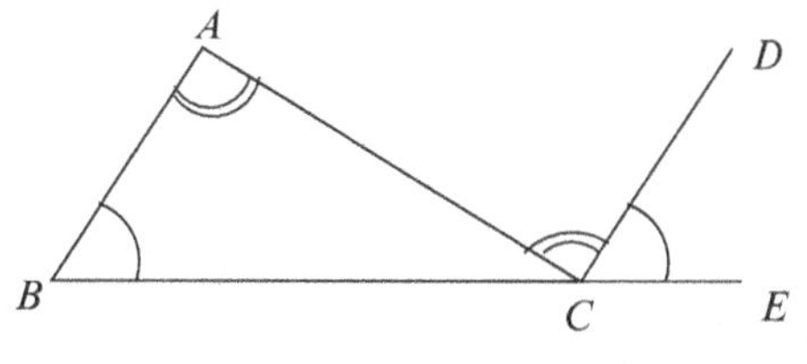

Then there is a *correct* proof that fills in all the nonintuitive gaps of this standard intuitive proof. The latter is unfortunately very subtle as well as very boring. For *middle school* students, the correct proof is not worth the investment of time and effort. Thus, on page 316 of Wu (2016a) (which was written for mathematics educators and middle school teachers), the standard, intuitive proof is given, followed by the remark on page 317, (*loc. cit.*) to the effect that the steps in the given proof are

> essentially correct. Nevertheless, from a strictly *mathematical* standpoint, one can find fault with them for certain omissions in the details.

Then it goes on to say that insofar as

> our main purpose here is to acquire geometric intuition and make the first step towards the mastery of geometric proofs,

the intuitive proof will serve. Needless to say, a correct proof should be given when the occasion calls for it in the high school course on geometry (see Section 6.5 of the second volume, *Algebra and Geometry*, of Wu 2020).

Teaching is, among many things, the result of negotiations between what is correct and what is possible in the face of the reality in a classroom The purpose of our effort to get teachers ready to teach PBM is to provide them with the needed mathematical information so that they have the freedom to decide what is possible in a given classroom.

Learning About Definitions

It should be clear from the discussion up to this point that definitions and reasoning are essentially intertwined, but there are a few things about definitions that deserve to be discussed separately.

For teachers brought up in TSM, perhaps the most difficult thing to accept about PBM is that the definition of a concept ceases being something to be memorized for standardized tests and then cast aside, but is now the foundation for any reasoning about the concept. TSM has no reasoning to speak of, so definitions play no role in its version of "mathematics learning." Teachers with a TSM background therefore have difficulty getting used to the fact that PBM puts every definition to use for the purpose of reasoning. We have seen how the definition of a fraction and the definition of the addition of fractions are used, *literally,* to derive the formula for the addition of fractions (see **Example 11.3 Revisited** in Sect. 11.3.2), how the analogous derivation happens with fraction multiplication (see **Example 11.4 Revisited** in Sect. 11.3.2), and how Luis' walking problem in the early part of this subsection can be solved simply by the use of reasoning once a precise definition of *constant speed* is given. This point is stressed throughout all six volumes of Wu (2011a, 2016a,b), and 2020, but if personal experience is any guide, it still does not come easily to teachers.

One may speculate that things would go more smoothly in the ongoing effort to convince teachers about the importance of precise definitions if the education literature would also make such an advocacy. This is not happening, however, because TSM has held sway over many educators as well.[24] Consequently, some educators downplay the importance of precise definitions. In discussing ratio and rate, for example, Susan Lamon makes the following statement:

> Even if we could precisely define ratios and rates and the difference between them, definitions do not discharge the full meaning of the idea being defined. The nature and meaning of rates and ratios come from problem situations. (Lamon 1999, page 165.)

There is an obvious misunderstanding here about the role of definitions in mathematics: the mathematical definition of a concept is not required to "discharge the full meaning" of the concept so defined. All that it is obligated to do is furnish *all* the information that is needed for any reasoning regarding that concept, no more and no less. In this connection, what Polya has to say is very much to the point: "The mathematician is not concerned with the current meaning of his technical term.... The mathematical definition creates the mathematical meaning" (Polya 1957, page 86). As we mentioned in Sect. 11.3.1, the main virtue of presenting a precise definition of a concept is to tell the whole *mathematical* truth from the beginning to facilitate mathematics learning.[25] This no-hidden-agenda feature of mathematics is essential to making mathematics learnable because, above all else, it establishes a sense of trust between the learner and the mathematics. It tells the learner that all the cards are now on the table, so just look closely at what you have got! The failure of TSM to develop mathematics according to precise definitions has so far resulted in gnawing suspicions and distrust from learners at the outset. This is no way to make mathematics learnable.

Having argued for the need to make teachers see the importance of precise definitions in PBM, we also want to supplement the argument with the remark that by no means are we advocating for the unmotivated (highhanded) presentation of definitions that we sometimes see in advanced mathematics. Any PD must also pay attention to the art of persuasion in giving definitions. This is why the definition of a fraction in terms of the number line in Wu (2016a), is preceded by eight pages explaining why a precise definition of a fraction is necessary (pp. 3-10, loc. cit.). Likewise, the definition of slope on page 66 of Wu (2016b) is preceded by almost five pages of discussion about the intuitive meaning of the concept of slope and how this intuition may be captured in a precise definition. There are many other such suggestions about how to present a precise definition in those two volumes as well as in Wu (2011a). Teachers should be aware of the many pedagogical flexibilities in making precise definitions an integral part of their teaching.

Finally, let us consider one more objection to the recommendation that there be precise definitions for every concept in school mathematics, to the effect that

---

[24]When all is said and done, mathematicians have to be held mainly accountable for the deterioration of the content component in mathematics education (see Wu 2011b).

[25]This view of the role of definitions in mathematics in general is a product of the twentieth century, as a result of many trials and errors to make mathematics more transparent and less prone to obscurantism. The attitude towards definitions of the mathematicians in centuries past was actually remarkably close to that of Lamon's (cf. e.g., Quinn 2012).

students should not be *told* the definition of a concept because a definition is something that one formulates at the end of an exploration. So the theory is that we should let students explore a new concept until they themselves come up with something resembling a correct definition. This viewpoint about how to teach mathematics in general, and teach definitions in particular, is part of an old pedagogical debate about the Moore Method vs. direct instruction. It will not be productive to wade into this debate here except to point out that, however, the exploration approach is done, it is very time-consuming and one must take the availability of instructional time into account (see Section 3 of Wu 1999c). Moreover, after a precise definition has been formulated at the end of the exploration, one must bring mathematical closure by *retracing the steps of the exploration to let students see how the relevant mathematics can be developed on the basis of the precise definition*. This will further decrease the available instructional time. Thus far, discussions in the education literature seem to be oblivious to the need for retracing the steps to show students how mathematics is developed using definitions and the need for additional instructional time to make room for the retracing. This is but one manifestation of the tendency in the education literature to make an advocacy without also meticulously enumerating the possible detrimental side effects (see Section 4 of Wu 1999c for a fairly comprehensive discussion).

Other Issues

For teachers transitioning from TSM to PBM, a minor—though a significant—issue to be confronted is that they will have to learn how to teach certain topics in a different logical order. We have already remarked in Sect. 11.3.2 that the concept of mixed numbers can no longer be taught right after fractions are introduced (as in TSM) but must wait until after the discussion of the addition of fractions (**Example 11.5 Revisited**). In the same subsection, we also remarked that finite decimals can no longer be taught in a separate track independent of fractions but must be taught as a special kind of fractions (**Example 11.6 Revisited**); this profoundly changes the teaching of finite decimals because we can now *explain* the algorithms for decimal addition, subtraction, and multiplication (for division, see pp. 81-86 in Wu 2016a). Perhaps the most prominent change of this kind is the teaching of the slope of a line in middle school. In PBM, the definition of the slope depends on having available the angle-angle criterion for similar triangles so that slope has to be taken up *after* a serious discussion of the concept of similar triangles (Wu 2016b, Section 4.3). Now, although similar figures in TSM are those "with the same shape but not necessarily the same size," PBM will insist on a precise definition of similarity. This, in turns, requires that the school curriculum pave the way for such a precise definition. Teachers must therefore be prepared for these massive changes in their internal conception of school mathematics. Since such a change in the teaching of slope is also part of CCSSM, I received an email from an indignant teacher in the state of Washington right after the release of CCSSM in June of 2010. He wrote:

> After 13 years of teaching high school algebra, I wonder why you see similarity as critically important to Algebra I mastery—that certainly never occurred to me as a teacher of algebra.

> ... What makes you say that a student needs to understand similar triangles in order to write the equation of a straight line between two points?

Clearly, teachers have to be willing to keep an open mind to learn PBM.

Among the fundamental principles of mathematics, the longitudinal coherence of school mathematics may be one of the hardest things for teachers to appreciate. The only way they can learn it is to be exposed to several grades' worth of PBM over a long stretch of time. For elementary school mathematics, we can be more specific. Take, for instance, the coherence of the four standard algorithms for whole numbers. We already pointed out that all four revolve around a single idea: if we can compute with single-digit numbers, then we can compute with any numbers no matter how large (see Sect. 11.3.1). It would be impossible to see such coherence unless one knows how to prove this fact for each of $+$, $-$, $\times$, and $\div$, and this may not be so easy especially for the long division algorithm. (Is that algorithm even a theorem? And if it is, what does it say in the first place? see Sections 7.3–7.5 of Wu 2011a.) Thus a PD session devoted to making elementary teachers see such coherence has to first give the *detailed* proofs of four separate theorems, and then has to give them time to digest them so that they can step back and take note of the similarity in these proofs. It should take no less than three full days. Teaching elementary teachers about fractions in a way that enables them to see the coherence between the four arithmetic operations on whole numbers and fractions should take at least another five full days. And so on.

It takes time to learn PBM.

11.5.2 The Inservice PD We Need

Our tentative conclusion is that, for real improvement in school mathematics education to materialize, we will need a massive investment in long-term, content-based, inservice PD to get our teachers ready to teach PBM. But can we put our trust in PD to get this done? The answer is unfortunately not straightforward.

First of all, we have to provide more details about this proposed long-term, content-based, inservice PD. By *long-term*, we have in mind a long stretch of time of 1 week to 3 weeks during the summer. One may believe that, for example, instead of 1 week in the summer, we can parcel out the 40 h of PD into 20 two-hour sessions during the school year, with one session per week. The problem with breaking up 1 week in the summer into 20 sessions in the school year is that teachers have too many obligations during the school year to remember what they learn from week to week. If we have any design on impressing the coherence of PBM on teachers, these 2 h sessions will not be the answer. Moreover, learning mathematics requires serious mental concentration. Given how teachers already have to multi-task all through the school year, summer may be the only time they can summon this kind of concentration necessary for learning. Creating a learning environment for teachers by holding the PD in consecutive days in the summer is thus the only viable option. Holding the PD in consecutive days also tends to yield the pleasant dividend of promoting collaboration among teachers. Beyond the summer session, we should also help them retain—or remind them to apply—the new knowledge in the following school year. To this end, some Saturday review sessions throughout the school year would also be advisable.

About the *content* of the content-based PD, we hope no argument is necessary at this point that this content refers to the content of PBM and not TSM. We note in passing that, because most professional developers were brought up in TSM, getting and vetting competent providers for the needed PD will be a nontrivial problem.

"Long-term" and "content-based" are certainly not qualities one normally attributes to most of the PD currently provided by school districts. Thirty years ago, Judith Warren Little wrote about the PD system in California and showed that, instead of providing learning opportunities for teachers, PD had too often devolved into a series of uncoordinated rituals (Little 1989). To those working in the trenches with the firsthand knowledge of mathematics PD in the past decades, not much has changed since the appearance of Little's article (see, e.g., Wu 1999b and U. S. Department of Education 2009, p. 95). This is by way of saying that if we are committed to using PD to help teachers learn PBM, we must be ready to fight for *long-term* and *content-based* as nonnegotiable requirements. By a happy coincidence, the 2017 publication of *Effective Professional Development* (Darling-Hammond et al. 2017) also comes to a similar conclusion. The work of Darling-Hammond et al. set out to discover the common features of effective PD in all fields and is not exclusively about mathematics, so it has little overlap with the present article. Nevertheless, it too concludes that effective PD must be "content focused" and "of sustained duration."[26]

In addition to these intellectual concerns, there is also a practical matter that is no less important. Learning mathematics is almost never a fun activity in the everyday sense of "fun"; it is *hard work*, and the PD cannot succeed without teachers' willingness to work hard. The only way to ensure that the teachers will put in the hard work is to pay them generously for their daily attendance. This therefore means that the desired PD will not only be of sustained duration and PBM-based, but also *expensive*.[27]

We should also address the role of pedagogy in PD. Let it be noted that our concept of "content" already takes pedagogy into account because PBM addresses not mathematics but *school mathematics*, i.e., how mathematics should be taught in schools. In addition, while specific pedagogical issues will inevitably arise in any PD, we must be careful to keep the amount of purely pedagogical discussions in the PD in check. I took note of this fact for the first time when I had the opportunity to observe other people's PD in California back in the late 1990s (Wu 1999b). What I found was that when content knowledge was only one of many topics of concern in

[26]We also agree with Darling-Hammond et al. on the need to "support collaboration," as we shall see in the next subsection.

[27]Though a generous stipend by itself will not guarantee hard work or success.

PD, it would not get the attention it deserved, *nor would it inspire the needed effort on the part of teachers to make foundational changes in their content knowledge.*

It remains to point out that while we have been talking about PBM as if it is a common entity that is easily accessible, the fact is that it is not. The PD providers for the PD under discussion will have to create their own materials because the PD literature is almost completely immersed in TSM. If people want to take a look at PBM, however, they can always look up the six volumes of Wu (2011a, 2016a,b), and (2020).

Having pinned down the general parameters of the PD we need, we are now ready to go to work—except that, as we mentioned at the beginning of Sect. 11.5, there is one more hurdle to overcome: recent research by the Institute of Education Sciences (**IES**) asserts that even "teachers who received the best of the best PD" are unlikely to "see large, lasting improvements in their practice, knowledge, or student learning" (Hasiotis 2015). More precisely, the two studies by Garet et al., one in 2011 (conducted over a 2-year period in 2007–2009) and another in 2016 (conducted in 2013–2014), raised serious doubts about the ability of content-intensive PD for inservice *mathematics* teachers to raise student achievement. Fortunately, a closer examination of these two studies—which will be given in the next section, Sect. 11.6—reveals serious flaws in the design of their PD that might have led them astray. Thus the jury on the alleged *in*effectiveness of PD is still out, and we all have every reason to proceed with the PD that has just been carefully outlined above. The next subsection gives a slight nudge in this direction.

11.5.3 An Inservice PD Program

The kind of PD proposed in the preceding subsection did not come out of the blue. It is based on the author's personal experience and it serves the modest goal of providing *one* data point that attests to the possible validity of such an approach to PD.

Each summer, from 2000 to 2013, I gave 3-week PD institutes for mathematics teachers of K–8, mostly in Berkeley, CA, and sometimes more than once a year. The goal of these institutes was for inservice teachers not only to learn PBM, but to also achieve long-term retention of the new knowledge. The institutes met 5 days a week (M–F), about 8 h a day (including lunch), with homework assignments every day. I would lecture to the whole group for 4–5 h each day, and the day would always end in small group meetings lasting 60–90 min, led by my three assistants. Each PD institute was followed by five Saturday follow-up sessions (one every 2 months) in the following school year to review the new content knowledge and to discuss the progress teachers were making putting it into practice in their classrooms.

Each year we put the word out about these institutes and asked teachers to apply to participate. Every participating teacher who attended all 3 weeks of the institute received a stipend of \$1500, i.e., \$100 a day.[28] For every follow-up Saturday session, each participant received a stipend of \$100. We actively encouraged group applications by teachers from schools in the same district because we believed that being able to consult with colleagues about mathematics would ensure better learning as well as better retention of PBM. It is gratifying to report that we did witness many cases of collaboration during and after the summer institutes that have continued to this day.

There were three kinds of institutes:

(1) **Elementary Institute:** whole numbers (4 days), elementary number theory (2 days), fractions—including decimals—and their arithmetic (6 days), percent, ratio, and rate (3 days). (Reference: Wu 2011a.)
(2) **Pre-Algebra Institute:** Review of fractions, percent, ratio, and rate (4 days), rational numbers (3 days), experimental geometry (3 days), geometric vocabulary, congruence, and similarity (4 days), length and area (1 day). (Reference: Wu 2016a.)
(3) **Algebra Institute:** use of symbols (2 days), linear equations in one and two variables, including a correct definition of the slope of a line (3 days), simultaneous linear equations (1 day), laws of exponents, exponential functions and their graphs (4 days), quadratic functions and their graphs (3 days). (Reference: Wu 2016b.)

Although these institutes were unapologetically devoted to the dissemination of PBM, pedagogy also played a role. In the first few years, I arranged for a short session of about an hour on pedagogical discussions at the end of each day, but it turned out that almost all the teachers were so absorbed in learning the mathematics that essentially *nobody wanted to broach the subject of pedagogy.* So I stopped making time for the discussion of pedagogy in the institutes thereafter. Instead, I made sure that pedagogy and content were given equal emphasis in the Saturday follow-up sessions. Teachers were asked to share personal stories about their own attempts to integrate the newly acquired knowledge into their classrooms, and their pedagogical strategies were then openly discussed and critiqued. It seemed quite clear that the teachers began to understand the mathematics on a deeper level when they tried to build their pedagogy on a foundation of correct school mathematics. Year after year, teachers would tell me that it was usually not until the fifth (and last) Saturday follow-up session, 9 months after their first exposure to the new content, that they *began* to feel that they owned the material. Apparently, it took the combination of the *intensive three-week immersion in content* (120 h) and the leisurely 9 months of gestation and fledgling attempts at classroom applications for them to begin making *foundational changes* in their content knowledge.

Naturally, I was interested in whether these institutes had any effect on the participating teachers' performance. Together with a school district's Director of Instruction, we applied twice to federal agencies, in 2012 and 2013, for a grant to teach fractions to elementary teachers and then (1) videotape the PD sessions and make them available online, and (2) follow each participating teacher's student scores for 3 years to examine their value-added measurements. However, our

[28] In the last six or so institutes, I tried to raise the daily stipend to \$125, but there was insufficient funding to do it.

applications were rejected. This explains why I have no hard data to report. Nevertheless, a few individual teachers have privately contacted me to share their personal successes, and these are recorded in Section 2.4 of Wu (2018). The anonymous evaluations by the teachers of the PD institutes from 2009 to 2013 are also available on request: please write wu@berkeley.edu.

11.6 The IES Impact Studies

In this section, we will closely examine the study by (Garet et al. 2011) on the impact of content-intensive PD on inservice *mathematics* teachers' ability to raise student achievement. Some attention will also be given to the impact study of Garet et al. (2016).

First of all, these two studies differ from earlier ones in their built-in methodological credibility: they used a rigorous experimental design and, in addition, had large meaningful sample sizes: the 2011 study involved some 150 teachers and the 2016 study 221 teachers. The bleak conclusion of the 2011 study was that

> after two years of implementation, the PD program did not have a statistically significant impact on teacher knowledge or on student achievement in rational numbers (page 53 of Garet et al. 2011).

The outcome of the 2016 study was that "the PD did not have a positive impact on student achievement" but "the PD had a positive impact on teacher knowledge" (page 40 and page 35, respectively, of Garet et al. 2016).

Together with an earlier impact study on *early reading instruction* (Garet et al. 2008), the 2011 impact study of Garet et al. led to the 68-page report published by TNTP, *The Mirage* (TNTP 2015), on the ineffectiveness of inservice PD in general, not just in mathematics or reading. The main conclusion of *The Mirage* was:

> In short, we bombard teachers with help, but most of it is not helpful—to teachers as professionals or to schools seeking better instruction. We are not the first to say this: In the last decade, two federally funded experimental studies of sustained, content-focused and job-embedded professional development have found that these interventions did not result in long-lasting, significant changes in teacher practice or student outcomes. (TNTP 2015, page 2)

The two "federally funded experimental studies" referred to above are the 2008 and 2011 studies of Garet et al. These studies, together with *The Mirage* and the 2016 impact study of Garet et al., were written up in the popular press (Layton 2015, and Loewus 2016), and the perception began to take hold that inservice PD as a means of achieving education improvement is a blind alley (see, e.g., Dynarski 2018). Were the conclusions of these impact studies valid, we would be wasting our time here talking about the use of PD to help teachers learn PBM.

For all these reasons, a critical examination of (at least) the 2011 impact studies on mathematics is overdue. Although the 2016 study is about fourth grade teachers (rather than seventh grade teachers as in the 2011 study), the flaws of the two studies are essentially the same from a broader perspective—both ignore the crippling effect of TSM on school mathematics in their PD design. For our purpose here, we will simply concentrate on the 2011 study.

11.6.1 The PD Program of the 2011 Impact Study

To better understand the PD program of Garet et al. (2011), we begin with a brief description of its design. It was a 2-year program for 7th grade mathematics teachers on the following topics: "fractions, decimals, percent, ratio, rate, and proportion." In the first year, teachers were given:

a 3-day summer institute on content instruction (18 hours)
5 one-day follow-up seminars during the school year (30 hours)
10 days of coaching (20 hours)

Of the *8* days of content instruction and seminars, 4 were devoted to *fractions and decimals* and the other 4 to ratio, rate, proportion, and percent. In the second year, teachers were given:

a 2-day summer institute on content instruction (12 hours)
3 one-day follow-up seminars during the school year (18 hours)
8 days of coaching (16 hours)

Of the *5* days of content instruction and seminars, 4 were devoted to ratio, rate, proportion, and percent, and 1 day to fractions and decimals. (By the second year, the PD organizers realized that teachers were having *real* trouble with ratio, rate, proportion, and percent, and they adjusted accordingly.)

The report states that the PD provided by the impact study made use of the number line for the discussion of fractions and emphasized precise definitions, but the report also states that *it was not designed to improve teachers' content knowledge* (page 21 of Garet et al. 2011). Rather, the focus of the PD was on pedagogical enhancements such as developing their ability to

> identify and address persistent student misconceptions... The pedagogical techniques that received the most attention were eliciting and responding to student thinking, using charts to keep track of particular student misconceptions... (page 21 of Garet et al. 2011)

Since students' most serious misconceptions regarding ratio, rate, proportion, and percent stem from TSM itself—which we will demonstrate in the next subsection—it is not clear how this PD could "address" these misconceptions without identifying and uprooting TSM and replacing it with PBM. We are not aware of any pedagogical strategy that can transform TSM into PBM.

In greater detail, this impact study refers to "the knowledge of topics in rational numbers that students should ideally have after completing the seventh grade" as **CK** (*common knowledge*), and to "the additional knowledge of rational numbers that may be useful for teaching rational number topics" as **SK** (*specialized knowledge for teaching*). Keep in mind that the "CK" as stated consists of nothing but TSM.

Of the total number of 13 (= 8 + 5) days devoted to content instruction and seminar in the PD program,

> the focus of the presentation in both years was on SK, and instruction in common knowledge of mathematics content CK was mainly implicit. ... the PD was not presented to teachers as an opportunity to improve their understanding of rational number content. (p. 21 of Garet et al. 2011)

Overall, we may summarize the PD program of the impact study as maintaining teachers' knowledge of TSM at the level of students' grade 7 textbooks and empowering them with better pedagogical techniques.

We note that such a PD program, favoring SK over CK, is not compatible with a recommendation from the National Mathematics Advisory Panel, to the effect that "teachers be given ample opportunities to learn mathematics for teaching. That is, teachers must know in detail and from a more advanced perspective the mathematical content they are responsible for teaching and the connections of that content to other important mathematics, both prior to and beyond the level they are assigned to teach" (Recommendation 19 on page xxi of National Mathematics Advisory Panel 2008). In the next subsection, Sect. 11.6.2, we will explain why *without a far better content knowledge than the grade-level TSM they possess at present, teachers cannot hope to become more effective in teaching ratio, rate, proportion, and percent.* Granting this, the conclusion of the impact study would have been more accurately described as follows:

> The study results are consistent with the expectation, as of 2019, that *a mathematics PD program that does not replace teachers' content knowledge of TSM with PBM will not be likely to have a statistically significant impact on student achievement.*

11.6.2 Ratio, Percent, and Rate in TSM

Recall that the stated goal of the PD in the impact study was to raise *teachers'* content knowledge to the level of what "students should ideally have after completing the seventh grade," and its main focus was on improving the *teaching* of ratio, percent, rate, and proportion. Simply put, what the IES impact study aspired to do was *equip teachers with the best knowledge base that TSM had to offer students in the 7th grade curriculum,* with the hope that these teachers would then be able to boost student achievement by better pedagogy alone.

This strategy was bound to fail because percent, ratio, and rate are among the most feared topics by middle school teachers and students. *Why fear?* Because:

(1) Understanding ratio, percent, and rate requires a fluent knowledge of fractions in the first place. Since the knowledge of fractions for most students is precarious, they are handicapped before they begin.
(2) These three topics come *after* the division of fractions, and we have to remember: *"ours is not to reason why, just invert and multiply."*
(3) The presentations of *percent*, *ratio*, and *rate* in TSM are seriously flawed.

Let us elaborate on (3).

Ratio and Percent

We first consider *ratio* and *percent*. How are they defined for students in TSM? Percent and ratio are words in our daily conversation, and we all have a vague idea what "the ratio of Democrats to Republicans in this gathering is about 2 to 3" means: if there are about 200 Democrats in the gathering, there will be about 300 Republicans there. But what if there are about 2365 Republicans, can students still figure out roughly how many Democrats there are? Mathematics is a discipline of precision, so students need a precise meaning of "ratio." In TSM, ratio is "defined" in a variety of ways, but the following are typical:[29]

- A ratio is a comparison of two numbers, a and b, written as a fraction $\frac{a}{b}$. You can write a ratio in three ways.

$$1 \text{ to } 45 \quad \text{or} \quad 1:45 \quad \text{or} \quad \frac{1}{45}$$

  You can write a ratio to compare two amounts—a part to a part, a part to the whole, or the whole to a part.
- A ratio is a comparison of two numbers. It may be written in three different ways. The ratio of the number of people who picked "hazardous waste material" (18) to the number of people who picked "greenhouse effect" (9) [in the preceding survey] can be written as:

$$18 \text{ to } 9, \qquad 18:9, \qquad \frac{18}{9}$$

  . . . If you think of a ratio as a fraction, then $\frac{18}{9} = \frac{2}{1}$. They are **equal ratios.**
- Ratios are encountered in everyday life. For example, there may be a 2-to-3 ratio of Democrats to Republicans on a certain legislative committee, a friend may be given a speeding ticket for driving 69 miles per hour, or eggs may cost 98 cents a dozen. Each of these illustrates a **ratio**. Ratios are written $\frac{a}{b}$ or $a : b$ and are usually used to compare quantities.

  A ratio of 1 : 3 for boys to girls in a class means that the number of boys is $\frac{1}{3}$ that of girls, that is, there is one boy for every three girls. Notice we could also say that the ratio of girls to boys is 3 : 1, or that there are three times as many girls as boys. The ratio of 1 : 3 for boys to girls in a class does not tell us how many boys and how many girls there are in the class. It only tells us the relative size of the groups.

Does any one of these abstruse "definitions" tell students clearly *what a ratio is?* In the first bullet, for instance, it suggests that a ratio is a comparison of two numbers that is written as a fraction. Since a fraction is a number, is ratio therefore also a

[29] These bulleted statements are taken directly from textbooks.

number? If so, why say it is a "comparison," which suggests that it is some kind of an "action"? But wait: a fraction $\frac{a}{b}$ implies that, by definition, a and b are both whole numbers. Does this mean we can only "compare" two whole numbers? Since people also use "ratio" to compare fractions—the ratio of flour to sugar in a recipe is $1\frac{3}{4}$ cups to $\frac{2}{3}$ cups—shouldn't the definition have been replaced by something like the following?

> A ratio is a comparison of two fractions, a and b, obtained by dividing the fractions: $\frac{a}{b}$.

In addition, the last sentence of the first bullet talks about comparing "two amounts" instead of two numbers. What is an "amount," and what is "a part to a part, a part to the whole, or the whole to a part" all about?

We can go further. For example, the statement that "the ratio of Democrats to Republicans in this gathering is about 2 to 3" has a well-known interpretation of "to every 2 Democrats there are 3 Republicans." Every student needs to know (1) what it means to say "to every 2 Democrats there are 3 Republicans" and, more to the point, (2) how does the definition of ratio lead to this interpretation? TSM does not even pretend to address these questions. (see Chapter 22 of Wu 2011a for some answers to the preceding questions.)

A *mathematical* definition must at least address these basic, mundane issues, but the first bullet does not. It is therefore easy to see why, given such a variety of "definitions" of a ratio, it is very difficult for learning to take place. This is why students fear ratios.

Without a definition of *ratio* that makes sense, there can be no reasoning to speak of (see the discussion of *learning about definitions* in Sect. 11.5.1). Similar comments can easily be made about the other two bullets.

We have thus seen that ratio is a confused concept both within TSM and in daily life. To introduce school students to ratio, we must find a definition that provides an entry point into such a vague concept that is *correct, simple, and therefore learnable.* If we introduce the concept of a *complex fraction* as the division of two fractions (see Chapter 19 of Wu 2011a), then a ratio of two fractions A and B can be simply defined to be the complex fraction $\frac{A}{B}$. All the standard ratio problems can then be easily solved *by the use of reasoning* (see Chapter 22 of Wu 2011a).

The sins of TSM on the subject of ratio run deeper. Implicit in the preceding "definitions" is the belief that *ratio* is an ineffable concept, so that even a verbose description will not "discharge its full meaning" (Lamon 1999, page 165). Yet, later in the school curriculum, TSM has no hesitation in defining a ratio simply as a division. For example, the *ratio* of the circumference to the diameter of a circle is the number π, the *ratio* of "rise over run" of a straight line in the coordinate plane is *one* number, the slope, and the *ratio* of the opposite side of an acute angle of t radians in a right triangle to the hypotenuse of the right triangle is also *one* number, $\sin t$. But TSM never addresses the disconnect between this and its earlier, inscrutable "definition" of a ratio,[30] and every middle school teacher is aware of that. This is why teachers, too, fear ratios.

Next, *percent*. Here are some typical TSM "definitions."[31]

- Percent is a special kind of ratio in which the second quantity is always 100.
- Remember that percent means "out of 100." You find a percent by first dividing to find a decimal.
- Percent: part of 100, or per hundred.

Do these "definitions" tell us what "percent" means? If students already have trouble understanding what a "ratio" is, how is the first bullet going to help them understand "percent"? Is "out of a hundred" or "of each hundred" a number, and if so, what number is it?

TSM is apparently indifferent to these concerns, but it will tell you how to solve "percent" problems by laying down some "rules." For example, the second bullet already has a built-in rule: divide to find a decimal. What is the *reasoning* that leads from "out of 100" to "divide to find a decimal"? None that we can see. It is an arbitrary rule dictated by TSM.

On the basis of the second bullet, the suggested way to solve "*what percent of* 80 *is* 45?" is this: $45 \div 80 = 0.5625$, "which when rounded to the nearest 100th" is 0.56. So the answer is: about 56%. Notice that we are using the fact that one can obtain the decimal representation of a fraction by long division to get the answer to this simple question. This fact about "dividing numerator by denominator to get a decimal" is in fact very difficult to prove at the level of school mathematics (for a preliminary explanation, see pp. 81-86 of Wu 2016a, and for the full proof, see Section 3.4 in the third volume of Wu 2020), and is in any case only taught by rote in TSM. Thus, to find an approximate answer to "*what percent of* 80 *is* 45?", students have to follow an arbitrary rule, *and* also use a difficult fact they could only memorize by rote. *No reasoning in any case.* What are we teaching and learning here except how to follow rules?

For another example, starting with the third bullet ("Percent: part of 100"), the answer to *what is* 45% *of* 80? can be found by one of two "rules," according to TSM. First rule: set up a proportion: $\frac{45}{100} = \frac{n}{80}$. By the CMA (cross-multiplication algorithm), $45 \times 80 = 100n$ and therefore $n = 36$. Second rule: multiply the number by the percent: $45\% \times 80 = 36$. TSM is silent about how to start with "part of 100" and arrive at either of these solution methods *by the use of reasoning*.[32] TSM is also silent on whether there is any connection between the two methods. As far as TSM is concerned, it is enough that both methods seem *intuitively* related to "part of 100," and that both methods yield the correct answer. This is why, in TSM, percent is not learnable as mathematics, and this is why teachers and students both fear it.

---

[30] A blatant example of incoherence (see the fundamental principles of mathematics in Sect. 11.3.1).

[31] These bulleted statements are taken directly from textbooks.

[32] See Nike's trademarked slogan: "Just do it."

One can consult Chapter 20 of Wu (2011a) for a simple definition of *percent* and see how—on the basis of this definition—reasoning leads to straightforward solutions of all such standard problems.

Rate

A great deal can be said about how the concept of *rate* is abused in TSM (see Section 7.2 of Wu 2016b), but we will limit ourselves here to discussing only *continuous rates*[33](those related to, e.g., motion, water flow, lawn-mowing). We already touched on the most troubling aspect of TSM's treatment of rate problems when we discussed Luis' walking problem in Sect. 11.5.1: its reliance on the fictitious concept of *proportional reasoning* to solve problems. But there are other issues.

TSM should make clear at the outset that *rate*, as it is understood intuitively, cannot be taught *as mathematics* in K–12 (it requires the use of the derivative and is therefore a calculus concept). But this message has never gotten out. As we noted in Sect. 11.5.1, what *can* be taught in K–12 are the concepts of *average rate* (over a fixed time interval) and *constant rate*, but TSM shows little or no inclination to define either precisely or teach either seriously. Instead, TSM serves up the following brew of "definitions" for *rate*:[34]

- A **rate** is a ratio that involves two different units. A rate is usually given as a quantity per unit such as miles per hour.
- A **rate** is a ratio that compares two quantities having different units of measure.
- A rate is the quotient of two quantities with different units. A quantity whose unit contains the word "per" or "for each" or some synonym.

Again, these "definitions" are not informative because, this time around, none of them even makes any pretense at trying to give students any *usable* mathematical information. In two out of three cases, these definitions are built on the concept of *ratio*. Since TSM never explains clearly what ratio is, how can students use what little they know about ratio to find out what "rate" is?

In Summary

The many mathematical flaws in the way percent, ratio and rate are taught in TSM make these topics *unlearnable*, and that is why they are feared by one and all.

When the 2011 IES impact study considered teachers' content knowledge to be adequate for teaching ratio, percent, rate, and proportion if it was equal to "the knowledge of topics in rational numbers that *students* should ideally have after completing the seventh grade" (page 21 of Garet et al. 2011), it *failed to recognize the damage TSM had done to teachers and to school mathematics education as a whole.* This is tantamount to saying that *TSM is good enough.* If the foregoing

---

[33]The concepts of what is "continuous" and what is "discrete" are well understood in mathematics, but for the case at hand, the *ad hoc* explanation of "continuous" given in the parentheses is sufficient.

[34]These bulleted statements are taken directly from textbooks.

analysis in this subsection means anything at all, it is that TSM is *not* good enough.[35] By not providing teachers with an improved content knowledge base, the IES impact study in effect forced them to present the standard unlearnable TSM to their students. *Was it the teachers' fault that they turned out to be ineffective in raising student achievement?*

In this light, the IES impact study—while its good intentions are undeniable—cannot serve as a reliable gauge of whether PD can improve teachers' effectiveness or student achievement. It would make sense to do a similar study with a PD program that teaches PBM to teachers in the first place.

11.6.3 Final Thoughts

We want to come to a real appreciation for why the 2011 IES impact study failed to raise student achievement.

Our first thought is that grade 7 is probably not the best grade to choose for the purpose of an impact study, for the following reason. We saw in the preceding subsection, Sect. 11.6.2, that ratio, rate, and percent are concepts that have been made more difficult by TSM than they actually are. Any attempt to improve teachers' mastery of these concepts must begin by providing teachers with precise definitions for these concepts and showing teachers—on the basis of these definitions—how to use reasoning to solve all the standard problems with ease. In other words, the PD must help teachers *relearn* these topics from the standpoint of PBM. There is a catch, however. Teachers cannot learn the new content knowledge about ratio, rate, and percent without first acquiring a new foundation for fractions such as that presented in Sect. 11.3.2. To help teachers of 7th grade to better teach ratio, rate, and percent, it is necessary to first revamp their knowledge of fractions. We are thus suggesting that the PD program of the 2011 impact study should have been more comprehensive in terms of content, and it should also be held over a longer time duration. A future impact study may try to avoid grade 7 and work with teachers in an earlier grade.[36]

If we accept the foregoing explanation that teachers' faulty knowledge of fractions and ratio, percent, and rate, based on TSM, cannot be used in a school classroom, then it immediately raises the question of whether the impact study's emphasis on SK (*specialized knowledge for teaching*) at the expense of CK (*common knowledge*) was a good decision (see Sect. 11.6.1). The impact study

---

[35]This conclusion is also consistent with the fact that, by the second year of the impact study, it became apparent to the researchers that the teachers were having trouble with the 7th-grade topics of ratio, percent, and rate.

[36]We may note that the 2016 impact study of Garet et al. (2016) did avoid this pitfall and chose to work with 4th grade teachers. However, the content component of the 2016 study has some issues of its own, including a lack of awareness of the damage TSM has done to teachers.

made the decision to essentially leave teachers' knowledge of TSM for 7th grade intact and empower them with some new pedagogical techniques, hoping that the same teachers would nevertheless raise student achievement. Having the benefit of hindsight, we would put the emphasis of the PD program on CK to help teachers eradicate their knowledge of TSM and replace it with PBM. Mindful of teachers' need for intensive content immersion, we would also increase the number of summer institute days[37] from 3 to 8, expand the length of the day from 6 to 8 h (including lunch), and give out daily homework assignments. (There is nothing like doing exercises to improve one's understanding of mathematics.) Of the eight summer institute days, a tentative suggestion would be to spend 5 days on the definition of fractions, the comparison of fractions and the arithmetic operations on fractions. Middle school teachers really need a firm foundation on fractions because almost everything they teach rests on this foundational knowledge. The remaining 3 days could then be devoted to a thorough discussion of *complex fractions* (see Chapter 19 of Wu 2011a), ratio, percent, and rate. We should add that complex fractions are absolutely essential for any discussion of ratio, percent, and rate. One of the reasons that TSM cannot make sense of these three concepts is precisely its neglect of the *concept* of complex fractions (see Wu 2016a, Section 1.7).

It goes without saying that the *content* of the eight summer institute days would be PBM rather than TSM (Sect. 11.3). Given the present lack of usable PBM materials (cf. the discussion in Sect. 11.4), allow me to suggest Chapter 1 of Wu (2016a), *without the long Section* 1.10 *on probability*, as a reference. (Wu 2016a was actually written explicitly for this kind of PD for middle school teachers.) A slightly different suggestion would be Part 2 of Wu (2011a), *without Chapters* 23 *and* 24.

The 10 days of coaching in the original PD design of the impact study is an excellent idea, but having to send so many coaches to different school districts poses a problem of getting qualified coaches. This also brings up a similar issue of how to get enough PD facilitators for the summer institutes to be given in the participating school districts. As we have indicated all along, most PD providers are themselves products of TSM. For them to be effective in helping teachers learn PBM, they themselves will have to undergo training to learn PBM first. Therefore, before teaching teachers, we will have to teach coaches and PD facilitators. There is no getting around this difficulty, the fact that the preparation for such an impact study—in addition to the usual logistical issues—would have to take place months before the study itself for the purpose of creating a corps of qualified coaches and facilitators. The content-intensive training for this purpose is similar to the content-intensive PD for the teachers, except that there will be less room for failure because the coaches and facilitators will be responsible for the PD after all. We will not go into the details of this training because much of it will take us far away from our main concern with PD for *teachers*.

[37]If my own experience in PD is any guide, 8 days of content instruction do not really count as "*intensive* immersion." But we do have to compromise with reality.

In summary, this discussion may give a better idea of why one should not jump to conclusions about the non-effectiveness of inservice PD for improving school mathematics education on the basis of one or two impact studies. We hope there will soon be a follow-up impact study with a better design.

11.7 Miscellaneous Remarks

(1) As of 2019, elementary teachers are generalists, and it is impractical to ask all generalists to teach PBM. The reality is that learning and teaching PBM will take more dedication and time than a typical generalist can afford. Any real improvement in elementary mathematics education will inevitably require the creation of *mathematics teachers*—often called *math specialists*—to teach the mathematics of elementary school (see Wu 2009). These math specialists will certainly need the content-intensive training in PBM mentioned above.

(2) If we want more effective teachers, we cannot talk only about *inservice* PD because we must teach PBM to all preservice teachers so that the new teachers coming out of the pipeline will help solve the TSM-infestation problem rather than adding to it. Now preservice PD is understandably a different beast from inservice PD, and we will only *lightly* touch on a few of the major issues that complicate the preservice picture.

First of all, few colleges are willing to offer a mathematics course for teachers because teaching such a course on PBM will likely require extensive cooperation between the school of education and the department of mathematics. Given the often frosty relationship between these two units on many campuses, this obstacle can be overcome only by a leadership with intellectual vision and dedication to social justice. On top of that, there is the obvious problem with textbooks because an overwhelming majority of the available preservice PD textbooks are mired in TSM. The presence of these books is part of the reason that TSM is continually recycled in the world of education. There is also a less obvious personnel problem, as we now explain.

In year 2019, those with the requisite mathematical knowledge to teach PBM—regardless of grade level—are overwhelmingly found in mathematics departments. A typical mathematician is, however, ill-equipped to teach a course on correct school mathematics, for several reasons. Such a course is "elementary" in the sense of the usual mathematical hierarchy and will therefore be treated like calculus, and it is sad but true that calculus is usually taught as TSM. Moreover, teaching a course on PBM requires mathematical sophistication on a level with teaching an upper division course like introductory analysis (ϵ's and δ's) or abstract algebra (groups, rings, and fields). But teaching the former like an upper division course for math majors would be unfair to future teachers and ill-equip those teachers to teach their future students. Very often, the proofs (explanations) in a course for teachers—if done correctly—would be the most intuitive, not the shortest possible. Short proofs tend to be mathematically sophisticated and, therefore, *generally* not appropriate for school students. We would prefer that preservice teachers learn something close to what they will have to teach. For example, the explanation that $(-a)(-b) = ab$ for rational number a and b is something that bedevils most middle school students along with many of their teachers. In Wu (2016a), the explanation of this fact takes a full five pages (middle of page 166 to the middle of page 171). The usual three-line mathematical proof, expanded to half a page, is finally given in the lower half of the last page, page 171. A typical mathematician, approaching such a course for teachers as one in pure mathematics, would surely scoff at such verbosity as a waste of time. There are many such *mathematical* issues that can potentially reduce the relevance of such a course on PBM to future teachers. Ideally, one can smooth over such bumps on the road if there is a good working relationship between the School of Education and the Department of Mathematics, but such a spirit of cooperation is currently in short supply. Getting the right people to teach such a course will probably be a thorny issue for a long time to come.

(3) Finally, we have advocated for sustained PD to teach teachers correct school mathematics. While this seems not to be happening yet, several education centers around the country have been offering PD for mathematics teachers that focusses on solving hard problems or doing mathematical research. Any effort at raising the content knowledge of mathematics teachers is welcome, so there is no doubt that these centers are doing something right for a certain population in school mathematics education. Nevertheless, we must not lose sight of the fact that we have to raise the *general level* of content knowledge of the average teacher if better school mathematics education is our goal. If those centers that teach problem solving or helping teachers do research could convey the message that the problem solving and the mathematical research are means to an end, the end being the replacement of TSM by PBM, they would be making a major contribution to the cause of better school mathematics education for all.

Acknowledgments I wish to thank R. A. Askey for his direct and indirect advice, Yeping Li and Bill McCallum for supplying some needed information, and Larry Francis for his indispensable help with corrections and suggestions. I also wish to acknowledge the many expository improvements that came from the reviewers' suggestions.

References

Aharoni, R. (2015). *Arithmetic for parents*. Singapore, Singapore: World Scientific Publishing.

Armstrong, B., & Bezuk, N. (1995). Multiplication and division of fractions: The search for meaning. In J. T. Sowder & B. P. Schappelle (Eds.), *Providing a foundation for teaching mathematics in the middle grades* (pp. 85–120). Albany, NY, USA: State University of New York Press.

Askey, R. (2018). What content knowledge should we expect in mathematics education? In Y. Li, J. Lewis, & J. Madden (Eds.), *Mathematics matters in education. Essays in honor of Roger E. Howe* (pp. 219–231). Dordrecht, The Netherlands: Springer. https://tinyurl.com/y7bfhgqf

Baldridge, S. (2013). *Eradicating Textbook School Mathematics (TSM)*. Blog posted Retrieved February 6, 2013 from: https://tinyurl.com/yaegv9cn

Bezuk, N., & Kramer, K. (1989). Teaching about fractions: What, when and how? In P. R. Trafton & A. P. Schulte (Eds.), *New directions for elementary school mathematics*. Reston, VA, USA: National Council of Teachers of Mathematics.

Billstein, R., Liebeskind, S., & Lott, J. W. (2007). *A problem solving approach to mathematics for elementary school teachers* (9th ed.). Boston, MA, USA: Pearson-Addison Wesley.

Bingea, A. (undated). *Adding and Subtracting Rational Numbers on the Number Line*. Retrieved on October 30, 2019 from: https://teachers.yale.edu/curriculum/viewer/initiative_16.05.01_u

Common Core (2010). *Common Core State Standards for Mathematics*. Retrieved from: http://www.corestandards.org/Math/

Common Core (2012). *K-8 Publishers' Criteria for the Common Core State Standards for Mathematics*. Retrieved from: http://tinyurl.com/bpgx8ed

Common Core (2013). *High School Publishers' Criteria for the Common Core State Standards for Mathematics* (2013). Retrieved from: https://tinyurl.com/y8upgtcr

Cuoco, A., & McCallum, W. (2018). Curricular coherence in mathematics. In Y. Li, J. Lewis, & J. Madden (Eds.), *Mathematics matters in education. Essays in honor of Roger E. Howe* (pp. 245–256). Dordrecht, The Netherlands: Springer.

Darling-Hammond, L., Hyler, M. E., & Gardner, M. (2017). *Effective teacher professional development*. Palo Alto, CA, USA: Learning Policy Institute. https://tinyurl.com/y9qbdfmf

Douglas, A. (2015). *The Creation of a Mathematics Education Program at City Tech, CUNY*. Retrieved from: https://tinyurl.com/y8wn3rz4

Dynarski, M. (2018). *Why is accountability always about teachers? Brookings: Series–Evidence Speaks*. Retrieved February 25, 2018 from: https://tinyurl.com/y9d9zff4

Education Week (2014). *Researcher Isolates Common-Core Math Implementation Problems*. Retrieved November 14, 2014 from: https://tinyurl.com/y7hme29c

Garet, M. S., Cronen, S., Eaton, M., Kurki, A., Ludwig, M., Jones, W., ... Sztejnberg, L. (2008). *The impact of two professional development interventions on early reading instruction and achievement* (NCEE 2008-4034). Washington, DC, USA: National Center for Education Evaluation and Regional Assistance, Institute of Education Sciences, US Department of Education. https://ies.ed.gov/ncee/pdf/20084030.pdf

Garet, M., Wayne, A., Stancavage, F., Taylor, J., Eaton, M., Walters, K., ... Dolittle, F. (2011). *Middle school professional development impact study: Findings after the second year of implementation* (NCEE 2011-4024). Washington, DC, USA: National Center for Education Evaluation and Regional Assistance, Institute of Education Sciences, US Department of Education. https://ies.ed.gov/ncee/pubs/20114024/pdf/20114024.pdf

Garet, M. S., Heppen, J. B., Walters, K., Parkinson, J., Smith, T. M., Song, M., ... Borman, G. D. (2016). *Focusing on mathematical knowledge: The impact of content-intensive teacher professional development* (NCEE 2016-4010). Washington, DC, USA: National Center for Education Evaluation and Regional Assistance, Institute of Education Sciences, US Department of Education. https://ies.ed.gov/ncee/pubs/20164010/pdf/20164010.pdf

Hasiotis, D. (2015). What Does the Research Say on Professional Development, Anyway? *TNTP Blog*. Retrieved July 28, 2015 from: https://tinyurl.com/y8lbybkb

Jensen, G. (2003). *Arithmetic for teachers*. Providence, RI: American Mathematical Society.

Keeghan, A. (2012). *Afraid of Your Child's Math Textbook? You Should Be*. Blog posted on February 17, 2012. Retrieved from: https://tinyurl.com/yxo2a97h

Lamon, S. J. (1999). *Teaching fractions and ratios for understanding*. Mahwah, NJ, USA: Lawrence Erlbaum.

Layton, L. (2015). Study: Billions of dollars in annual teacher training is largely a waste. *Washington Post*. Retrieved, August 4, 2015 from: https://tinyurl.com/y7zzr8t8

Little, J. W. (1989). District policy choices and teachers' professional development opportunities. *Educational Evaluation and Policy Analysis, 11*(2), 165–179.

Loewus, L. H. (2016). Gates Chief Acknowledges Common-Core Missteps. *Education Week*. Retrieved May 23, 2016 from: https://tinyurl.com/yd5juay5

Loewus, L. H. (2016). Study: Improving Teachers' Math Knowledge Doesn't Boost Student Scores. *Education Week*. Retrieved September 28, 2016 from: https://tinyurl.com/zh6hhez

Loewus, L. (2017). Majority of Teachers Say Reforms Have Been 'Too Much'. *Education Week*. Retrieved December 19, 2017 from: https://tinyurl.com/y7yw8lyd

Moynahan, J. (1996). Of-ing fractions. In D. Schifter (Ed.), *What's happening in math class?*. New York, NY, USA: Teachers College Press.

Namkung, J. M., & Fuchs, L. S. (2016). Cognitive predictors of calculations and number line estimation with whole numbers and fractions among at-risk students. *Journal of Educational Psychology, 108*(2), 214–228.

National Mathematics Advisory Panel (2008). *Foundations for success: The final report of the national mathematics advisory panel*. Washington, DC, USA: US Department of Education. Retrieved from http://tinyurl.com/hrzm3uu

National Research Council (2001). J. Kilpatrick, J. Swafford, and B. Findell (Eds.), *Adding it up.*. Washington, DC, USA: National Academy. https://tinyurl.com/y6mtcr5k

Otten, S., Herbel-Eisenman, B. A., & Males, L. M. (2010). Proof in algebra—Reasoning beyond examples. *Mathematics Teacher, 103*(7), 514–518.

Polya, G. (1957). *How to solve it; a new aspect of mathematical method*. Garden City, NY: Doubleday.

Poon, R. C. (2014). *Principle-Based Mathematics: An Exploratory Study*. Dissertation at University of California, Berkeley. Retrieved from: http://escholarship.org/uc/item/4vk017nt

Quinn, F. (2012). A revolution in mathematics? What really happened a century ago and why it matters today. *Notices of the American Mathematical Society, 59*, 31–37. https://www.ams.org/notices/201201/rtx120100031p.pdf

Romberg, T. A., & Carpenter, T. P. (1985). Research on teaching and learning mathematics: Two disciplines of scientific inquiry. In M. C. Wittrock (Ed.), *Handbook of research on teaching* (3rd ed.). New York, NY, USA: Macmillan.

Sawchuk, S. (2016). Are teachers getting the right kind of common-core PD? *Education Week*. Retrieved Febuary 1, 2016 from: https://tinyurl.com/y8238pzj

Siegler, R., Carpenter, T., Fennell, F., Geary, D., Lewis, J., Okamoto, Y., ... Wray, J. (2010). *Developing effective fractions instruction for kindergarten through 8th grade: A practice guide* (NCEE 2010-4039). Washington, DC, USA: National Center for Education Evaluation and Regional Assistance, Institute of Education Sciences, US Department of Education. https://tinyurl.com/y66ffyec.

TIMSS 1995 Results (1995). *National Center for Education Statistics*. Retrieved from: https://nces.ed.gov/timss/results95.asp

TNTP (2015). *The mirage: Confronting the hard truth about our quest for teacher development*. New York, NY, USA: TNTP. https://tntp.org/assets/documents/TNTP-Mirage_2015.pdf

U. S. Department of Education (2009). *State and local implementation of the no child left behind act, volume VIII—Teacher quality under NCLB: Final report*. Washington, DC, USA: Office of Planning, Evaluation and Policy Development, Policy and Program Studies Service.

Wu, H. (1998). *Teaching fractions in elementary school: A manual for teachers*. Retrieved from: https://math.berkeley.edu/~wu/fractions1998.pdf.

Wu, H. (1999a). *Some remarks on the teaching of fractions in elementary school*. Retrieved from: https://math.berkeley.edu/~wu/fractions2.pdf

Wu, H. (1999b). Professional development of mathematics teachers. *Notices of the American Mathematical Society, 46*, 535–542. http://www.ams.org/notices/199905/fea-wu.pdf

Wu, H. (1999c). The joy of lecturing–With a critique of the romantic tradition of education writing. Appendix to *How to Teach Mathematics*, S. G. Krantz (pp. 261-271, 2nd ed.). Providence, RI, USA: American Mathematical Society.

Wu, H. (2002). *Chapter 2: Fractions (Draft)*. Retrieved September, 2002 from: https://math.berkeley.edu/~wu/EMI2a.pdf

Wu, H. (2006). How mathematicians can contribute to K-12 mathematics education, *International Congress of Mathematicians. Volume III.* (pp. 1676–1688). Zürich: European Mathematical Society. http://math.berkeley.edu/~wu/ICMtalk.pdf

Wu, H. (2009). What's sophisticated about elementary mathematics? *American Educator, 33*(3), 4–14. https://math.berkeley.edu/~wu/wu2009.pdf
Wu, H. (2011a). *Understanding numbers in elementary school mathematics*. Providence, RI, USA: American Mathematical Society.
Wu, H. (2011b). The mis-education of mathematics teachers. *Notices of the American Mathematical Society, 58*, 372-384. https://math.berkeley.edu/~wu/NoticesAMS2011.pdf
Wu, H. (2011c). Bringing the common core state mathematics standards to life. *American Educator, 35*(3), 3–13. http://www.aft.org/pdfs/americaneducator/fall2011/Wu.pdf
Wu, H. (2016a). *Teaching school mathematics: Pre-algebra*. Providence, RI, USA: American Mathematical Society. *Its **Index** is available at:* http://tinyurl.com/zjugvl4
Wu, H. (2016b). *Teaching school mathematics: Algebra*. Providence, RI, USA: American Mathematical Society. *Its **Index** is available at:* http://tinyurl.com/haho2v6
Wu, H. (2018). The content knowledge mathematics teachers need. In Y. Li, J. Lewis, & J. Madden (Eds.), *Mathematics matters in education. Essays in honor of Roger E. Howe* (pp. 43–91). Dordrecht, The Netherlands: Springer. https://math.berkeley.edu/~wu/Contentknowledge1A.pdf
Wu, H. (2020). *Rational numbers to linear equations, algebra and geometry, and pre-calculus, calculus, and beyond*. Providence, RI, USA: American Mathematical Society.

Author Index

Y. Li et al. (eds.), *Developing Mathematical Proficiency for Elementary Instruction*, Advances in STEM Education, https://doi.org/10.1007/978-3-030-68956-8

Subject Index

Y. Li et al. (eds.), *Developing Mathematical Proficiency for Elementary Instruction*, Advances in STEM Education, https://doi.org/10.1007/978-3-030-68956-8

The manufacturer's authorised representative in the EU is Springer Nature Customer Service Centre GmbH, Europaplatz 3, 69115 Heidelberg, Germany. If you have any concerns regarding our products, please contact ProductSafety@springernature.com

Printed and bound by CPI Group (UK) Ltd, Croydon, CR0 4YY
28/09/2026
02224091-0013